BIOLOGICAL RHYTHMS
IN HUMAN AND ANIMAL PHYSIOLOGY

by
Gay Gaer Luce

DOVER PUBLICATIONS, INC., NEW YORK

ACKNOWLEDGEMENTS

This monograph derives substance and shape from the gracious contributions of many scientists and other professionals, but mainly from the continuous suggestions and criticism so generously given by Dr. Alain Reinberg, Docteur en medicine, Docteur des Sciences, Fondation A. de Rothschild, Laboratorie de Physiologie, Paris.

Guidelines and consultation during the development of this report were provided by Dr. Ira Black and Dr. Charles F. Stroebel. Melvina Hindus and Margaret Mullins assisted in the research, and Jane Tucker in manuscript editing. The report was coordinated by Mrs. Dorothy Freed.

Published in Canada by General Publishing Company, Ltd., 30 Lesmill Road, Don Mills, Toronto, Ontario.

Published in the United Kingdom by Constable and Company, Ltd., 10 Orange Street, London WC 2.

This Dover edition, first published in 1971, is an unabridged and unaltered republication of the work originally published by the U. S. Department of Health, Education, and Welfare and the National Institute of Mental Health in 1970 as Public Health Service Publication No. 2088 under the title *Biological Rhythms in Psychiatry and Medicine*.

This report was prepared for the Program Analysis and Evaluation Branch, Office of Program Planning and Evaluation, National Institute of Mental Health.

International Standard Book Number: 0-486-22586-0
Library of Congress Catalog Card Number: 76-169785

Manufactured in the United States of America
Dover Publications, Inc.
180 Varick Street
New York, N. Y. 10014

TABLE OF CONTENTS

Page

FOREWORD

From the moment of conception until death rhythm is as much part of our structure as our bones and flesh. Most of us are dimly aware that we fluctuate in energy, mood, well-being, and performance each day, and that there are longer, more subtle behavioral alterations each week, each month, season, and year.

Through studies of biological rhythms, many aspects of human variability - in symptoms of illness, in response to medical treatment, in learning, and job performance - are being illuminated. Already some of our changes of mood and of vulnerabilities to stress and illness, our peaks of strength and productivity, can be anticipated. Moreover, by the end of this decade, much that is still considered unpredictable in health and human performance may become foreseeable through research into the nature of biological time cycles. As a result, timing promises to become an important factor in preventive health programs and medicine. For example, since the effects of drugs depend in part upon the time of administration, timing may be used as a critical aspect of treatment. Evidence now suggests, too, that X-ray treatments, surgery, and even psychotherapy are influenced in their outcome by timing. Research on biological rhythms also promises to have an impact on problems of work performance - including accidents and absenteeism; a new concept of scheduling as part of health may one day influence the determination of work-shifts among transportation and communication personnel, and members of various professions.

No corner of medicine - from the laboratory testing of new drugs and procedures to clinical and public health programs - is likely to remain untouched by the new explorations into biological rhythms. These studies are being made by scientists working with support from the National Institute of Mental Health, who have now joined forces with members of various disciplines ranging from biology to entomology and mathematics. The results of their work are provided in this comprehensive report. Contained here is compelling evidence that man is constructed not only of matter, but that he is temporally organized - and that this organization carries with it significant implications for man's mental and physical health.

The National Institute of Mental Health has been fortunate to support some of the very few researchers who could lead the way in this new field, thus taking an initiative in an area that is still not widely acknowledged. It is hoped that this report - part of a continuing effort to analyze and evaluate the Institute's programs - will draw the attention of both scientists and laymen to an exciting and promising area of scientific activity.

Bertram S. Brown

Bertram S. Brown, M.D.
Director
National Institute of Mental Health

There is a season for everything
And a time for every purpose under the heaven:
A time to be born, and a time to die;
A time to plant and a time to reap. . . .

Ecclesiastes

Chapter I. Introduction: Rhythms, Their Scope and Influence

Invisible rhythms underlie most of what we assume to be constant in ourselves and the world around us. Life is in continual flux, but the change is not chaotic. The rhythmic nature of earth life is, perhaps, its most usual yet overlooked property. Though we can neither see nor feel them, we are nevertheless surrounded by rhythms of gravity, electromagnetic fields, light waves, air pressure, and sound. Each day, as earth turns on its axis, we experience the alternation of light and darkness. The moon's revolution, too, pulls our atmosphere into a cycle of change.

Night follows day. Seasons change. The tides ebb and flow. These various rhythms are also seen in animals and man. We, too, change, growing sleepy at night and restlessly active by day. We, too, exhibit the rhythmic undulations of our planet.

Circadian Rhythms

In concert with the turning earth, plants and animals exhibit a very pronounced daily rhythm. Often external cues synchronize living organisms into an exact tempo. However, when men or animals are isolated from their usual time cues, they do not keep to a precise solar day (24 hours) nor even a precise lunar cycle (24.8 hours). Nonetheless, isolated creatures do show rhythms that do not deviate very much from the 24 hours. This daily rhythm is denoted by the popular term "circadian." It means "about a day," from the Latin, *circa dies*.

Mollusks, fish, cats, marigolds, baboons, men—indeed, most living organisms show a circadian rhythm of activity and rest. Time-lapse photography has captured the circadian dance of plant life, showing how leaves lift and drop, open and close every 24 hours. Man, no exception to this daily ebb and flow, may be unaware that his body temperature, blood pressure, and pulse, respiration, blood sugar, hemoglobin levels, and amino acid levels are changing in a circadian rhythm. So, too, do the levels of adrenal hormones in his blood and concentrations of essential biochemicals throughout his nervous system. Urine also shows the influence of the circadian cycle. We excrete rhythmically, not merely according to the time of liquid intake. As a decade of laboratory researches has disclosed, there is a rhythmic fluctuation in the contents of the urine, along with almost every physiological function, from the deposition of fat or sugar in the liver to the rate at which cells are dividing.

In health we have an appearance of stability that cloaks a host of rhythms, hormonal tides, intermeshed with surges of enzyme activity, production of blood cells, and other multitudinous necessities for life. Our smoothness of function seems to rest upon a high degree of integration among these circadian production lines, and they in turn may act as timekeepers for us, guiding us in our periods of energetic endeavor, or rest, acting as distributors of our dreams and the tidal motions underlying our ever-shifting moods. Although we appear constant from the outside, we can feel inside that we are not really the same from one hour of the day to the next.

The corollary of all this circadian change is very dramatic, for a creature's strength and weakness also vary, depending upon biological time of day. Life and death may hang in the balance of timing. Mortality has been decided, experimentally, not by the amount, but by the time of day that a rodent received X-rays or was injected with pneumonia virus, bacteria, or drugs. Exposure to loud noise affects a rodent little during his period of rest, but may hurtle him into a frenzy, into convulsions, and even death if it occurs during his activity period. A volume of anecdotal literature attests that man, too, must be different at different biological times of day. Now, these casual observations are being researched. People do perform differently on psychological and physiological tests at different hours. We may soon learn why more pregnant women go into labor during the night or early morning hours than afternoon, and why doctors receive so many of their calls from patients with coronaries during these same hours. Deaths and symptoms of diseases do not seem to be distributed evenly around the clock. The pain of glaucoma and symptoms of certain allergies or of asthma seem to occur mostly at certain hours. As we should expect, drugs also affect us differently according to biological time of day. Biological time of day does not necessarily jibe with the local clock time. For example, a person who works at night and sleeps by day is likely to be 180 degrees out of phase with the daytime workers of the world. He sleeps when they are awake. His temperature is falling as theirs is rising. His adrenal steroid levels are low when theirs are high. The positions of various body functions indicate "time of day" in the body—biological time of day.

Subjectively, people do notice that they are changing during the 24 hours of the day. Some people express this by their preference for afternoon or morning work, or may notice that they are emotionally resilient at certain hours and irritable at others. Like mechanical clocks, we synchronize our activity with the imperatives of society around us, and squeeze ourselves into the 24-hour schedule of modern life. But unlike mechanical clocks, our bodies will not instantly adjust. Jet travel has forced many of us to realize that internal time must be respected. A person is not the same at 4 a.m. as at 4 p.m., a fact that is unpleasantly palpable to the traveler who flies from Moscow to New York and finds himself trying to make mid-afternoon conversation in an office, while his body feels that it is not quite dawn. Two weeks later, long after he feels adjusted to local time, his body may be showing signs that adjustment is still not complete.

Relevance to Research

The fluctuations that continuously transform us each day are small, and superficially some of them may seem random. Nonetheless, biological cycles are pertinent to experimentation in many fields of psychology, pharmacology, and medicine. In the past, experimenters in these fields have made delicate manipulations to hold the experimental conditions constant but have not realized that their animal or human subjects were changing right under their hands. More sophisticated researchers may carefully take their measurements at the same time each day, but unless they have controlled many aspects of their subjects' environment this precaution may be futile. Time, indeed, may be a factor in some of the embrangled debates that arise when one scientist cannot replicate the experiment of another. Was something amiss with the initial study, or did the second scientist fail to take precisely the same steps? Even if the replication were conducted at precisely the same clock hours, this does not mean that animals or volunteers had all been synchronized to exactly the same schedule and were at the same biological time of day. The experiment may not reproduce the original findings because the subjects were at different points on their cycle.

Monthly, Seasonal, Annual Cycles

Although the day is our most important social unit, there are other cycles that alter a person. A woman differs at the various points of her menstrual cycle, before, during, and after. These monthly hormonal tides influence not only her physiology, but her emotional responses, performance, and as some slight evidence suggests, even the content of her nightly dreams. Psychological ramifications of the monthly cycle are most visible in women with premenstrual tension, a common yet often overlooked symptomatology now being studied by Dr. Oscar Janiger, at the University of California in Irvine, and by a team of National Institute of Mental Health grantees, at Stanford University Medical School. Men also exhibit a few signs of monthly change. Sanctorious, a 17th century doctor, used a fine scale to weigh healthy men over long periods of time and he reported that normal men underwent a monthly weight change of about 1 or 2 pounds.

Dr. Christian Hamburger, a renowned Danish endocrinologist, kept daily records of his urinary hormones (a group known as the 17-ketosteroids) for some 16 years. When the daily hormonal fluctuations were later analyzed by Dr. Franz Halberg, at the University of Minnesota Medical School, they showed a roughly 30-day rhythm in addition to a pronounced daily rhythm. Because few physicians are cycle conscious, and few individuals have the patience to accumulate serial records of their own changes, such rhythms are often caught only obliquely. A 28-day rhythm was observed in the drifting bedtime of a physicist during a year's research in Antarctica. He would go to bed later and later for 28 days, and then revert to his original time of retiring, a pattern discovered by Drs. Jay Shurley, at the University of Oklahoma Medical School, and Chester Pierce, at Harvard Medical School.

Advances in data analysis will probably reveal many slow rhythms in man, cycles with long periods that have not yet been documented. Even known seasonal and annual rhythms are difficult to study. We secrete something known as "summer hormone," a thyroid product that helps to reduce body heat, but we still know little about it and how its secretion is triggered so as to anticipate the hot summer months. Presumably, there are other subtle changes in our metabolism, consonant with seasonal change.

Annual changes in tissue have been observed in laboratory animals kept under standardized conditions of food, temperature, and humidity, as reported by Dr. H. von Mayersbach and his colleagues, at the University of Hannover, Germany. They found changes in the structure of liver cells and in the amount of a sugar, known as glycogen, that makes energy available to the animal. The seasonal changes were enough to make experimental results seem unstable.

Recently, an even more startling and controlled study has been performed by Drs. Erhard Haus and Franz Halberg. Their mice lived in a controlled environment where lighting was standardized, so as to exclude the major cues to seasonal change. Nonetheless, these inbred, antiseptically treated laboratory animals showed annual cycles in their blood levels of an adrenal hormone, corticosterone. Even without cues, their bodies registered an annual rhythm.

Man's own annual rhythms were largely a matter of anecdotal reports, or very sketchy statistics, until the last several years. Rhythms in the number of suicides and suicide attempts and in deaths from arteriosclerosis do seem to exist. Dr. Halberg has analyzed statistics from the Minnesota Department of Health and has seen that, at least, in the temperate zone, there are several distinguishable rhythms. There is a peak in the number of deaths from arteriosclerosis around January, a peak in suicides around May, and accidental deaths peak in July and August. Some of these rhythms may relate to social customs, such as summer vacations, while others, indeed, may relate to seasonal changes within us, and may help us to explain certain medical and psychiatric disorders.

Very recently, a venturesome researcher has found what may be clues to the well-known, but mysterious winter madness, the "arctic hysteria" reported among Eskimos. This sudden experience, like psychosis, may last for a few hours or days in winter. Dr. Joseph Bohlen, at the University of Wisconsin, has found that there is an annual physiological rhythm among Eskimos living at Wainwright, Alaska. Aided by his wife, Bohlen studied ten individuals around the clock for ten days during each of the four seasons.

It took something of a heroic effort, for the Bohlens had to visit their subjects' houses every two hours to gather urine samples and record oral temperature, blood pressure, pulse, and make tests of hand-grip and eye-hand coordination. They found that the Eskimos' body temperature and the amount of potassium excreted in urine were rising and falling in precisely 24-hour cycles, although they had no circadian cardiac rhythm. The urinary calcium excretion cycle was of greatest interest. Calcium, like other electrically-charged elements in the body, has a profound influence upon the functioning of the nervous system since it appears to be necessary in the transmission of nervous messages. Every Eskimo in Bohlen's study showed a striking annual rhythm in calcium excretion. They excreted eight to ten times as much calcium in winter as they did in summer.

Eskimos seem to suffer from an unusual amount of emotional illness, and it has been postulated that their symptoms are a kind of periodic illness. In Dr. Bohlen's study, we can see that some of the keys to mysterious disorders, such as arctic winter madness, may lie buried in the time cycles of man's physiology. Studies of these cycles may begin to offer explanations for certain social customs and strange human talents.

Time Sense

People who can "set" themselves for a night's sleep or a 20-minute nap and awaken "on time" are envied by all. Yet, many creatures measure time as accurately as if they contained a timer. We have recently learned about the marvelous time sense of bees. In one experiment bees were placed in a salt mine in constant light and temperature, where they learned to feed at certain dishes only at certain times of day, despite the fact that there were no time cues from the sun. In another experiment, bees were trained to collect sugar water from 8 a.m. to 10:15 a.m. in France. The hive was flown to the Eastern United States. Now the bees arrived for feeding at 3 a.m., which was precisely 24 hours later. It would have been the correct time in France. Bees in similar experiments have adjusted to local time within a period of about a week.

Ants have learned to show up for food at various intervals, as precisely as if carrying a watch, and the eyes of crayfish turn light every 12 hours, even in constant darkness.

Migrating fish, mammals, and birds orient themselves in part by relying on their inner time sense. They maintain a particular angle to the sun (or to polarized light when the sun is not visible). Their problem is analogous to that of a person who is dropped, blind-folded, into the Sahara Desert and told to walk east. One would need to know whether it was 10 a.m. or 2 p.m. to know from the sun's position which direction was east. In studies of time sense, birds, bees, crabs, and flies have been flown around the world and exposed to many experimental situations. Few such studies have been done with man. Indeed, the physiological bases for human-time sense are only beginning to be researched.

When a person awakens at the precise time he has elected beforehand, it almost seems as if he is responding to biological signals inside. Perhaps there are subtle signals, of which most of us remain unaware, signals that function as a delicate stimulus to give a person time information. During the last several years experiments with animals have demonstrated a kind of "trainable" yet unconscious time sense. Experimental creatures have exhibited abnormal symptoms that might be called "psychosomatic" at the precise time of day they had previously experienced unpleasant experimental manipulations. Can time of day become a conditioned stimulus? Perhaps within the "clock" of the brain remain echoes of past events in the glands and circulatory system. Is this how certain illnesses evolve? Are there periodic symptoms at the hour of a trauma, out of awareness but remembered by the body? Perhaps this could illuminate the so-called "anniversary symptoms" seen by psychiatrists. Such questions multiply from new studies of man's rhythms that promise a dramatic departure from traditional research on time sense.

Subjective Time Sense

The experience of time and the personal evolution of time sense have been more explored by philosophers, anthropologists, and historians than by physiologists and other medical scientists. Time sense and rhythm develop before language. Before a baby speaks he will drum on his crib and show a love of music and rhythm. Some infants rock rhythmically. Psychiatrists have speculated that the newborn infant loves to be cradled and rocked partly because it orients him in the way his mother's heartbeat may have oriented him in the womb. Out of an environment of heartbeat and protection, one is born into seeming chaos. Maybe rhythms do serve to comfort and reorient babies after the lonely confusion of being born. Surely young children show greed for dance, poetry, marching, and rhythmic repetition. However, some of our earliest time measures are not like heartbeat, but mark off stages of growth or other intervals.

There seem to be critical periods in the life cycle. There are critical moments in which a microsecond of light in a constantly darkened experimental room will pull an entire population of larval flies into synchrony to be born. There are critical periods for humans, during which a malnourished baby may incur permanent mental retardation. Many researchers think there are critical periods for learning language.

The passage of time also seems to be experienced differently by the child than by the adult. Drs. Jean Piaget and Paul Fraisse have explored this difference in their psychological studies. Until about age eight, time generally feels very expanded or seems to pass slowly. Most parents have heard their children make such statements as, "If I ever live to be nine..." in tones implying that it would take about a century.

Through repeated experience, adults have learned to measure units in estimating time and to apply reasoning. In young children, each day is its own universe. Tomorrow is a barely comprehensible concept. Moreover, until a certain age children do not seem to have a sense of the ordering of events or classification of order, even after adults have tried to teach them about it. Piaget reports one typical instance:

> "How old are you?"
> "Seven years old."
> "Do you have a friend who is older than you?"
> "Yes, this one next to me is eight years old."
> "Very good. Which one of you was born first?"
> "I don't know. I don't know when his birthday is."
> "But come on, think a little. You told me that you are seven years old and that he is eight, now which of you was born first?"
> "You'll have to ask his mother. I can't tell you."

Children do not seem to be able to comprehend certain concepts when they are young and are often notoriously subjective in their perceptions of things. In their feelings about the duration of time, or at least, the reason for their subjectivity, may be a physiological system that functions at a rate different from that of the adult.

The speed of the transactions of neurons in our brains and bodies must be calibrated to the rhythms of the physical world we perceive. Our brains must respond at a certain rate to hear sounds of certain frequencies. We do not have receptors to respond as fast as those of the dog, for instance, which means that our pets can hear ultra high sound but we cannot. We can see certain colors, but others are beyond our vision. We can see discrete movement, until it becomes so fast that our brains no longer separate the frames of a moving picture and, instead, see continuous motion, or flashes of light occurring so rapidly that we perceive only a beam.

The role of physiological speed in perception has fascinated many scientists.

> The relativity of our reference point (writes Roland Fischer in *The Voices of Time*) can be demonstrated by taking a moving picture of a plant at one frame a minute, and then speeding it up to thirty frames a second. The plant will appear to behave like an animal, clearly perceiving stimuli and reacting to them. Why then, do we call it unconscious? To organisms that react 1800 times as quickly as we react we might appear to be unconscious. They would in fact be justified in calling us unconscious since we would not normally be conscious of their behavior.
>
> Our restricted range of consciousness can be demonstrated by exposing a man to a sequence of similar or nearly similar stimuli, such as a succession of frames on television. If the frequency of these stimuli is increased to about fifteen cycles per second, previously distinct images cease to be separated and a spatialization or fusion of events takes place.

We all know that time interval perception is relative. Albert Einstein once wrote, "When you sit with a nice girl for two hours you think it's only a minute. But when you sit on a hot stove for a minute you think it's two hours. That's relativity."

Time Estimation

Does our varying sense of time have a plausible explanation? Accurate estimates of short-time intervals depend upon some internal pacemaker. Perhaps, in a monotonous situation, the stimulus input is too low, while in exciting situations the input is excessive relative to an internal pacemaker. But metabolic changes also may be involved.

Dr. Hudson Hoagland of the Worcester Institute of Experimental Biology, once asked himself, "If clock time perceived requires motion, and we judge time with our brain, would the velocity of chemical metabolism alter our time sense?" If metabolism and temperature were high, two minutes of brain time might pass in only one minute of clock time—and we would think that time was dragging. As Hoagland relates in *The Voices of Time* (Fraser, J., ed.), he began thinking about time sense, body temperature, and chemical kinetics in the early 1930's. Around that time his wife had a bad case of influenza. He recalled that she had a temperature of 104 F. one afternoon and asked him to go to the drug store. He was gone 20 minutes, but when he returned she insisted he had been gone for hours. Since she was usually very patient, this intrigued Hoagland. He took a stopwatch, and without telling her why, asked her to count to 60, at a speed she thought was about one per second. His wife, a trained musician, had a good time sense. But she counted to 60 in many fewer seconds. Throughout her illness, and later when she was well, Hoagland repeated the test some 25 times. It was consistent. When her body temperature was high she counted fast; when her body temperature was low, she counted slowly. Hoagland also raised the body temperatures of student volunteers by diathermy and tested their perceptions of short-time intervals before and during raised temperatures. Like his wife, the students counted faster when their body temperatures were high, suggesting that perception of short-time intervals may be modulated by a metabolic-chemical pacemaker system in the brain. In the progression from childhood to maturity, there is a decline in the rate at which one consumes oxygen and, indeed, a slowing of the metabolism. Does the higher metabolic rate of the child suggest why time seems to move so slowly for him, while time passes swiftly for the elderly person?

If Hoagland's physicochemical hypothesis is correct, with reference to short-time intervals, time should, indeed, be a variable experience for snakes and lizards and other cold-blooded animals. On warm summer days the time should seem to pass slowly, while on cold days it might seem to pass rapidly. There are, of course, more factors in general time perception, and in the experience of short intervals, than could be explained by metabolism. Indeed, there must be a multiplicity of factors in both time estimation and sense of time, doubtless governed by different loci in the nervous system. For example, the pacemaker governing heartbeat is probably located in other brain centers than those we use in "feeling" or sensing our heart beat.

A Rhythm in Time Estimation

A considerable literature on time estimation by psychologists shows how variously people have explored the questions asked by Hoagland. In 1962, the French speleologist, Michel Siffre, lived two months in a cave, isolated from all normal time cues. Several times during each waking period he estimated two minutes, counting to 120 via telephone to a base-camp operator above ground. These data were later analyzed by Drs. M. Engeli and Franz Halberg, at the University of Minnesota, and showed that he fluctuated—counting too fast or too slowly—in a rhythmic manner. That same year Dr. D. H. Thor predicted that time estimation and time perception both would depend on time of day, varying according to a person's period of maximal alertness. Thor was working at the Johnstone Training and Research Center in Bordentown, New Jersey.

Thor and his associates asked 450 people to tell them the time of day without looking at a clock. They asked the question at six different points between 8 a.m. and 6 p.m., and averaged the estimates at each time. The subjects were most accurate between 8 a.m. and 10 a.m., and again around 4 p.m. During midday, people typically misjudged the time, thinking it earlier; while at the end of the day they assumed the hour was later than it was in actuality. There was no doubt that their subjective sense of the clock hour varied in a predictable way, following a diurnal rhythm.

In 1968, Dr. Donald Pfaff, at Rockefeller University, showed that the estimation of short time intervals varied in a way that correlated with body temperature. Pfaff's subjects were consistent and when asked to denote a five- or eight-second interval, they made the interval smaller when their body temperatures were at their peak and longer when temperature was lower—much as Hoagland would have predicted. Thus, studies of time sense must, themselves, encompass a reference for physiological time of day if they are to be precise.

If Michel Siffre instigated interest in the circadian fluctuations of individual subjective time perception, he also exemplified one of the strange misperceptions of time that seems to occur during isolation. He spent 60 days in a cave, and emerged thinking he had been there only 35 days. Many factors influence the perception of both long and short time intervals.

Altered Mental States and Drugs

Under hypnotic time regression, some extremely susceptible subjects relive past experiences, and grown men act like infants, capable of poking an inquisitive finger into a flame, unable to handle matches.

People demonstrably change their sense of time when they are under the influence of such hallucinogenic and excitatory drugs as LSD or psilocybin. If asked to tap on a Morse key at a self-chosen rate and instructed to tap as evenly as possible, they will tap fastest at the peak of the drug state, says Roland Fischer in *Interdisciplinary Perspectives of Time*. At the time of increased tapping rate, they experience "a flood of inner sensations" or time contraction. In such states, people arrive too early for appointments if they do not consult their watches. At the same time they experience an expansion of nearby space, and this is evident in their enlarged handwriting. An opposite effect occurs under tranquilizers, which are used as antidotes for hallucinogenics, and under these drugs people are often late for appointments and their handwriting shrinks in size.

Distortions of time sense are reported among many psychotic patients, and the bleak distortions of many schizophrenics have invited tests that might define them operationally. Testing psychiatric patients in hope of finding some unique distortion of time estimation that would correlate with a diagnosis has proven exceedingly difficult. In their studies of schizophrenic patients and normals, Dr. William T. Lhamon, at the New York Medical College, and Dr. Sanford Goldstone found that methods and sensory modality influenced the results: it mattered greatly whether one asked patients to estimate the duration of a sound or a visual display.

Drs. F. T. Melges and C. E. Fougerousse, Jr., have been studying psychiatric patients at the University of Rochester School of Medicine. Recently, they studied a sizeable group of patients when they first entered the hospital, and again when they had improved and felt better. When asked to produce a specified short-time interval, patients in unhappy emotional states gave inaccurately brief intervals, as if their nervous systems were running at an accelerated pace. Patients in pleasanter states gave more accurate intervals. Psychotic patients seemed to have a more distorted time sense than depressed or neurotic patients, and their time judgments became more accurate as they improved.

Dr. Robert Ornstein, at the Langley Porter Neuropsychiatric Institute in San Francisco, has reviewed the literature on time experience in his book, *On the Experience of Time*. He concludes that our sense of duration cannot be correlated with any physiological "clock." Time sense, as he sees it, is more psychological than biological. Time durations seem long to us if we are conscious of a great deal of information. Ornstein performed nine ingenious experiments that demonstrated a role of awareness and memory in time experience. Subjects reported feeling the assigned intervals as increased in length when the amount and complexity of the stimuli within the measured interval were increased. People judged intervals as comparatively longer when they recalled the contents than when they did not, suggesting that our sense of duration may be related to memory as well as to information processing. A variety of experiments and studies of people under hallucinogenic drugs has begun to reveal that subjective time experience must be related to the elusive phenomenon we call "conscious awareness."

Cultural Concepts of Time

If neural firing rates and altered biochemistry may lie behind some of the time distortions observed in psychiatric illness, environment also must play a role. What we consider healthy or abnormal in the general perception of time is probably modulated by cultural

The tremendous effect in the West of the rejection of reincarnation by the early, Christian thought-makers.

attitudes. Temporal attitudes pervade a culture to such an extent that they are almost invisible. Probably they are more influential than we like to imagine. In subtle but powerful ways, cultural concepts of time have helped to mold the history of civilizations. Ancient Egyptian culture was exceedingly involved in natural cycles of the sun and the moon, for instance. However, dedication to a philosophy involving eternal life gave Egyptian existence dimensions that had been absent among their African predecessors, who were also guided by moon cycles.

Time concepts may help to explain the accuracy of early Chinese histories. These are unrivalled. Not only did the Chinese document events from earliest antiquity, but they expressed an orderly respect for family tradition and rules of human conduct, qualities that appear to have been generated by a philosophy embodying respect for time cycles of considerable magnitude. Naturalists and astronomers saw that the cycles of the sun and moon were reflected in life, and this in turn influenced their philosophy: "The sun at noon is the sun declining; the creature born is the creature dying." Ancient Taoism was cycle oriented. Time was divided into seasons and eras, considered part of an infinite chain of duration—past, present, and future.

The calendar was an inevitable focus of interest in an agrarian society, and there was almost no Chinese mathematician or astronomer who did not work on calendars. Between 370 B.C. and 1742 A.D., there were about a hundred different calendars, each one embodying astronomical events with ever greater accuracy. Early in Chinese history, astronomers belonged to the history department of the ruler. The Chinese were so historically oriented that they meticulously dated all objects and inscriptions. Twenty-five dynastic histories (beginning about 90 B.C.) offer a wealth of material on science, government, and customs.

In the first century A.D., Buddhism and a philosophy of continual metamorphosis spread through China. Naturalists and philosophers observed evolutionary transformations in living organisms. Some 16 centuries before Darwin they expressed a kind of evolutionary naturalism that embodied a succession of phylogenetic unfoldings rather than the concept of a single train of evolution with which Western science began. The complex concepts of time held by the Chinese led to remarkably sophisticated theories that included accurate perception of astronomical change, views on the nature of fossils, and explanations of the unity of these vast time cycles in the development and personal history of each man. In the Chinese view, man's place was a humble and appropriate part of the time cycle. Morally and biologically, he fit into the expanse of nature and history. Thus, he saw his world and his future very differently than does any modern Westerner.

In the 13th century, the Chinese *Book of Changes* gave an estimate of phases in the evolution of life covering about 130,000 years. At that time the Chinese were calculating astronomical periods in millions of years. Western attitudes of that era are primitive by contrast. Judeo-Christian perception of time was linear. The flow of time was believed to begin with some specific point in space time. Indeed, in 17th century Europe, people piously followed Bishop Usher's calculation of the date of the Creation of the Universe—as October 6, 4004 B.C. After all, time had to begin with some significant event.

This simple linearity dictated much of Western thought, custom, and philosophical egotism. It encouraged a self-centered concept of our place in the universe, our hustling individuality, and our philosophies of cause and effect. These notions have indeed been instrumental in the development of Western science.

If time were linear, one could ask ultimate questions about "the beginning," but other cultures with different concepts of time have shown them in their development.

Westerners measure time by action. How different are the Hindustani. For instance, India has never produced a written history. After all, why would people trouble to make detailed chronological records of their national development if they lived in a time domain characterized by a changeless sense of everbecoming? By contrast with Westerners, Indians may seem lacking in urgency. Their universe, world, and social order are eternal; personal life is only a sample of a succession of lives, repeating themsleves endlessly. Transmigration of souls and perpetual rebirth make meaningless any quantitative view of a particular period of time. Life, infinitely recycled, makes history less significant, and an individual's biography is merely a transient moment in the process.

The Japanese Buddhist concept of the transience of the physical world has very different consequences: it has led to intuitive, sensitive admonitions that if all things are transient, one must appreciate but not cling to them. "Time flies more swiftly than an arrow and life is more transient than dew." Thus, the ultimate reality is what one sees and experiences. The urgency of Western action is a new phenomenon in Japan.

In Japanese sensibility, time is not an absolute, nor an objective set of categories, but a process. It is the change of nature. Man is part of that change and able to appreciate it, feeling transience to be part of the eternal loveliness of the universe rather than a threat to the ego (as Western man sees mortality).

Even in a brief and oversimplified sketch, it is apparent that cultural concepts of time have a pervasive influence upon individuals and upon major social developments. In medicine, for instance, one can see why ancient peoples might have accepted notions of biological rhythmicity, connecting human life with natural cycles—and why there might be resistance to such ideas within our own culture.

Ancient Medicine

Ancient medicine took periodicities into account. Early Western medicine inherited a residue from ancient moon cults, from fertility rituals calibrated to the phases of the moon. In the early 18th century, Richard Mead, an extraordinary British physician, wrote a treatise on the relationship between illness and the movements of the sun and moon. His "Discourse Concerning the Action of the Sun and Moon on Animal Bodies" is filled with examples.

> The Girl, who was of lusty full Habit of Body, continued well for a few days, but was at Full Moon again seized with a most violent Fit, after which, the Disease kept its Periods constant and regular with the tides; She lay always Speechless during the whole Time of Flood, and recovered upon the Ebb. . . .

Early Greek therapies involved cycles of treatment, known as metasyncrasis. Patients were not fed the same food and herbs or given the same exercise every day, but in rotations of three at seven days. *Caelius Aurelianus on Chronic and Acute Diseases* (translated by I.E. Drabkin) describes these treatments:

> And [in these remission periods] give the patient food of the middle class, and then apply the regular restorative cycles, but in such a way as to avoid the necessity of sudden changes. That is, in the first cycle let the patient take food of the middle class, in the second fowl, in the third the flesh of field animals, and in the last pork. In the next stage two successive cycles may be joined together and finally all. . . .

Over 2400 years ago, Hippocrates had advised his associates that regularity was a sign of health, and that irregular body functions or habits promoted an unsalutary condition. He counseled them to pay close attention to fluctuations in their symptoms, to look at both good and bad days in their patients and healthy people. Around 300 B.C., Herophilus of Alexandria is said to have initiated measurements of biological periodicity by counting the human pulse with the help of a water clock. Ironically, the early 20th century saw some ancient medical precepts revived in a fashion that was to become a fad—a fad that is popular even today.

Mythology

In 1887, Whilhelm Fliess published his formula for the use of biological rhythms. He asserted that everyone was bisexual. The male component (strength, endurance, courage) was keyed to a cycle of 23 days. The female cycle (not menstrual, but a cycle of sensitivity, intuition, love, and other feelings) had a period of 28 days. Both cycles, he claimed, are present in every cell and play a dialectic role throughout life—from birth on. They manifest themselves in the ups and downs of vitality, physical and mental, and eventually determine the day of one's death. In the 1930's, a teacher at Innsbruck added a 33-day creativity cycle of mental acumen and power.

Fliess and his magnum opus, *The Rhythm of Life: Foundations of an Exact Biology*, might have faded into obscurity had he not been a close friend of Sigmund Freud. For many years Freud credited Fliess with a great breakthrough in biology. Fliess' "breakthrough" linked 23- and 28-day cycles with changes in the mucosal lining of the nose. He related nasal irritation to neurotic symptoms and sexual abnormalities. He diagnosed ills by inspecting the nose and applying cocaine to what he called "genital cells" in its interior. He operated on Freud's nose twice. Fliess' blatantly unsophisticated understanding of simple mathematics is evident in his formula, which is transparent junk. Yet, every year it is offered to the public in new books on "biorhythms" that promise a reader the ability to chart his own cycles of physical or emotional vulnerability and strength in advance. By predicting his strength and weakness, much as a woman might put menstrual dates on a calendar, a person is supposed to be able to plan his activities for maximum advantage. If the formula is childish, the underlying idea may not be so far-fetched.

Normal Emotional Rhythms

Long undulations in emotion and cycles of subtle psychological change may occur, unnoticed, even among normal, healthy people. In 1929, for instance, Dr. Rex B. Hersey reported a curious observation on industrial workers. He had arranged to follow and question a group of workers in their plant. The men were "average guys," quietly haunted for many weeks by Dr. Hersey and his checklist of questions. By the end of the study, his data showed signs of four-to-six week cycles of emotional change. In some men the shift from an easygoing amiability into a period of tension and irritability came imperceptibly. Most of the mood swings were so gradual and moderate that the men themselves did not notice.

Until recently, only very dramatic alterations in mood, such as manic-depressive psychoses, have been considered worth recording. Today, however, it may be possible to look at the less perceptible slow rhythms of emotional transition. Some psychiatric hospitals, following the lead of the Institute of Living, Hartford, Connecticut, are beginning to analyze day-to-day staff notes on patients by computer, and here the data show periodic mood shifts that had passed unnoticed before. Perhaps the same techniques will be used to discover whether healthy people also undergo minor cyclic shifts in mood over periods of weeks or months. Computer analyses of rhythms have utilized an agreed-upon language, which specifies the various aspects of a cycle.

Basic Terms: Frequency, Phase, Period, and Amplitude

Any recurring event, whether it is mania, sleep, or a rise in body temperature, can be described as a cycle, and conceived as a circular process that starts from a point of origin and returns to it. The time it takes to complete the cycle is the interval known as the period, whether this be an hour, day, microsecond, or year. Period is represented by the Greek letter (τ): This period may be the time that elapses between the peaks of body temperature, or nightly bedtime, an event that is conveniently described as if it occurred by the completion of a 360-degree turn. The period of a day is about 24 hours.

The frequency of a rhythm is the reciprocal of the period: ($\frac{1}{\tau}$). Adults usually sleep for a few hours every 24-hour period, and so sleep occurs with a frequency of 1:24. This is the frequency at which body temperature and many of our physiological functions fluctuate.

Frequency and period are not sufficient to describe an individual's rhythmicity. Indeed, until now, most of medical research has concentrated upon the levels and the degree of fluctuation to be found in body temperature or adrenal steroids. The amount or extent of daily change in body temperature is greater in some people than in others. This is a difference in the amplitude (C) of the rhythm. One person may show a change of 2.1 degrees, while another changes by only 1.5 degrees.

Besides differences in amplitude, two individuals may not show their peaks and troughs of body temperature at exactly the same clock time. Thus, they may be more or less out of phase (Φ) with reference to some external marker. If one person usually works at night and another works by day, their body temperature rhythms are likely to be about 180 degrees out of phase, or roughly opposite. Phase describes the time location of some part of the 360-degree cycle, usually the peak or the trough of the cycle with reference to some external point, such as local time, or the beginning of sleep. However, one could also describe the peak phase of one body substance with reference to the peak of another internal cycle.

Most researchers presently identify the phase of a rhythm by its peak and trough.

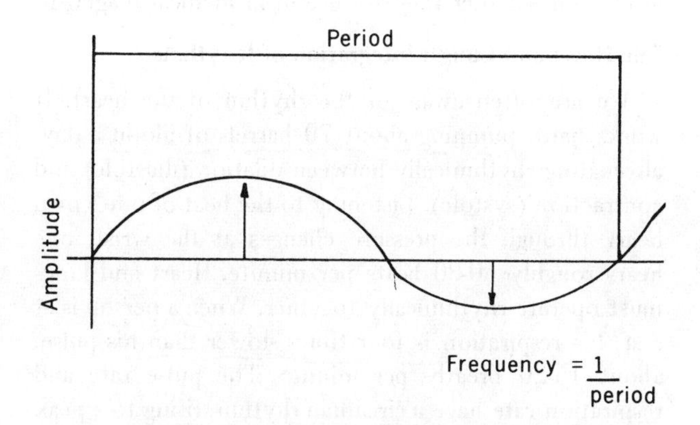

The spectrum of biological rhythms is enormous. Some are of exceedingly high frequency, such as certain brain-wave rhythms, and some have periods of a year or longer.

At the most basic level, metabolic rates and their limits are being explored. For instance, Dr. Britton Chance, at the University of Pennsylvania, has studied high-frequency cycles of activity in certain enzymes by measuring small changes of fluorescence that indicate the increase and decrease of essential biochemical intermediates in a single yeast cell.

The properties of neural pacemakers have been probed by quite different methods. Dr. Felix Strumwasser, at the California Institute of Technology, has tickled the scientific community with his elegant studies of nerve cells from a sea slug found in tidal pools. Although this creature moves at a snail's pace, it does have a locomotor cycle related to the alternation of light and dark, a rhythm that is comically apparent if one compresses 24 hours of film into a few minutes. Then the proverbially slow slug appears to be jumping and darting around during his activity period. When kept in the laboratory, away from the usual time cues of nature, its energy output shows a drift like that of isolated man toward a long day, a 26-hour cycle. Strumwasser, in search of the cells responsible for the activity rhythm, discovered one giant nerve cell with a pattern of self-sustained bursts that increased at dawn and at dusk. By way of microelectrodes implanted into the cell, Strumwasser began to tease it with tiny pulses of electricity. It was a first look at the neural coding of a cell that might be a pacemaker for the mollusk's circadian activity rhythm.

Microscopic rhythms, such as these, are being studied in efforts to understand the pacemakers that regulate other, more gross, continuous rhythms in the body, such as those of the heart. The heartbeat, in turn, is part of the overall rhythm of the body. One laboratory in Marburg, Germany, is attempting to discover how several of the body's major rhythms may intermesh at several levels and whether they are useful in medical diagnosis.

The Heart and Lungs: Integration of Rhythms

We are often aware of the rhythm of the heart. It works hard, pumping about 70 barrels of blood a day, alternating rhythmically between dilation (diastole) and contraction (systole). Listening to the beat of one's own heart through the pressure changes at the wrist, one hears roughly 60-80 beats per minute. Heart and lungs must operate rhythmically together. When a person is at rest, his respiration is four times slower than his pulse, about 15-20 breaths per minute. The pulse rate and respiration rate have a circadian rhythm, rising to a peak by day and falling to a low point during sleep.

Dr. Gunther Hildebrandt, at the Institute of Physiology in the University of Marburg, Germany, has studied the relationships among several body rhythms and concluded that a heartbeat to respiration ratio of 4 to 1 is a sign of health. In Germany, many people with cardiac or other chronic symptoms recuperate in health spas. Balneology is something more than immersion in natural spring waters or carbon dioxide baths. In a spa the patient follows a rigid regimen of baths, treatments, food, and sleep—perhaps forcing him to regain an overall circadian rhythm of body function. In studies of patients at one spa, Bad Orb, Hildebrandt saw that some patients had a heart-respiration ratio of 6 to 1 before carbon dioxide treatment, but a 4 to 1 ratio following treatment.

The circulatory system is a complicated, closed, hydraulic system. A number of German scientists have searched for relationships between heart rate and athletic activity for causes of arrhythmia. Dr. Hildebrandt has observed that people who do not feel well, but whose electrocardiograms appear "normal," often have exhibited abnormal ratios of pulse and respiration or other signs of internal discoordination during late sleep. One way in which internal rhythmicity has been studied is by stressing a person with heat, cold, or exercise and examining aspects of heart function.

At another level, the laboratory has been studying blood flow through muscle tissue. A rhythmic decline and rise in flow volume that suddenly doubled its wavelength was observed when a drop of adrenaline (epinephrine) was injected. Many aspects of long and short rhythms of the body are being assayed for new criteria in judging health and illness, new measures of physiological efficiency that may prove useful in diagnosis and treatment.

Some of these body rhythms are rarely noticed. For instance, skin temperature, particularly the temperature of the hands and feet, changes in a circadian rhythm. Ordinarily, during sleep, temperature is higher on the left side of the body, and during the day, slightly higher on the right. This alternation in skin temperature has gone as unremarked as the alternation we experience in breathing. Anyone who has suffered with a bad head cold will surmise that we do not breathe evenly through both nostrils at once: we alternate. The yogic masters have long known about these alternations. A normal man will breathe mainly through one nostril for three hours, while the tissues of the other are slightly engorged. Then, in a three-hour exchange he will breathe predominantly through the other nostril.

Alternating functions, from left to right, and a rise in body temperature that corresponds with a cooling of extremities, are relatively unstudied. By looking at these rhythms will we understand certain asymmetries in a body that is thought to be predominantly symmetrical? Do we, indeed, alternate in the use of our "duplicate" organs, our double-sided brain?

Quite a few scientists now suspect that the long and short biological cycles probably have an overall integration. Certain rhythms are clearly determined by a creature's physiological characteristics. For instance, heart rate depends upon an animal's size, his surface-to-volume ratio, and circulatory efficiency. But whether the rhythm is as rapid as a discharge of nerve cells, or as slow as the menstrual cycle, various types of rhythmic phenomena are bound to interact and to influence one another. They are also bound to be influenced by a person's habits, as Hippocrates remarked so long ago.

A healthy person lives in harmony with his environment. Only when he takes a jet trip across time zones and feels uncomfortable for a day or a week does he become aware of interior schedules. Now he is sleepy and hungry at "inappropriate times." What occurs inside might be likened to a major production line adjustment within the world's most complex factory. The analogy of a factory may indicate how important rhythmicity may be for our health.

No factory could produce an item as simple as a shoe unless each step were part of an orderly, well-timed sequence. Parts must be ready before they are needed on the assembly line, yet they cannot be overstocked so as to burden storage facilities. Analogously, man's infinitely more complicated system must rely upon an orderly sequencing of millions of events so that a person functions smoothly. When production is well meshed, the clockwork is scarcely perceptible, but when it runs amuck, the result may be illness.

In 1960, Dr. Erwin Bünning, of the University of Tübingen, Germany, opened a symposium in Cold Spring Harbor by suggesting that there must be a rhythmic harmony between cell behavior and the whole organism in health, probably organized around the unit of a day. When a person travels across several time zones, his internal rhythms adjust to the new local time, each at its own rate. Therefore, some functions of the body have adjusted within three days, but others have not adjusted at the end of a week. This may mean internal cacophony. As Bünning stated, "a glandular tissue may be in the phase of hormone production while another organ, being in another phase, cannot make use of the hormone; or an enzyme may be very active in a particular tissue at a time when its substrate is not available. Every transatlantic air traveler knows the physiological discomforts that may arise from such a lack of cooperation. More important, however, are graver illnesses, even cancer, which may be shown to be evidence for this. . . ."

At the level of the cell, disordered rhythms in division and multiplication may be symptomatic of malignancy. Studies by Dr. Franz Halberg and his many associates have begun to find a contrast in the rhythms of healthy and cancerous tissue. Halberg has also shown that disordered rhythms of cell behavior are reflected, on occasion, in abnormality of gross rhythms. The daily body temperature rhythms of certain cancer patients have proven to be extraordinarily disordered, obviously differing from those of other sick patients.

As electronic techniques have been developed that enable us to trace a variety of human physiological functions for hours at a time, researchers have become aware of rhythmicity as an aspect of "normalcy." In sleep laboratories, where healthy adults exhibit predictable rhythms of cyclic change in nightly sleep, a deviation in rhythm (from fever, mental illness, or drugs) is often the first warning signal of the symptoms to follow. Many abnormal rhythms are known by the names of illness. A common name for a disordered sleep rhythm, for instance, is insomnia.

It is abundantly clear that healthy living things are not only internally rhythmic; they are also synchronized with their environment. Is environmental change the source of a creature's circadian rhythms? Some have suggested that they are genetic, reflecting an inbuilt rhythmicity that evolved some 180 million years ago as an adaptation to this turning earth. Others think they are imprinted upon the organism early in life by the external periodicities of environment and parents. Still other theorists speculate that rhythms are caused by geophysical events to which living things constantly respond anew. We could, after all, be very sensitive cosmic receivers, acting in accord with rhythmic changes not only in light and temperature but in cosmic rays, barometric pressure, ionization, magnetic fields. The most generally accepted theory suggests that inherited oscillators allow creatures to respond to recurrent geophysical stimuli.

Research on biological rhythms began in earnest in 18th century botany. Leaves and flowers were observed to show circadian movement, which continued even when plants were isolated from time cues, as in caves. The suspicion that such a rhythm might be inherited was stated and bolstered by the work of Dr. Erwin Bünning. Experiments with plants in a constant laboratory environment showed that their circadian rhythm was not precisely 24 hours, yet never more than about three hours shorter or longer. Moreover, each plant species showed a typical circadian period in isolation. One plant might show a 24.2-hour cycle while another would have a 25.6-hour rhythm. Hybrids were found to have intermediate periods, strongly suggesting that the rhythm had a genetic basis.

As zoologists entered the field, experiments with unicellular organisms, insects, and mammals began to delineate the complexity of endogenous rhythms. Dr. Colin Pittendrigh, now at Stanford University, has experimented with a variety of organisms, from flagellates and fruit flies to mammals. He concluded that the circadian rhythm must originate in some basic molecular mechanism—a clocklike mechanism with a periodicity of 23 to 25 hours that shows a sustained oscillation in the absence of external synchronizers. Dr. Pittendrigh feels that this inheritance is an evolutionary adaptation to a strongly rhythmic planet, in which sun and moon create periodicities of 24 to 25 hours.

As physiologists experimented with birds, mammals, and man, the importance of environmental influences suggested that circadian rhythms might be to some extent acquired. Living things seem to use time cues to orient themselves. Dr. Jürgen Aschoff, and his associates, have been working on the problem of time cues at the Max-Planck Institut für Verhaltensphysiologie in Erling-Andechs, Germany. They have shown, with birds and man, that light is a most prominent synchronizer in the hierarchy of time cues, but that a hungry animal would adapt his activity and rest to the timetable of available food.

For rodents, as Dr. Franz Halberg amply demonstrated, light and darkness are the most important synchronizers of the rhythm of activity, and of physiological rhythms, such as the level of adrenal hormones. If the light schedule were shifted a few hours, the adrenal rhythm would follow, adjusting to the new schedule within about four days. If the light schedule were inverted, the adrenal rhythm might need nine days to adjust. If mice were blinded, however, their adrenal hormone rhythms would begin to deviate from the precise 24-hour day, so that they showed rhythms with a period between 23 and 25 hours.

Developmental studies with mammals have suggested that circadian rhythms might be acquired. Both the social and physical environment may play a major role in the evolution of these rhythms. Human infants, for instance, only slowly develop circadian rhythms of sleep and waking, of urinary excretion, and other functions. Of course, the fact that patterns do not appear at the start does not necessarily mean they are not inherited. Certain genetic illnesses, such as Huntington's Chorea, are inherited but do not appear until late in life. There are other evidences suggesting that inherited propensities must be very much influenced by environment.

Dr. Mary Lobban, working at the National Institute for Medical Research in Hampstead, London, has found that urinary excretion rhythms in Arctic Eskimos differ from those of people in temperate zones. Eskimo excretion patterns, nonetheless, exhibited some circadian rhythmicity despite the lack of a pronounced day-night difference during most of the year.

It is probably fair to say that many of the scientists now studying biological rhythms believe that inherited mechanisms give earth organisms the propensity to synchronize themselves with certain periodicities in the environment. A roughly 24-hour oscillator within living cells may help the creature to survive by acting in tune with its changing environment.

In attempting to understand the rhythmicity of life forms, there are some scientists who propose that creatures could be a kind of cosmic receiving station, responding to a continuous inpouring of information from rhythmic geophysical changes of the environment. Dr. Frank Brown, of Northwestern University in Evanston, Illinois, has been the most active proponent of the cosmic theory, suggesting that an independent internal timing system for rhythms is not necessary to life since the environment is always generating rhythmic signals. These include variations in terrestrial magnetism, electric fields, and background radiation and other signal sources, such as gravitation from which no creature on earth can be completely isolated. Only experiments in space can isolate life from these cycles. If animals and plants comprise a multiplicity of very sensitive receivers, they could utilize this continuous influx of information from external pacemaker signals.

Using a variety of life forms in which he has ascertained rhythmic changes, Dr. Brown has sought geophysical periodicities that might account for them. His point of view is worth expression for it underscores what we do not yet know about biological rhythms and suggests that man may be a more diversely sensitive creature, more subject to the invisible influences of the surrounding universe than Western science usually considers him.

The Cosmic Receiver

Many forms of earth life exhibit both solar and lunar rhythms. Fiddler crabs, for instance, show a rhythm of color change, darkening and lightening each day in exactly solar 24-hour rhythm, even in isolation. On the other hand, they show an activity cycle that is the length of a lunar day, 24.8 hours.

In studying daily cycles of metabolism in crabs, Dr. Brown fortuitously found they were the mirror image of average daily fluctuations in cosmic radiation. Dr. Brown's contact with physicists in the mid-1950's was fortunate, indeed. At the time, cosmic radiation cycles underwent unusual modifications of form, including cycle inversion. The oxygen consumption cycles of potatoes, carrots, shellfish, and rats also varied with the cosmic ray changes so as to suggest that living organisms might be responsive to cosmic radiation and, perhaps, also to geomagnetic influences.

Actually, our atmosphere, and especially its magnetic field, protects us from primary cosmic rays. These must enter the ionosphere, one of earth's protective layers, which is composed of electrically charged particles (ions) of gaseous matter. The ionosphere is held close to earth by the geomagnetic field, shielding us from harmful rays from outer space. However, the ionosphere is not steady. It attenuates at night as it draws away from earth and increases in density by day, when it draws close to earth.

In his laboratory, Dr. Brown interfered with the normal activity of cosmic rays by manipulating lead shields. He created a cascade, or serial flow, of the rays so that fiddler crabs could be exposed to intensified cosmic ray showers every other 12 hours. In doing so, he saw that he amplified a normal 24-hour solar rhythm—the darkening and lightening of their surface color, which occurs every solar day—24 hours.

The amount of primary cosmic rays entering earth's atmosphere is the inverse of the geomagnetic strength. Since geomagnetic strength is a factor that also fluctuates, Brown wondered if living creatures were sensitive to earth's magnetic field as well. To investigate geomagnetic influences, Brown performed experiments with planaria (worms) and mud snails. He used as his experimental focus the direction in which the creature veered. Worms and snails tend to veer in given compass directions at certain times of day, month, and year. By simply rotating a bar-magnet beneath the experimental grid on which the creatures rested, he could change their tendency to take a certain compass orientation. They reoriented themselves according to very weak changes in magnetic field. Using essentially the same kind of test, he showed that these organisms were very responsive to the natural ambient electrostatic fields in the atmosphere. When tested with weak gamma fields, they also responded both to the strength and direction. Clearly, if organisms could sense these geophysical properties, there was no such thing as placing a creature in "constant" conditions. Even in the deepest, most shielded underground chamber, such factors as tides of neutrinos, or rhythms of geomagnetism, would continue to convey some of the periodicity of the cosmos.

Geomagnetism, Brown conjectured, might be an important kind of compass cue. Other cues might be related to earth's electrostatic and electromagnetic fields. At any moment in time, the geophysical field varies with geographic direction, and for any geographic direction or location, there is a continuous variation of earth's field in time. The same cues that might offer directional information could also be used to sense time. Worms, for instance, were given an easterly light source, to which they repeatedly responded with a characteristic veering, when they were directed north. Even later when the worm had been turned south with a light source from the west, it appeared to recall the formerly easterly light source. Memory for light appeared to be associated with the directional magnetic fields at the time. If one assumed great geophysical sensitivity, as Brown reports, one could also account for distortions in daily cycles that occur in supposedly isolated subjects in supposedly constant conditions, distortions that Brown has correlated with weather changes in some instances.

Brown believes that all living organisms gain information about time and orientation in space from weak electromagnetic fields in the environment. In his terms, these fields are used by the central nervous system and provide a kind of medium enabling the bioelectric activity of the brain. Presumably, the full use of the atmospheric media could give man unusual "sense" information, including abilities to detect weather changes. Animal lore is rich in this respect. Fiddler crabs have been reported to disappear into inland burrows two days before an oncoming hurricane. Foresters in the Pacific Northwest sometimes predict snow by the behavior of the elk, which begin to gather in the shelter of trees two to three days before a blizzard.

The question remains—Is man sensitive to magnetic field changes? No one has produced experimental demonstration that man responds to magnetic changes comparable to those of the environment. Some people have suggested that the discomfort of crossing time zones may be generated by crossing bands of magnetic force. The earth is a huge magnet, layered from north to south like an onion, with bands of magnetic force, whose lines of changing field strength lie between the North and South Poles.

During the last few years there has been some evidence that human beings may be sensitive to very strong electric fields. In one of the Max-Planck laboratories near Munich, a research team led by Drs. Jürgen Aschoff and Rutger Wever, has been studying human volunteers in underground bunkers, one of which is heavily shielded against ordinary electromagnetic fields. By introducing artificial electric fields into the bunker, the experimenters influenced the circadian rhythms of volunteers, shortening the period of their cycles. A number of experiments, described in another section, showed that slight effects could be observed after using very strong fields, but these do not prove that human beings would respond significantly to the corresponding natural fields. Although electromagnetic sensitivity is not one of the receptor modalities man customarily enumerates among his senses, it may be that certain individuals are unusually sensitive to exceedingly small changes in magnetic field strength. In attempting to explain the mysterious capability of the "dowser," the person who can divine underground water, the French physicist, Dr. Y. Rocard, at the Universite de Paris, Paris, followed dowsers on their sensing missions. They would hold their arms straight and taut, balancing before them a long hickory stick. Rocard followed with a magnetometer, and found that they were responding to tiny (3-5 milligauss) changes in the earth's magnetic field strength. Since they held their arms with a certain degree of muscle tension, he reasoned that the small changes affected bioelectric transmission in their arms. He then planted electric coils underground to create changes in magnetic strength similar to those of nature. By reinforcing people as they went over the test ground, he was able to condition the ability to detect .3-1.0 milligauss changes in magnetic field.

Dr. Brown suggests that variations in earth's magnetic field may form a space-and-time-grid for living creatures, and that even man may unconsciously use a variety of subtle cosmic cues rather than depending upon inherited clocks for his physiological rhythms. Perhaps a major value of this point of view is that it suggests the possibility of overlooked inputs into man's nervous system, inputs relevant to understanding human behavior.

Sunspots and Behavior

Periodically, in mental hospital wards, aggressive patients show intense surges of activity: individuals become hostile, excitable, even violent. Such outbursts occur only sporadically but are striking enough to have invited study. At Douglas Hospital in Montreal, continuous, round-the-clock observation studies of patients over periods of several months did, indeed, show a picture of such periodic outbursts. Correlations between increased aggression and staff on duty, changes in menu, medication, or visiting days were too weak to explain the group behavior. Barometric pressure, temperature, humidity, and other environmental factors were juxtaposed against the hospital calendar of aggressive behavior. When no explanation could be found, Dr. Heinz Lehmann compared his hospital data against data from the U.S. Space Disturbance Forecast Center in Boulder, Colorado. There appeared to be a correlation between solar flare activity (sun spots), geomagnetic disturbances, and excitement on the ward. It seemed unlikely but the study continues. Since sun flares are bursts of gaseous material, high energy particles that influence the ionosphere, causing changes in magnetic fields on earth, a relationship is not impossible. Sun storms sometimes cause a noticeable deflection in a compass needle. Perhaps, since the brain is at least as sensitive as a fine compass, it also responds to large magnetic disturbances.

We know that rats are sensitive to X-rays, which act as aversive stimulation. The human brain may also respond to such inputs. The possibility of an expanded "sensory" range has been raised by studies of biological rhythms.

Time Structure

Timing in the heart, in muscle contractions, and nerve cell transactions has been an issue for research for some time. Until recently, however, nobody considered the possibility that the entire body and brain might depend upon an integrated time structure. The human being is often treated as if he were—or should be—a constant system with homeostatic balance, capable of great flexibility in dealing with exigencies outside.

There is growing evidence that all mammals are predictably changing from one hour of the day to another. Until the last decade, it was possible to imagine that these changes were random fluctuations. Now, when a fluctuation is called "random," it often means that its variation in time has not been carefully studied. Over the years experimenters have searched the anatomy for the origin of circadian rhythmicity. They have removed stomachs, yet the animal's activity rhythms remained. They have removed adrenals, but the circadian rhythms persisted. Mice born without pituitary glands are rhythmic, as are rodents from which the thyroid has been removed. Isolated organs also beat with circadian rhythmicity. Dr. Erwin Bunning once removed sections of hamster intestine, and their contractions sustained a circadian rhythm for three days. Dr. G. Edgar Folk, Jr., and his associates took isolated hearts and adrenals from hamsters, and the hearts continued to beat with a circadian rhythm and the adrenals continued to secrete steroids in a circadian rhythm.

Since light and darkness perform such an important role in synchronizing mammal activity rhythms, animals have been blinded, or left in constant light or darkness; but their rhythms of activity have deviated only a little from the precise 24-hour schedule. Indeed, as Dr. D. H. Thor and his associates have shown, when rats are given control over their own lighting they choose the stimulation of darkness and light in a circadian rhythm. Attempts to deflect animals from a circadian activity, or temperature pattern, have involved brain lesions, stress with drugs, scheduled lighting changes, changes in feeding, and physical injury. These many experiments have reinforced the notion that the circadian rhythm is persistent and that different body systems may exhibit this rhythmicity independently of each other. Perhaps one definition of health is a state in which these rhythms are integrated in certain phase relationships. Surely, as the experiments of Dr. Franz Halberg and his associates demonstrate, timing is an inescapable part of our structure, and essential to well-being.

If human beings and animals change as they follow multitudinous cycles each day and night, subtle diagnostic tests (psychiatric, endocrine, etc.) will evoke different responses at different points on the person's daily cycle. As psychiatric experiments now demonstrate, the relation between rhythmicity and mental health is not coincidental. Experiments at the Institute of Living indicate that biological time influences learning, memory, and the rate at which an animal can unlearn fear. Stress has been followed by psychotic and neurotic behavioral reactions, accompanied by characteristic changes in circadian physiological rhythms.

The study of biological rhythms offers a first step toward deciphering the time structure of other living creatures and of ourselves. Temporal structure, like the structure of matter, must be an important aspect of the anatomical and functional ability of organisms. Without temporal structure, indeed, we could not exist.

"For though the outward man perish,
yet the inward man is renewed day by day.

Corinthians II

Chapter II. Sleep and Dreaming

In consonance with the alternation of light and darkness on earth, living beings alternate between rest and activity. From the elephant to the butterfly, the primitive opossum to the chimpanzee, earth creatures all suspend their activity for a prolonged rest once every 24 hours. If they are nocturnal, like the owl, mouse, or rat, they sleep during the hours of light and forage for food in protective darkness. Sometimes, predators or scarcity force an animal to shift his hours of activity, but the basic cycle does not change. Animals still alternate between rest and activity in a period of about 24 hours.

This recurrent experience is the most tangible of circadian rhythms. Whether it is impressed upon us by habit or ancient heredity, the routine of nightly sleep is so strongly a part of man's custom that it is difficult to alter. Indeed, when a person abruptly and spontaneously exchanges night for day, waking by night and sleeping by day, it can be a sign of serious illness, such as encephalitis, which often damages important loci in the brain. Without sleep, man performs poorly at his work, cannot sustain concentration, becomes irritable, and eventually begins to act deranged. He not only needs repose, but he must have it every 24 hours.

Physiologists have asked themselves whether this unit of a "day" is really immutable. Dr. Curt P. Richter, of Johns Hopkins University, Baltimore, Maryland, tried without success to alter the 24-hour cycle of rat activity by assaulting animals with shock or drugs, exposing them to prolonged danger, by freezing them, stopping the heart, removing certain portions of the brain. Even after blinding, animals showed a cycle that differed little from 24 hours in period. The circadian rhythm of activity and rest seemed profoundly inherent in the animal.

Dr. Nathaniel Kleitman, while at the University of Chicago, used himself as a subject. He tried to live a 28-hour day during a month's sojourn in the isolation of Mammoth Cave, Kentucky. A student, younger by some 20 years, did manage to adapt, but Kleitman found that he was always out of kilter with their artificial "day" and "night." He was sleepy when he should have been alert and not hungry at meal times.

Sleep and waking help act as synchronizers for many internal functions. When a person lives on a regular routine of daytime work and nighttime sleep, his physiological functions follow a related rhythm. Temperature rises and falls once every 24 hours, as do plasma concentrations of certain hormones in the brain and blood and depositions of an important carbohydrate in the liver. Some functions or chemical concentrations reach their peak in the morning while others reach their peak in afternoon or night, maintaining particular phase relationships. When certain phase relationships are disturbed a person may notice it. For instance, instead of urinating mostly in the morning, he has to get up out of bed in the middle of the night. Or he may feel tired precisely at hours he usually does his best work. To some extent we are analogous to exceeding complex watches, with millions of intermeshing gears inside. Like watches, we are capable of being reset by external means. We do, indeed, resynchronize ourselves when we travel, by shifting our hours of sleep and waking in accordance with local time.

Just as sleep forms the "dark" side of our circadian rhythm of behavior, sleep itself is a rhythmic phenomenon. Nightlong sleep undulates in repeated cycles through levels of consciousness, physiological, and neural change. Its striking rhythmicity was first noticed in infants, who were observed to undergo a quiescent then active state of sleep with considerable eye movement. Dr. Eugene Aserinsky, as a graduate student with Nathaniel Kleitman at the University of Chicago, first noted these eye movements in babies and then began watching adults sleep.

Aserinsky and Kleitman began to monitor nightlong sleep, using brain-wave tracings, better known as EEGs.

Small sensors (electrodes) on the face and skull transmitted changes in bioelectrical polarity from facial muscles and from the brain to the amplifier system known as the electroencephalograph machine. The pens of this EEG machine move up and down with changes from positive to negative charge, across continuously moving graph paper, so that the resulting record appears in the form of waves. During deep, slow-wave sleep, the changes in polarity occur slowly, and the voltage may be as high as 300 microvolts. As Aserinsky and Kleitman monitored sleep, they found recurrent episodes of rapid-eye-movements coinciding with an EEG pattern of very rapid irregular changes of very low voltage. They had a hunch that their subjects might be dreaming. Systematically, they awakened volunteers from different phases of sleep; out of the rapid-eye-movement (REM) state, people almost always reported dreaming. More rarely did they recall dreams when awakened from other sleep. A decade of subsequent research was stimulated by this finding. Today, we know that dreaming, or some kind of describable mentation, can occur in every stage of sleep. But REM sleep has remained a focus of interest, in part, because it provides the greatest recall and certainly the most intense emotionality.

Dr. William C. Dement, now at Stanford University Medical School, pursued the discovery of REM sleep, monitoring nightly slumber in volunteers for months. REM sleep seemed a bridge to the dreams and thereby the unconscious of humankind, and Dement began to see that episodes of REM activity occurred in everyone, every night, lasting anywhere from ten minutes to an hour. Moreover, these REM periods recurred regularly, about every 90 to 100 minutes. When Dement looked at the sleep of cats he found EEG rhythms resembling those of man. Perhaps, the rhythmic occurrence of REM sleep indicates a universal biological rhythm, at least among all mammals.

Other scientists soon began to examine sleep in human volunteers and many animal species, and found that the REM state was accompanied by irregular pulse and breathing, a complete slackening of skeletal muscles, and changes in brain temperature. People and animals can be difficult to awaken from this state, like creatures lost in concentration, yet the brain waves resemble those of lightest sleep. In describing animals, indeed, this stage is aptly called paradoxical sleep.

The Nightly Rhythm

Sleep research over the last 15 years has shown us that we spend each night of our lives in an unremembered voyage, shifting, and bobbing on a recurrent tide of colorful dreams and strange states of being. Each night, in much the same routine, we plunge into these levels of consciousness in a predictable manner. We fall asleep, loosening our hold on the world, eyes beginning to roll, temperature dropping, muscles relaxing, breathing growing regular. In lightest sleep (Stage I), the EEGs almost resemble the swift pinched scrawl of waking—EEGs showing slightly new bursts of activity as sleep deepens very slightly (Stage II). Then the EEG slows into larger, even waves (Stage III and IV) as deepening sleep blankets us profoundly. About ten minutes after falling asleep, a person may be in light sleep with images and reveries that could be mistaken for waking fantasy. After roughly 40 minutes, he is almost out of reach in the bottomless depths of Stage IV sleep. By the end of about 90 minutes, he has returned up through lighter stages to the surface—Stage I. There, if he is healthy and undrugged, he will not awaken, but his eyes will begin darting as if tracking some action beneath the closed lids; visual portions of the brain will emit volleys of signals in short bursts. Breathing and pulse become irregular, muscles flaccid. If awakened, one almost always recalls dreaming. This is also the time when a person with peptic ulcers shows abnormal nighttime secretions of gastric acid. All males, from infancy to old age, have penile erections around or during REM periods, perhaps a sign of activation within drive centers in the brain.

Four or five times a night, a normal person goes through a cycle lasting 85 to 110 minutes, in which he descends to the depths of slow-wave sleep and rises again to REM sleep. Slowly, over the night, the quality of sleep changes. Toward morning, sleep grows lighter, body temperature begins to rise, and certain hormone levels increase as the body becomes ready for waking. Before this rhythm, itself, attracted sleep researchers, they attempted to discover the function of the different stages of sleep. REM sleep and Stage IV were dramatically different and invited the most exploration.

REM Sleep and Stage IV

Motivated by the Freudian concept that dreaming might be a safety valve for sanity, Drs. William Dement and Charles Fisher, at Mt. Sinai Hospital in New York, began to deprive volunteers of nightly REM sleep. They awakened them each time they began to show the eye movements and EEGs of REM sleep. All other sleep was permitted. As the night wore on, volunteers dropped into the REM state more and more often as if they had built up some "pressure" or need for it. When finally allowed a night of uninterrupted sleep, they fell, with unusual alacrity, into REM sleep, and seemed to "compensate" for their previous loss. Some showed signs of emotional disturbance by day. Interruptions at other times of night produced no such reactions. Initially, this suggested there might be a nightly minimal requirement for REM sleep (and dreaming), a quota which would be about 20 percent of an adult's nightly sleep time.

Very deep, quiescent, oblivious Stage IV sleep normally occurs early in the night, by contrast with REM sleep which is most prominent toward morning. It is often called Delta sleep (one name for high amplitude, slowly changing EEGs) and is a stage in which sleepwalking occurs, as well as most episodes of bedwetting and nocturnal terrors. When deprived of sleep altogether, a person first makes up for loss of Stage IV sleep by spending long periods in that bottomless state, and should he cut his normal sleep to about two hours, he will spend much of that time in Stage IV. Drs. Wilse B. Webb, Robert Williams, and Harmon Agnew, at the University of Florida, have deprived volunteers of Stage IV sleep by nudging them with electric shocks into lighter sleep. After a few nights, these subjects complained of bodily malaise, apathy, and depression.

Stage IV sleep can be enhanced by daily exercise. This deep sleep state appears to play a physiological and restorative role quite essential to well-being, but very different from that of REM sleep.

Dr. Dement and his coworkers, at Stanford University, have reported that depriving cats and rats of their equivalent of REM has led to abnormal animal behavior. Deprived tomcats would aggressively stalk others, mounting dead or anesthetized animals—an unheard of breach of normal cat behavior. These cats would also exhibit strangely voracious and uncontrolled hunger. The studies suggested that normally REM sleep may give periodic discharge to mounting excitement in drive centers. Without such a release, the animal's brain-wave responses to repeated clicks indicated abnormal excitation. Left alone, they compensated for some of their deprived paradoxical sleep.

Rats showed their abnormal response to deprivation by falling into convulsions after minor electroshock. After shock, they did not compensate much in their sleep for lost REM time. Had shock discharged some brain excitability in the manner of paradoxical sleep? Perhaps, this will help explain the mechanism of electroshock therapy with some mentally-ill patients.

REM and Learning

Recently, REM sleep has been viewed as a cycle of brain activity that is important to learning, memory, and adaptation. Dr. Johann Stoyva and associates, at the University of Colorado School of Medicine, saw increases in REM sleep among subjects adapting to distortion goggles they wore during the day. Dr. Ramon Greenberg, and his associates at the Boston V.A. Hospital, saw increases in REM sleep among aphasics who were improving in speech with coaching, but found no increases among patients who failed to improve. Dr. Jimmy Scott observed increases in REM sleep in subjects when they were required to adapt to a slightly new experimental situation such as a day's isolation, or even a half day's isolation. Dr. Howard Roffwarg and his coworkers, at Montefiore Hospital in New York, have found that people wearing red goggles all day reported "red tinted" dreams when awakened from REM sleep.

On the whole, when people have been deprived of REM sleep by repeated awakenings they have grown irritable. Some of them become anxious, and, occasionally, a subject has become truly disturbed. By contrast, deprivation of Stage IV sleep seems to have left subjects with a feeling of lethargy, of depression, and apathy. Perhaps, this is because REM and Stage IV sleep have different hormonal accompaniments.

Hormonal Accompaniments of REM and/or Stage IV

Dr. Elliot Weitzman and his coworkers, at Albert Einstein College of Medicine and Montefiore Hospital, New York, have found that the nightly output of certain hormones from the adrenal glands occurred in intermittent bursts during the latter third of the night's sleep, when REM sleep is occurring at its maximum. Such hormones are part of a class of compounds known as steroids which have a particular kind of carbon-ring system. The hormones in which Dr. Weitzman noticed changes in output were from the outer layer or cortex of the adrenals, a group known as the 17-hydroxycorticosteroids and commonly abbreviated 17-OHCS. Dr. Weitzman and his coworkers developed a method of putting a small catheter into the arm of a volunteer, so that blood samples could be obtained at short intervals throughout sleep without disturbing the subject.

Recently, a number of other researchers have refined a similar catheter system to sample blood for levels of growth hormone at 20-minute intervals. Growth hormone is released from the pituitary gland and stimulates bone growth. Its action is inhibited by sex hormones. Recently, this hormone has been associated with deep, slow-wave sleep which is far more predominant in the nights of young children than in adolescents or adults.

Drs. Y. Takahashi, D. N. Kipnis, and W. H. Daughaday, at the Washington University School of Medicine in St. Louis, have studied eight young adults for some 38 nights. Forty minutes after the person fell asleep, roughly the time he should be in Stage IV sleep, there was a peak of growth hormone in the blood. The level remained high in the first three-and-a-half hours of the night, then dropped to a base line at the time of waking. This corresponds to the distribution of slow-wave sleep. On some nights the subjects were awakened after they had slept for only a couple of hours; then after two to three hours awake, they were permitted to return to sleep. In each of these subjects, there were two peaks of growth hormone coming early in each separate period of sleep.

Others have looked at hormone release in normal nightly sleep, and also after volunteers had reversed their sleep time from night to day. Drs. J. F. Sassin and D. C. Parker, and their colleagues, at the U.S. Navy Medical Neuropsychiatric Research Unit in San Diego, found that the peaks of growth hormone seemed to coincide with slow-wave sleep, adapting to its new schedule rather than occurring at a particular time of day. Hormone levels seemed to be associated with the rhythm of sleep, itself, rather than appearing in the blood in an independent circadian rhythm, a finding corroborated by Dr. Weitzman and his associates.

The rhythmic alternation of neural and biochemical activity in sleep are only beginning to find medical applications. It may be possible to diagnose and treat the quality of sleep and sleep disorder from a hormonal point of view.

Good and Poor Sleep

Until recently, the quality of sleep was assessed largely by subjective description or a count of body movements. In the early 1960's, Dr. Lawrence J. Monroe, while a graduate student at the University of Chicago, compared a group of self-proclaimed "good sleepers" with a group of people who admitted they slept poorly. Each man slept in the laboratory for two nights, under identical conditions, wearing EEG electrodes on his head, a temperature probe rectally, and sensors to detect pulse and blood pressure.

On physiological measures, the poor sleepers differed noticeably from the good sleepers. They distributed their deep sleep less evenly over the night. They had higher rectal temperatures before and during sleep, and their temperatures did not drop as low as did those of the good sleepers. Among other physiological signs of "arousal," they had phasic constrictions in the peripheral blood vessels of their fingers, showed more body movements, and higher pulse rates in sleep than did "good" sleepers. Judged by their responses to psychological tests, the "poor sleepers" were far more anxious and more prone to somatic distresses and neuroticism than were the "good sleepers."

Body temperature curves suggested that the "poor" sleepers might be people who were out of synchrony with the 24-hour day. The temperatures of "good" sleepers began to show a marked decline at bedtime, falling steeply to a low point in the wee hours of the morning. About an hour or two before they awakened in the morning, their temperatures began to climb. Their temperatures were up to "normal" by the time they got up. By contrast, the temperatures of the "poor" sleepers declined less and were still declining when they arose in the morning. Perhaps, one of the characteristics of the "poor sleeper" is that his physiological rhythms are not in synchrony with the local clock time. He may be perpetually wishing he could go to bed later and could rise later in the morning than is socially convenient, for his body time actually may lag behind clock time, showing a longer period than 24 hours.

Insomnia and Depression

Various disturbances of the nightly sleep rhythm are now used to identify more serious forms of insomnia. One of the classic tortures of insomnia is an inability to stay asleep. Some people drop off easily, only to awaken in the dark, cold hours before dawn, unable to sleep again. This "early morning awakening" is extremely common in severe depression.

In 1961, Dr. William W. K. Zung and his associates, at Duke University Medical School, thought this form of insomnia might be due to an oversensitivity to sounds during early morning sleep. When they played sounds to normal and depressed patients who slept all night in the sleep laboratory, they did find that the depressed people were more easily affected by the noises. In addition to Dr. Zung, a number of researchers were then taking a close look at the poor sleep of depressed patients (Drs. S. Gresham, H. Agnew, and R. Williams at the University of Florida; Dr. Frederick Snyder at NIMH; and Dr. David Hawkins at the University of Virginia). Although the patients were not a homogeneous group and consequently exhibited a range of sleep troubles, several findings were common to all of the studies. These patients lacked the regularity of the "normal" sleep rhythm. Many of them showed little or no deep Stage IV sleep, and others spent abnormally little time or sometimes abnormally much time in the REM state. Most of them complained that they found their sleep unrefreshing, often punctured by waking, and easily disturbed by the slightest noise.

The first researcher to examine disturbed sleep for its rhythm was Dr. Zung. In 1963, he began to compare the nightlong sleep records of normal and depressed persons, evaluating them by a mathematical method known as Markov chain analysis. For each EEG stage of sleep, he and his coworkers derived the probability with which they could predict the person's shift to the next stage of sleep. Would a sleeping person in Stage II—light sleep—be more likely to drift upward into REM sleep, or downward into deep Stage III? The probability was derived empirically by computer, after evaluating many successive shifts from one stage to another on many nightlong EEG records. Zung and his colleagues thus derived conditional probabilities allowing them to predict a normal person's passage through the stages of a night's sleep. Normal sleep showed consistent organization and rhythmicity. By contrast with normal people, the sleep of some very depressed patients showed EEG patterns so disorganized they could hardly be interpreted.

According to most of the nightlong studies in major sleep laboratories, depressed patients seem to shift quickly from one stage to another, often awakening from REM sleep. A normal person takes 70 to 90 minutes after falling asleep before he reaches REM sleep, but a very sick patient may fall into REM stage shortly after closing his eyes, in the manner of someone who has been forcibly deprived. Whereas the normal person generally expends most of his deep Stage IV sleep in the first half of the night, a depressed person may sleep fitfully, lie awake for a few hours, and then lapse into a long period of Stage IV slumber abnormally placed in the early morning hours. The varieties of insomnia suffered by these people might be described as a disorganization of the pattern of nightly sleep.

One aspect of this fragmentation was described by Dr. Zung through all-night studies of patients who slept in the laboratory just after entering the hospital and again just before discharge. At first they showed abnormally short cycles of sleep. Instead of undulating gently through four or five cycles of about 90 to 100 minutes, which is the normal course in a night, they had eight or ten cycles within a short night of sleep. When patients began to get well, sleep improved, and they no longer shifted so rapidly from one EEG stage to another.

Although the relationship between sleep cycles and mental health is unknown, a clear relationship can be seen between health and rhythmicity in depressed patients. Moreover, the drugs that seem to lift a depression also improve the rhythm of sleep by lengthening the cycle, as Dr. Zung has shown with compounds known as tricyclic antidepressants. Typically, one woman who felt miserable—an insomniac who slept fitfully upon entry to the hospital—showed normal sleep patterns after four weeks on the tricyclic compound, desipramine. Zung tested the same drug on normal volunteers between 20 and 40 years old. It increased the amount of deep Stage IV sleep, slightly reduced the amount of REM sleep, and lengthened the cycle from 90 minutes to about 120 minutes. The potency for protracting the sleep cycle, thus normalizing deviant rhythms, may be an important property of this drug. Other drugs, such as phenelzine, have been found to have the opposite effect: they shorten the sleep cycle. Zung's work suggests that one way to evaluate the probable efficacy of a drug may be to look at its effect upon the length of the sleep cycle.

Phase Shift and Depression

A depressed person who shows disordered rhythms in sleep may reflect the disorder in his biochemistry as well. Ordinarily, a person excretes water in a rhythmic fashion, more by day than by night, more in early morning than evening—if he lives an ordinary schedule of sleeping at night and actively going about his business by day. The urine contains such elements as sodium, potassium, and calcium, as well as metabolites of steroid hormones produced by the adrenal cortex. The relative concentrations of these also vary in a rhythmic orderliness around the 24 hours.

Recently, a study of the urinary steroids of depressed patients was analyzed by Dr. Franz Halberg. Using a method of mathematical analysis with a display of data known as a cosinor, he showed that normal people had their peak excretion at a different time of day than did the patients. More precisely, when plotted over 360 degrees, the peak phases of the normal and depressed patients were 100 degrees apart. Following treatment, however, the depressed patients showed a pronounced phase shift so that their peak excretion of 17-hydroxycorticosteroids and 17-ketosteroids overlapped in time with the peaks of normal persons.

In attempts to find possible hormonal differences between normal and depressed people, most researchers have been taking serial blood samples at longer than hourly intervals. Recent work by Dr. Howard Roffwarg and his associates, at Montefiore Hospital, New York, suggests that blood cortisol rhythms defined by infrequent samples may be misleading. Cortisol levels in the blood rise and fall precipitously in episodic bursts. By taking blood from an indwelling catheter at 15- to 20-minute intervals throughout the night, the Montefiore group looked at the cortisol rhythm in depressed patients before and after treatment. In general, their circadian cortisol rhythms resembled those of normal people. However, when a patient suffered from insomnia during the early part of the night, the base hormone level shifted and there were sharp cortisol peaks, generally associated with awakenings. The episodic appearance of cortisol suggests that its secretion into the blood may be related to the restlessness of depressed patients. During the early morning hours, the sharp peaks and nadirs of cortisol result in such variability that even hourly samples are bound to be inadequate, since cortisol spurts into the blood and drops like a fountain. One must know where, on the fountain, one obtained one's level. Studies of the circadian rhythm of plasma cortisol based on infrequent samples may need reevaluation.

Patterns of hormone secretion may eventually shed light on the insomnia of depressed patients, but in the meantime, the insomnia, itself, has become a useful early warning signal of biochemical and mood disorder.

Narcolepsy

One of the most interesting and instructive of sleep disorders has been known as narcolepsy. This is sleep that breaks into the normal waking portion of the day, intruding upon the unwary victim, sometimes causing him to collapse into a seeming faint, with abnormally timed symptoms of REM sleep. Laughter, anger, or tears—the normal emotions of living—may catapult one kind of narcoleptic into an apparent seizure. Many narcoleptic seizures are now thought to be related to the mechanisms that produce REM sleep.

This alarmingly misplaced REM sleep has inspired quite a few investigations since Drs. William Dement, Allan Rechtschaffen, and George Gulevich first discovered that the EEGs during narcoleptic attacks corresponded to the patterns of REM sleep.

Dr. Pierre Passouant, at the Université de Montpellier in France, has watched ten narcoleptic patients around the clock in one of his studies. During the day a few of them had narcoleptic attacks, showing REM activity every two hours. Most of them had attacks every three-to-four-hours. When patients were put to bed and left in darkness every two hours (to facilitate sleep during the daytime), the REM periodicity became more regular and was comparable to the REM activity observed in normal people during the night's sleep. When these people lay in the dark for 24 hours, they exhibited an incredibly high number of REM periods, sometimes less than an hour apart. Their rhythm of sleep and waking, indeed, resembled that of a newborn infant.

Dr. Stephen E. Mitchell and his associates, at Stanford University Medical School, have attempted to trace the etiology of illness in over 50 narcoleptic patients. Usually the attacks began after a major life-stress, such as marriage, divorce, or military combat. Often these people got less than usual sleep at night, and typically would drop into REM sleep soon after they fell asleep.

Dr. James A. Lewis, at Stanford University Medical School, examined the electroencephalograms of 14 narcoleptic patients, doing a round-the-clock study on urinary excretion of adrenal hormones in some of them. Most normal people show a visible rise and fall of adrenal cortical hormones, the high point coming about the time of awakening in the morning, the low point at night before or just after sleep. However, the narcoleptic patients showed little rhythmic change in hormone levels. Perhaps their irresistible sleeping illness is associated with the lack of a well-defined circadian rhythm in hormonal output from the adrenal cortex. The resemblance between rhythms of REM sleep in narcoleptics and in newborn infants suggests that narcolepsy may imitate the sleep activity of the infant's brain. The lack of apparent hormonal rhythm also resembles an infantile stage of development.

Space

Many attributes of REM sleep have been illuminated by the abnormalities of illness, or the effects of drugs. Recently, manned-space flight has added information. The weightlessness of space flight might be expected to alter sleep rhythms in a number of ways. Experimental evidence has testified that the vestibular system, important in our sense of balance, plays an imposing role in the eye movements of rapid-eye-movement sleep. One of the first connections was discovered in 1965 by Drs. Ottavio Pompeiano and A. R. Morrison, at the University of Pisa, Pisa, Italy. The vestibular system consists of the semicircular canals in the ear, the utricle, and saccule of the inner ear and neural extensions. This system exerts reflex control over the eyes, permitting us to keep our eyes on an object even when we move our heads. Without vestibular control, the world would appear to rotate each time we turned our heads. Drs. Pompeiano and Morrison found that the destruction of nerve cells in portions of the vestibular system would abolish the rapid-eye-movements that occur during REM sleep. Recently, researchers have begun to wonder whether a loss of REM sleep might, in turn, impair a person's sense of balance or alter his usual response to acceleration.

During acceleration, people ordinarily show eye movements resembling those of REM sleep. However, in normally rested people, the frequency of eye movements decreases during continued exposure. Drs. James W. Wolfe and James H. Brown, of the Army Medical Research Laboratory in Fort Knox, Kentucky, wondered whether this decrease in eye movements could be related to functions occurring in REM sleep, and whether it would appear in men who had been deprived of REM sleep. Because it is difficult to awaken individuals at REM onset, they used total sleep deprivation instead. The 16 volunteers were accelerated while seated in a chair in a dark capsule, and their responses were recorded before and after a night of sleep deprivation. After a sleepless night the accelerations elicited more frequent eye movements although the men were awake and alert. Thus, changes in sleep might affect the vestibular responses of waking man. In space, indeed, weightlessness might incur REM deprivation since man is accustomed to the continuous vestibular stimulation provided by gravity. Dr. Charles A. Berry, at the NASA, Manned Space Craft Center in Houston, has described sleep-hungry astronauts attempting to orient themselves by wrapping an arm or leg around something while free-floating in space.

If the first Americans in space did not sleep soundly nor enough, it was partly due to poor scheduling that demanded too-brief periods of sleep, sometimes at hours when the men normally would have been awake. Soviet cosmonauts maintained their usual earth schedule while in orbit. In 1961, when Titov spent 25 hours in orbit, he easily fell asleep at his usual bedtime. He awakened earlier than he planned, only to see his arms dangling weightlessly and hands floating in midair.

> "The sight was incredible [Titov reported]. I pulled my arms down and folded them across my chest. Everything was fine—until I relaxed. My arms floated away from me again as quickly as the conscious pressure of my muscles relaxed and I passed into sleep. Two or three attempts at sleep in this manner proved fruitless. Finally, I tucked my arms beneath a belt. In seconds, I was again soundly asleep.
>
> "Once you have your arms and legs arranged properly [Titov added], space sleep is fine. There is no need to turn over from time-to-time as a man normally does in his own bed. Because of the condition of weightlessness there is no pressure on the body; nothing goes numb. It is marvelous; the body is astoundingly light and buoyant. . . . I slept like a baby."

When permitted to sleep on their usual earth schedule, some American astronauts are reported to have rested comfortably despite the eerie and unaccustomed state of weightlessness. No continuous brain-wave tracings are available from most of these sorties into space to tell precisely what happened to the rhythm of human sleep in weightlessness. We do know a little about changes in astronaut Frank Borman early in his seven-day orbit in the Gemini program. On his first night, he alternated between light sleep and arousal, receiving little slow-wave sleep. However, on the second sleep period in flight, he showed three descents into Stage IV sleep, perhaps compensating for his first day's sleep loss. A small amount of space data has been obtained as well from a macaque monkey sent into earth orbit on Biosatellite III.

A male monkey was launched into earth orbit on June 28, 1969, for an intended stay of 30 days, but because he soon deteriorated he was brought back to earth on the eighth day. During space flight, his EEG patterns differed radically from his base-line sleep records on earth. He shifted from stage to stage very rapidly, and during the day he showed moments of drowsiness, or sleep, and sudden awakening. Such rapid shifts during waking had been observed in the records of astronaut Frank Borman, during his second day of the Gemini VII flight. It has been thought that the lability of EEG states, accompanied by shifts in heart rate and other physiological measures in the monkey, was related to prolonged weightlessness.

No marked changes in EEG sleep patterns were recorded from the aquanauts of Tektite I, who spent 60 days in a nitrogen-saturated atmosphere some 50 feet below the surface of the water. If anything, the divers recorded by Drs. Paul Naitoh and his coworkers, at the Navy Medical Neuropsychiatric Research Unit in San Diego, slept slightly longer than usual, showing slightly more delta sleep, and tended to go to bed progressively later, arising later each day.

Isolation

It is difficult to identify the factors that produce specific changes in sleep. In relative isolation, for instance, on research missions in the high polar wasteland of the Antarctic plateau, men do not necessarily adhere to a precise 24-hour day, and the calendar of at least one physicist showed that he went to bed later and rose later in a drift toward a 24-1/2-hour day. Moreover, men on polar expeditions have often complained of insomnia which they call "the Big Eye." By the standards of temperate zone sleep laboratories, their sleep may not be normal. EEG tracings taken by Drs. Jay Shurley, at the University of Oklahoma Medical School, and Chester Pierce, of Harvard University Medical School, revealed that sleep in the Antarctic lacked Stage IV. Tracings of nightlong sleep of men on the U.S. research team showed no deep slow-wave sleep, despite the fact that the men did considerable physical work.

One man was studied intermittently after his return to the United States. He exhibited no deep Stage IV EEGs until he had been back for over a year. This startling and prolonged change raised new questions about the possible influence of the electromagnetic environment upon the nervous system.

Deep in the bowels of the earth in damp, chilly caves, isolation is a different experience from that at the poles. Since the early 1960's, a handful of people have voluntarily endured months in this hostile atmosphere, in a perpetual night lit only by their lamps, broken only by their own internal schedules. In 1966, a young man spent six months in a cavern deep in the Maritime Alps of France, connected by cable with a research station at the surface. He would, before going to bed, affix electrodes to his head and face and plug himself into the electroencephalograph. In the initial weeks underground, he was wakeful for about 33 hours, sleeping 14 hours on what seemed a 47-hour day. Then he returned to a roughly 24-hour schedule and, subsequently, into a period of long stretches of activity interspersed with long sleeps.

The EEG data, as reported by Dr. Michel Jouvet, of the Faculté de Médicine in Lyon, France, revealed no major changes in the distribution or quantity of Stage IV sleep, but there was an increase in the percentage of REM sleep (to 28 percent). Moreover, the man in the cave would rapidly drop from waking into REM sleep, sometimes in five minutes rather than in the usual 70 to 100 minutes. This rapid onset of REM sleep has been observed among narcoleptic patients and people in withdrawal from alcohol or drugs. More recently, two other men lived in caves for four-and-a-half months, portions of that sojourn on a 48-hour day. Their EEGs showed extremely long REM periods, sometimes as long as three hours, and they reported being drowsy much of the time. It has been conjectured that REM activity and dreaming might compensate for lack of stimulation in the lonely darkness of cave isolation. However, the shift of REM sleep toward the beginning rather than the end of the first sleep cycle also suggests REM sleep may be related to the former circadian period of sleep and activity. Spectral analysis may indicate whether the distribution of REM sleep bore the trace of a persistent circadian rhythm in a 48-hour activity cycle.

The Presumed Functions of REM Sleep

Within the brief but rich history of modern sleep research, the most flamboyant and abundant speculations cluster around the function of the REM period. Psychoanalysts have considered it the vehicle of dreaming and laboratory studies of REM dream reports gave some support to this view. As people noticed parallel physiological changes in man and animals, REM sleep seemed to be a time when drive centers discharged. Other functions have also been suggested. Dr. Arnold Mandell and his associates, now working at the University of California in San Diego, had analyzed urinary metabolites they obtained at short intervals throughout the night from sleeping, catheterized volunteers. The excretions of metabolites of neurotransmitters (VMA and VHA) suggested that metabolism and nourishment within the central nervous system might be one function of this phase of sleep. Some years before, Dr. Seymour Kety, of Harvard University, saw that brain-blood flow was faster in REM sleep than other sleep, also suggesting a period of higher brain metabolism. Many scientists have searched for the underlying brain mechanisms.

Since the 1950's, Dr. Michel Jouvet and his associates have tirelessly sought to locate the clusters of brain cells that control various aspects of REM, or paradoxical sleep. Experiments with cats have shown that a region of the pons, in the brain stem, a primitive region of the brain just above the neck, mediates paradoxical sleep. When this region is destroyed, animals sleep, but no longer enter the paradoxical state. When this pontine region is electrically stimulated through hair-fine electrodes, the cats soon droop in flaccid relaxation, paws and whiskers twitching, eyes moving in the familiar signs of paradoxical sleep.

Jouvet's work has shown that the dramatic loss of muscle tone in REM, which renders a cat as limp as a rag, is controlled by a specific locus in the pons—the nucleus coeruleus. This aspect of paradoxical sleep has seemed to be triggered by specific groups of nerve cells and may be related to body temperature. Are the other aspects of this state so specifically controlled?

Recently, Dr. Jouvet and Dr. William C. Dement have been exploring the meaning of bolts of nerve discharge from portions of the brain stem and the visual system, from the pons, the occipital region, and the lateral geniculate. While a sleeping cat's eyes have appeared to be tracking some inner action, these PGO (ponto-geniculo-occipital) spikes were recorded in volleys of 50 to 60 a minute. On a strict tally, a cat seems to show between 11,000-14,000 each day, but at a very low rate in slow-wave sleep. They begin about 35 seconds before the creature shows the familiar signs of paradoxical sleep, and strongly resemble EEG discharges that are recorded from the brain during moments of visual attention. When animals are deprived of paradoxical sleep, the spikes increase in frequency and appear in other states of consciousness. Recently, Dr. Allan Rechtschaffen, of the University of Chicago, has shown that the PGO spike phenomena might be measured in human beings through the activity of the extra-ocular muscle. The reduction in REM sleep that follows certain drugs, such as the monoamineoxidase inhibitor drugs given to depressed patients, may be accompanied by increases in the equivalent of ponto geniculate ocular excitation. It would be interesting to learn whether the neural networks that result in PGO bursts also influence that activity of the hypothalamus during REM sleep.

The hypothalamus, deep in the center of the brain, shows an increase in temperature during REM sleep, and a surge of activity may be the cause of the many events on the surface that have typified the REM state: erections in males, irregular pulse and respiration, changes in the conductivity of the skin, and the dilations of peripheral blood vessels.

It is no surprise to learn that there is a surge of discharge in hypothalamic cells during REM sleep, as discovered by Drs. Walter D. Mink, of Macalaster College in St. Paul, Minnesota, Phillip J. Best, and James Olds, at the University of Michigan. Using almost invisible electrodes deeply implanted into the brains of specially conditioned animals, this research team was able to compare neural cell activity in several brain regions during a variety of waking and sleep stages. Their rats had been trained to press a lever and then wait very quietly for food, giving the researchers a way of obtaining brain waves from a motionless yet alert and motivated creature. EEGs were also obtained during highly motivated active periods, slow-wave sleep, and paradoxical sleep. Hypothalamic cells, in direct contrast to cells in the hippocampus, fired more frequently (3-1/2 times) during REM sleep than in other sleep or waking conditions. This does not prove that there is an organizational integrity in REM sleep involving the network that controls PGO spikes and the hypothalamus. REM sleep may have many aspects.

One aspect of REM sleep appears to be adaptational. A number of scientists have hypothesized that a major function of this state is for filing new experience, so that it can be retrieved in the future; a time for memory, adaptation, and processing without interference from the outside world. This "cerebral aspect" of REM sleep suggests that the primitive brain regions of the pons and hypothalamus must control only certain of its features. Presumably, the REM state also involves the cortex (regions of the brain that are related to intellection), as well as subcortical regions of the forebrain, which have been shown to belong to a system of nerve centers that cause sleep of various durations and qualities.

Because of the dramatic properties of REM periods, sleep has often been described as if it were a binary process, an alternation between two very different states of consciousness—REM and non-REM sleep. Excitations of nerve cells in specific regions have been said to trigger REM sleep, as if it were normally turned on and off like a light switch. Described in this way, the REM state did not seem to be a phase that one entered and departed in the continuous manner of many physiological cycles but seemed almost discrete. It is no wonder that so many physiological phenomena—bed-wetting—for example, initially seemed to coincide with REM sleep but were later found not to correlate perfectly. Instead, enuresis and REM sleep overlapped like events that occur cyclically in phase—or almost precisely in phase. Had REM sleep been considered one of many related ultradian cycles—instead of the triggering state—researchers could have been pleased that such phenomena as bed-wetting occurred in cycles almost in phase with REM sleep. They might have searched for possible mechanisms behind both cycles, in hope of explaining abnormal nighttime urination.

The Language of Rhythms

During the early renaissance of sleep research, REM sleep and Stage IV were often described as if they were part of a hydraulic system. If deprived of either state, a person would seem to make up the loss. Normal people of a given age seemed to have a certain "nightly quota" of each state, and drugs that may have been delaying the first REM cycle were described as suppressing it. This hydraulic language seemed to suit the data; the concepts were simple, and everyone understood what was being said.

Today, another viewpoint and language are available. A good many basic questions about sleep can now be reexamined in terms of basic biological rhythms. Unfortunately, the terminology for discussing cycles has been ambiguous and as diverse as the beliefs and training of the many scientists in the field—they have included astronomers, entomologists, endocrinologists, paleontologists, doctors, biologists, mathematicians, agriculturists, psychologists, and psychiatrists. During the last decade, these various contributors have been brought together often in international symposia; and scientists, continents apart, have begun to cooperate in their research. Discussions that began sounding like the Tower of Babel are slowly leading to agreements on terminology. This review precedes an accepted vocabulary; and since it attempts to represent each author in more or less his own words, it is bound to be inconsistent.

There are, however, certain basic notations that have been generally accepted and are worth repeating: the period of a cycle represented by the Greek letter, τ, is the elapsed interval between two peaks or troughs, while the reciprocal, $\frac{1}{\tau}$, of this period is the frequency. The magnitude of the change, or amplitude, C, describes another dimension of the cycle. The phase ϕ or specific location on the cycle, is usually described with reference to some external time point.

Language from biological rhythms is beginning to infiltrate sleep research. Some investigators are manipulating sleep in order to discover how sleep stages relate to the circadian rhythm of roughly 24 hours. Others have been interested in the 90 to 100 minute sleep cycle. This rhythm has a higher frequency than a "day" and is often called "ultradian." "Ultradian" refers to cycles with periods shorter than circadian, such as an hour; while "infradian" refers to cycles with periods longer than circadian, such as weeks or months. Other sleep researchers are studying slow, gradual displacements in the stages of sleep. For example, data from the Antarctic has shown that a low-frequency rhythm of 28 days may occur in an individual's time of sleep onset in an "infradian" rhythm. Coined by Dr. Franz Halberg, the terms "circadian," "ultradian," and "infradian" are becoming popular, and, therefore, convenient for identifying a particular biological or psychological cycle.

Recently, a number of sleep researchers have wondered whether the ultradian rhythm seen in sleep is a cycle restricted to sleep or whether it may be present throughout the 24 hours.

A Basic Ultradian Rhythm?

Over 30 years ago, Dr. Nathaniel Kleitman began systematic observations of change in human behavior and physiology, in temperature, performance, rest, and alertness. Kleitman found hints of regularity and postulated a "basic rest and activity cycle." He concluded that such fundamental alternations between activity and repose were inherent in the nervous systems of all homeothermic animals—animals who maintain body temperature within a narrow range—whatever the climate outside. As Kleitman saw, the human infant exhibits a 60-minute basic rest-and-activity cycle, encompassing quiescent and active sleep. Actually, the activity was the REM part of sleep. Most sleep researchers have called this a sleep cycle. Outside the realm of sleep research, however, "sleep cycle" might be misinterpreted to mean the cycle of nightly sleep, so a more precise phrase, such as "paradoxycle" has been suggested by Dr. Daniel Kripke, a psychiatrist and researcher associated with Albert Einstein College of Medicine in New York. This is particularly apt for describing animals, in whom the equivalent of REM sleep is known as paradoxical sleep.

In man's sleep, the cycle period is measured from REM stage to REM stage. It encompasses a revolution from rapid, low-amplitude brain-wave activity, to slow, high-amplitude brain-wave activity, and back. Human babies sleep in cycles of 55 to 60 minutes, but adults take from 90 to 100 minutes. A newborn cat requires only eight minutes for a cycle, while an adult cat takes a half hour. Like the cardiac or respiratory cycles, the period of this cycle appears to be related to the size of the animal and, perhaps, is determined in part by its rate of metabolism as well as external conditions of security. A small creature, the rat, shows cycles of ten to twelve minutes in sleep, but the elephant has two-hour cycles.

In trying to understand the nature of this cycle and to ascertain whether it continues around the clock, several approaches have been used with human subjects. Scientists have looked for the occurrence of REM activity in sleep at different times of day and night, in people who were napping, undergoing sensory isolation, or on unusual schedules. The rhythm of napping and of narcoleptic attacks has been analyzed. If the 90- to-110-minute cycle represents an alternation of drive intensity, one might expect to see signs of it during waking. The eating behavior of people during daytime isolation has been interpreted as another clue. Studies of neural and neurochemical rhythms in animals have offered corroboration of an ultradian rhythm of attention that may continue in waking, although it is most visible in sleep.

Naps

Since the early 1960's, researchers have noticed that when they monitored volunteers during daytime naps, their sleep did not have the same pattern as night sleep. Was the distribution of sleep stages related to the person's phase on his circadian rhythm? Although they expressed it differently, this is what Drs. Wilse B. Webb and Harmon W. Agnew, Jr., at the University of Florida, found when they invited subjects to take three-hour naps in the laboratory at either 9 a.m., 12:30 p.m., 2 p.m., or 4 p.m. Stage IV did not appear in the early morning. However, it increased in the late afternoon. REM sleep hardly appears in late afternoon or the first part of the night's sleep, but took up 32 percent of the nap that began at 9 a.m. Stage II, light sleep, varied least, but even its distribution was circadian, and it occurred least during the first two hours of nightlong sleep, increasing in the morning. Thus, REM sleep predominated in the morning and Stage IV during afternoon and early night.

A somewhat different study of people with unusual schedules was performed by Dr. Gordon Globus, who is now at the University of California in Irvine. He concluded that REM sleep was highest in the morning, least in afternoon, and its distribution was related to clock time. In other words, the REM sleep, a unique storm of neural activity, may be a rhythm of the nervous system related to the individual's circadian rhythm. Thus, the placement and amount of REM sleep in a nap may depend upon a person's phase in his circadian cycle. These studies mark the entrance of an entirely new viewpoint in the exploration of sleep, with many implications for the design of experiments and the reinterpretation of conventionally obtained data.

One of the first studies to be conducted from a "rhythms" point of view was done at Holloman Air Force Base, New Mexico. Five volunteers gave interesting responses to isolation in an experiment conducted by Dr. Daniel F. Kripke. For the first 12 hours, each man adjusted to the isolation in the dimly lit, sound-proofed chamber. During this time, he was permitted to sleep for 8 hours, but he was asked to remain awake for the subsequent 24 hours.

In the boring, dark compartment, each man found the perceptual deprivation disorienting and terrifying and could not manage to stay awake. Since each one wore sensors, permitting a continuous record of EEG, pulse, muscle tone, eye movement, and activity, it was clear when the volunteers were asleep. Using a system of analysis developed by communications engineers, the researchers saw that certain physiological functions varied rhythmically. Ninety-minute rhythms (16 cycles per day) were visible whether the subjects were asleep or awake. When subjects were only intermittently awake, their napping was rhythmic. Although the naps contained no REM sleep, they tended to occur at 90-minute intervals.

A number of researchers have begun to inspect waking activity on the suspicion that the 90- to 100-minute cycle might be a rhythm of waking consciousness, a rhythmic waxing and waning of attention. Dr. Ekkehard Othmer and his coworkers, at Washington University in St. Louis, have seen evidence of the same cycle under gradations of isolation and sensory deprivation.

In this study, three coeds were recorded on a polygraph for 24 hours. After a normal night's sleep, the subjects were asked to keep their eyes closed and refrain from all activity excepting at the three meal times. Throughout the 24 hours, they showed an EEG pattern like drowsiness or light sleep followed by periods of REM activity. Neither changes in the level of illumination in the room nor the subjects' position (lying or sitting) changed this pattern.

Under one condition, the girls rested in bed but were awakened at the onset of light sleep. The researchers reported, "Since dream activity was not interrupted at the onset of sleepiness, observation showed that the subjects entered periods of dreaming even without preceding undisturbed sleep." Even when they were awake, they had periods of rapid-eye-movements accompanied by varying degrees of muscle relaxation. Interviews suggested that this state might have indicated daydreaming.

In another study of men students in isolation, Drs. Othmer and Juan C. Corvalan deprived volunteers of all REM activity for 24 hours. It was difficult to keep the subjects out of REM; yet, after a cluster of "attempts" at REM sleep, an interval of non-REM occurred. The researchers thought that perhaps the deprivation procedure did not lead to a real increase of REM onset, but rather a segmentation of the original REM period which might initiate a disruption of the entire REM cycle.

In the following 24 hours, the REM percentage increased by more than 100 percent over the base line. Such a marked increase has been observed only in studies that deprived volunteers of nighttime REM over several nights. The effectiveness of 24-hour deprivation further suggests an occurrence of REM activity at night and also during the day.

Eating—A Sign of an Ultradian Drive Rhythm?

A number of psychiatrists have speculated that the roughly 90-minute cyclicity in sleep might involve an alternation of drive accumulation and discharge. Survival impulses, drives, such as hunger, sex, and aggression, show themselves during REM sleep. The neural mechanisms underlying these behaviors lie beneath the cortex in brain regions that are old in an evolutionary sense. One can observe them in the REM sleep of infants without special equipment, for babies in this state are actively sucking, grimacing, having penile erections, and showing many expressions of emotion. Are these the signs of a biological tide that arises and ebbs roughly every 90 minutes in the adult as part of a life-long alternation between the "charging" and "discharging" of drive centers?

Tangential evidence for a 90-minute cycle of hunger was noted in 1922 when a group of subjects, who were kept in bed all day, showed a roughly 90-minute rhythm of body motion and stomach contractions. In 1965, Dr. Stanley Friedman, at Mt. Sinai Hospital in New York City, postulated that a drive cycle, if there were one, might be visible in oral behavior. If so, one might find cycles of oral behavior and eating throughout the day.

Working with Dr. Charles Fisher, Friedman created a comfortable den where each volunteer might spend eight to nine hours in isolation, reading, writing, away from the sound and fury of social activity. A refrigerator was stocked with food. An electric coffee pot kept a fresh supply hot and aromatic. There were magazines, cigarettes, and books. On the other side of a mirror sat the observer, Dr. Friedman. Each oral item was rated: for example, a cigarette was worth 3 points, a 9-ounce carbonated drink 5 points. The observer recorded exactly what a person put into his mouth, and noted the time.

When points were plotted against time, it was clear that the volunteers in seclusion did not go to the refrigerator at random times. There was a waxing and waning of the intensity of oral activity, with a range of 85 to 110 minutes and a mean periodicity of about 96 minutes.

Dr. Friedman subsequently replicated the study with mild, chronic schizophrenic patients, who had shown the same REM sleep rhythm as normal people. He anticipated that they would show the same oral rhythm. They were told that they would be observed as a test of relaxation during an isolated period during which they might do as they pleased. Occasionally, Dr. Friedman dropped in for a minute or so to relieve the boredom. He had found earlier that some normal subjects grew bored after about 6 hours and began to exhibit shorter cycles of eating and smoking intensity—cycles of roughly 60 minutes. One man who was studying for an examination, and who felt very tense on his second day in the observation room, showed an increase of 14 percent in oral activity over his first day level, and his average cycle shortened to about 65 minutes by comparison with the 95-minute cycle observed on his first experiment day. Another subject, observed after a night of REM deprivation, also showed 60-minute cycles of eating. Shortened cycles seemed to occur during or after stress and have been described as a kind of regression toward the roughly 60-minute cycle of infants.

The schizophrenic subjects, in general, showed oral behavior cycles of 90 to 100 minutes, in what may be interpreted as additional evidence of an ultradian drive activation during the day. More recently, similar evidence of an underlying drive rhythm with a period like that of the sleep cycle has been shown in animals.

Other Signs of Ultradian Rhythms

Recently, Drs. Daniel F. Kripke, Franz Halberg, Thomas Crowley, (at the University of Colorado), in Denver, and G. Vernon Pegram (at Holloman Air Force Base) have observed ultradian rhythms in the EEG and muscle tone of rhesus monkeys on a controlled lighting regimen of 12-hours light and 12-hours dark. The sleep cycle of the monkey is about an hour, a cyclicity that was observed in the electromyograph recordings of muscle tones during the day. Moreover, when EEG rhythms were analyzed around the clock, eight out of nine animals showed an ultradian cycle of theta, although this slow EEG rhythm was of smaller amplitude during the day than during darkness. The ultradian rhythms of the REM cycle, as seen by day, show a circadian frequency and amplitude modulation.

The possible nature of this continuous ultradian rhythm, showing the periodicity of the paradoxical sleep cycle, has been illuminated by Drs. M. B. Sterman and Dennis McGinty, working at the V.A. Hospital in Sepulveda, California, and UCLA. They had trained eight cats to generate a specific slow-wave EEG pattern for a food reward. This "sensorimotor rhythm" seemed related to the inhibition of phasic motor behavior, and the cats quickly learned to generate the pattern. The animals were allowed to work until satiated on a test day and then were monitored through several REM cycles. When their performance rate was measured, it became clear that peak performance occurred cyclically every 15 to 30 minutes, which corresponds with the period of the REM cycle. When another group of animals was allowed free feeding over a 24-hour period, they showed bursts of feeding activity every 15 to 30 minutes, in a waking cycle that was continuous with their REM cycle during sleep.

In further studies the researchers eliminated portions of the brain, and recorded the tone of the neck muscle, certain reflexes, and behavior. Cats with only a brain stem showed a cyclic change in muscle tone. The cycle, about 30 minutes, seemed to correlate with the cycle of body temperature. In normal intact animals, body temperature fluctuates in an ultradian rhythm of about 30 minutes. The frequency and duration of muscular flaccidity in the decerebrated cat followed the temperature cycle. Elevations of body temperature would increase the frequency, while lower temperatures had an opposite effect. The neural mechanisms causing the neck muscle to relax during REM sleep and cyclically during the day appear to be located in the lower brainstem, and to be influenced by metabolic processes, thus producing the ultradian rhythm of atonia seen in sleep and waking.

Evidence from the study of cats strongly suggests that waking behavior may be modulated in a manner similar to the REM sleep cycle. The rhythm may reflect a metabolic periodicity that influences the activity of the central nervous system, both in the higher brain regions and the brainstem. Dr. Sterman has postulated that this ultradian rhythmicity occurs continuously, like the cardiac cycle, throughout life in developed creatures.

The concept of ultradian cycles in the nervous system has a usefulness that was not foreseen ten years ago. As sleep researchers gathered normative descriptions of sleep from thousands of volunteers, they first saw the important differences between Stage IV and REM sleep. When people or animals were deprived of either stage, they tended to compensate by sleeping more in that stage later. Normal sleep was often described by the percentages of REM and Stage IV in a night's sleep. Experimenters wrote as if there were a nightly REM quota that was altered by drugs or illness. Then the latency of REM onset seemed important, along with the intensity of eye movements or certain brain-wave configurations. More recently, attention began to shift to the temporal attributes of the sleep rhythms. If sleep had been analyzed for ultradian and circadian rhythmicity at the start, some of the confusion of early sleep research would have been eliminated, but along with it, much of the initial excitement. Experimenters would have adopted more uniform methods and parsimonious terminology. Instead of counting the minutes from the onset of sleep to the first REM period, the percentage of REM in total sleep, the proportion of REM in the first half of the night, or length of individual REM periods, many separate temporal parameters might have been summed up in expressions of rhythmicity. Experiments, too, might have been altered. Drugs might not have been expected only to suppress the first REM period, but also to shift the phase. Although they used different terms, many scientists were, indeed, measuring phase shifts of a roughly 90-to-100-minute cycle, as it was advanced or delayed within a context of 24 hours. Others were documenting the shortening of the period, or its variability under stress, but expressing their findings in a number of terms.

A Binary Autocorrelation Method for Analyzing Sleep Rhythms

Dr. Gordon Globus was among the first of the sleep scientists to postulate that a roughly 90-minute cycle seen in sleep might have its counterpart by day, as invisible to casual observation as the rapid-eye-movement cycle in nightly sleep had remained throughout so much of man's history. With this possibility in mind, Dr. Globus began to wonder if the so-called 90-minute rhythm was actually a consistent cycle as normative descriptions had implied. How might one analyze hundreds of miles of EEG records from night-long studies so as to obtain a quick scan of sleep rhythmicity? He evolved an autocorrelation method that is simple and inexpensive. Measures, which are derived from volunteers, are related to the period (or length) of the cycle and to the consistency of the cycle. Both the average period of a cycle and consistency can be potent criteria for examining sleep data.

Dr. Globus has drawn upon all-night sleep recordings from a number of laboratories to test his method for abbreviating data. Using a program he developed for computer analysis, he found it possible to analyze the rhythmicity of a night's sleep in about seven seconds, a procedure that might cost many hours by hand analysis. He was able to show that the period of the cycle of sleep stages has a mode of about 100 minutes in the normal person.

Usually, when people volunteer for a sleep experiment, the researchers will discount records taken during the first several nights of sleep, since the adjustment to the laboratory situation sometimes disrupts sleep for two or three nights. In 20-night studies of individuals, some have taken a week to adapt. Among sleep researchers this "first-night" effect is well known, for most people react with slight tension to sleeping in a research bedroom and wearing electrodes around their faces and heads. Despite the comfort and privacy of these bedrooms, the first nightlong record will contain "abnormal" patterns, and, among the signs of anxiety, a reduced percentage of REM sleep. When nightlong EEG records were run through a rhythmicity analysis, Globus found that there was an obvious first-night effect. Later records were taken from the same individuals after they had grown accustomed to sleeping in the laboratory, and these nightlong records showed more consistent rhythmicity. Consistent cycle periods may be criteria for healthy, restful sleep in the laboratory.

Dr. Globus also analyzed records from people who had undergone severe sleep loss, and found relatively normal rhythmicity. He analyzed records from a group of hypersomniacs, over-sleepers, who had been studied in the laboratory of Dr. Allan Rechtschaffen, at the University of Chicago, and found that they, too, were typical by a rhythmicity analysis. He then evaluated records of daytime narcoleptic attacks in a patient studied by Dr. Passouant. These attacks showed a roughly 100-minute rhythm, a periodicity like that of REM sleep. Data on daytime naps also suggested that REM sleep might be an ultradian rhythm, locked into the circadian rhythm. In other words, the 90- to 100-minute cycle may be like a ripple superimposed on the slower, circadian cycle. Further studies should reveal whether motivation, stress, heat, cold, drugs, conditioning, or other manipulations can alter this rhythmicity.

Sleep Reversal: Hormonal Rhythms

Sleep reversal is a simple, if arduous, technique for exploring the nature of an ultradian rhythm within the 24 hours. A volunteer who ordinarily sleeps by night is suddenly switched to a daytime sleep schedule, temporarily making his activity cycle the inverse of many physiological cycles. By this taxing method, Dr. Elliot Weitzman and his associates have offered some insight into the complaints of workers on night shifts, who claim that they suffer from poor sleep and a sense of general discomfort. A part of their trouble may be hormonal.

Since the 1950's, it has been known that adrenal cortical hormones (or steroids) show a rhythmic change of concentration in the blood. Every 24 hours the levels of hydrocortisone (17-hydroxycorticosteroids) rise and fall predictably. Adrenal cortical hormones help to regulate metabolism and energy supplies in the body, and appear to influence the transmission of nerve impulses, thereby influencing sensory acuity as well. They come from the outer layer of the adrenal glands, the region known as the cortex. Until recently, it was assumed that the blood level of these glucocorticoids declined from about midday until midnight and then began to show a steady rise to a high point just before or after waking.

In 1965, Dr. Weitzman took samples of blood at short intervals throughout the night in sleeping subjects. When the levels of blood hydrocortisone were analyzed it was clear that the hormone did not steadily seep into the bloodstream. It increased in "puffs" that were usually very close to REM periods, suggesting a direct relationship. Today, the imperfect but striking coincidence in timing suggests that the steroid and REM rhythms might be separate but in phase so long as the volunteers are healthy and maintaining a steady schedule.

During early morning sleep, between 4 and 6 a.m., the blood level of adrenal steroids declines with astonishing rapidity. Dr. Weitzman wondered whether the rhythm of cortisol in the blood was an ultradian rhythm, linked to a roughly 24-hour rhythm. Could the spurts of blood cortisol during early morning sleep be a biochemical transition from sleep into activity, anticipating the demands of waking? Dr. Weitzman and his colleagues elected to explore the circadian rhythm of hormonal changes by sleep reversal. Initially, they studied five men for three weeks in the metabolic ward of Albert Einstein Hospital where food and water intake could be precisely measured. The volunteers slept in a laboratory bedroom where brain waves and other physiological functions were recorded. Blood samples were taken from catheter tubes through the wall between bedroom and control room – deftly done without disturbing their sleep. For the first week, the subject kept to the usual hospital routine of sleep from 10 p.m. to 6 a.m. Blood, urine, and temperature were taken every four hours except for the midpoint of the eight-hour sleep period. On the eighth night, the volunteer was kept awake and permitted to go to sleep at 10 a.m. For the next two weeks, he slept from 10 a.m. to 6 p.m.

During the first week in the hospital, all of the young men showed the usual adult sleep pattern. About 70 minutes after they first fell asleep, they floated up into their first REM experience which lasted about 10 minutes. Then they sank into a period of deep delta oblivion, rising again for a more protracted REM period after about 90 minutes.

Each man reacted to sleep reversal with a striking and prolonged change in this usual sleep progression. The volunteers lapsed rapidly into REM sleep in a manner resembling narcoleptics or people who had already been asleep for several hours. Four or five minutes after falling asleep some subjects would enter REM sleep. Despite plenty of sleep, this unusual pattern continued. For two weeks they fell into REM sleep with abnormal celerity. The sleep pattern looked especially strange because Stage IV (deep, slow-wave sleep) had rapidly adapted to the sleep reversal. From the start it appeared in the early hours of sleep, as usual. Throughout the last one-to-two-hours of sleep the subjects awakened intermittently, which effectively decreased REM in the very portion of the sleep period where it normally predominates.

Although these volunteers obtained what most people would call sufficient sleep, it was not very good sleep. Their diaries revealed that the volunteers were not feeling like themselves, and certainly were not vigorous or refreshed. They said they felt uncomfortable and listless.

Only toward the end of the second week did they begin to adapt to sleeping by day. Then their EEGs revealed that they were beginning to revert to their original sleep progression. One factor in their initial listlessness and discomfort may have been related to adrenal hormone levels. In some instances, the sleep reversal seemed to damp the release of adrenal steroids, and the excretion levels fell unusually low. This may be relevant to the well-being of many workers including city policemen.

Nightworkers sometimes mention their insomnia and fatigue. When Dr. Daniel F. Kripke and his coworkers at Holloman Air Force Base, studied ten medical corpsmen on night duty, seven out of the ten complained of insomnia and digestive problems. Although night shift was made inconvenient by certain social factors and customary foods were not available, their sleep and health improved as they stayed longer on night shift. Their sleep showed more REM sleep and Stage IV. Like the subjects in Dr. Weitzman's study, these corpsmen initially showed a disorganization of sleep rhythms, and REM sleep was interrupted by Stage II. Adaptation seemed to take several weeks.

After analyzing a series of observations, Dr. Weitzman and his coworkers concluded that two weeks had not been sufficient for complete reversal. This was corroborated in 1969-1970 in a longer, more extensive study. This time each individual was under surveillance for nine weeks.

For three weeks each of the men lived in the metabolic ward of the hospital, while base-line values were taken for EEG, urine, temperature, and blood constituents—sampled day and night. In addition, one day each week, every four hours, blood and urine samples were taken. These were analyzed for the adrenal hormone, cortisol; for the hormone insulin; and for the electrolytes; sodium and potassium. There are many metal particles known as electrolytes because of their electric charge. In charged or ion form, sodium and potassium can influence the transmission of messages among nerve cells and the amount of fluid contained in cell bodies. Electrolyte measurements in urine have become a standard, if indirect, manner of gauging the stability of body functioning. After three weeks of base-line study, the subject was kept awake all night and allowed to sleep only by day for the next three weeks. Then he returned to his normal schedule of sleeping by night for his last three weeks in the study. Although adaptation may be accelerated under the more vibrant conditions of ordinary life, in the artificial situation of the hospital laboratory it takes about nine weeks to study an individual as he adapts to a reversal of his sleep-waking schedule.

Body temperature did not move into phase with the new schedule until the end of two weeks. Full reversal (as seen in four volunteers) occurred only between the end of the second and third week. Three weeks seemed to be the outer limit for the volunteers, and all had shown temperature reversal by the time they were again switched back to sleeping by night.

When they were returned to their former sleep schedules, the volunteers showed an astonishing rapidity of adaptation. Body temperature returned to phase with activity—not in two-to-three weeks, but in-two-to-three days. This incredible alacrity suggested that perhaps the "reversal" had been superficial. Is there asymmetry in the body's circadian "clock"? Dr. Weitzman has conjectured that sleep reversal and milder shifts, such as those incurred by crossing geographical time zones, may resemble a revolving door. If one fails to exit at the proper moment, one has to go all the way around again.

Although there has not been sufficient data for generalization, both experiments indicated that body systems were shifting and adapting at different speeds. For instance, a person quickly shifts his pattern of urine excretion. If the greatest urine volume were excreted during sleep, it would cause a person numerous awakenings and disruption. In young volunteers, at least, the rhythm of water excretion seems to adapt almost immediately. However, body temperature may take some time, and a person initially finds himself walking around, trying to act alive and alert when his body temperature is at its nadir. Body temperature showed a complete reversal by the beginning of the third week. Deep sleep soon reappeared in the early part of the night, as if related to activity and exertion more than a circadian rhythm inherent in the nervous system. If body temperature took a long time to reverse, urinary creatinine rhythms reversed promptly. The slow adjustment of REM sleep, indeed, suggested that this state may be modulated by circadian metabolic rhythms.

The plasma cortisol rhythm was equally slow in adapting. Measurements were made every 30 minutes by sampling blood through an indwelling catheter in the arm of the sleeping volunteer, but the characteristic base-line cortisol rhythm did not occur during two weeks of reversal. Moreover, the typical urinary circadian cycle of 17-OHCS was not fully reversed during two weeks of sleep reversal. Measurements of growth hormone levels in the blood showed that this hormone responded differently to the 180-degree schedule shift. Growth hormone and Stage IV sleep shifted at once to the early part of the "night." It is interesting to note that these hormones, both products of two hypothalamic-pituitary systems, responded differently to the phase shift. Growth hormone release appeared to shift immediately while the adrenal cortisol pattern took several weeks.

Data in these studies indicate that people adapt very slowly to 180 degree phase shifts, yet world travelers appear to adjust comfortably to a new locale in less than three weeks. Dr. Weitzman has speculated that ordinary waking life, with its demands for activity, attention, and motivation, may give a different picture of phase shifts than does a hospital laboratory in which volunteers have spent their nights awake in relative isolation, reading, or watching television. In further studies, medical student volunteers will be studying hard and taking emergency-room duties during the nights awake. Although it seems likely that the quality of waking might accelerate the reversal of sleep stages, body temperature, and adrenal steroid rhythms people do not phase shift very rapidly on rotating shifts.

This was a problem for the New York City Police Department in 1970, when newspapers reported that many officers were "couping," pulling off the road so they could sleep in their patrol cars. The effects of rotating shifts were explained to the department by Dr. Weitzman who described the fatigue that attends the physiological transition.

Rapid shifts of the light-dark schedule have been studied in animals, and there have been at least two such studies of human beings. Dr. George Curtis, at the Eastern Pennsylvania Psychiatric Institute in Philadelphia, has observed healthy young volunteers adapt to a nearly random alternation of short periods of sleep and waking.

Dr. Weitzman and his coworkers have recently conducted a pilot study with a man whose sleep-waking cycle was three hours for the duration of the experiment. In part, because of the ultradian rhythm of the REM cycle, which is 85 to 110 minutes, and other ultradian cycles of hunger or restlessness observed in man, it has been conjectured that three hours may be the shortest feasible unit for waking and rest. The volunteer who lived this cycle for ten days was a 40-year-old former athlete, now working in construction. He spent seven baseline nights sleeping as usual in the laboratory. He then began to alternate two hours waking and one hour of sleep in the 2:1 ratio that typifies the usual human cycle of waking and sleep. Traces of his circadian temperature rhythm peristed. This meant that it was harder for him to sleep during midday than at night or in the morning, although his body temperature curve slowly began to flatten. Since he was partially sleep-deprived (50 percent), he was tired and reluctant to be awakened at the end of his allotted hour. Ordinarily an amiable and cooperative person, he showed the symptoms of hostility and irritation that commonly appear under cumulative fatigue. After a single night of uninterrupted sleep he remarked on his new sense of well-being. Ten days may not be long enough for adjustment, and this pilot study does not demonstrate that people cannot adjust to short cycles of activity and rest. It seems possible that in this dimension, as in others, men vary. Some may alter their activity rhythms considerably, while others suffer a great deal from relatively slight changes.

Sleep reversal studies should begin to yield a medically useful time chart, telling the rate at which circadian functions adapt to a change in schedule.

Adaptability may involve a number of factors as well as the inherited propensity of the individual, among them conditioning and alternations of light and dark. Drs. J. R. Mouret and P. Bobillier, of the Faculté de Médecine in Lyon, France, have done a provocative experiment in which they have conditioned rats to alter their circadian rhythm of sleep quite radically. By placing them on restricted feeding schedules that made food and water available only during the hours of light, they found that the nocturnal animals would shift their time of feeding and activity. Normally, rats eat, drink, and show the most activity during the hours of darkness. Within six days of the experimental regimen they began to exhibit about as much slow-wave sleep during the dark period as in their normal sleep period of illumination. Interestingly, paradoxical sleep did not stretch out across the 24 hours in a similar manner. Under the new schedule the rats seemed to show more than usual paradoxical sleep which occurred mostly during the dark period, thereby creating a total inversion of their usual cycle.

Drs. Charles H. Sawyer and James H. Johnson, at UCLA, had found that rats could be stimulated by darkness to begin paradoxical sleep. When living in a rapidly alternating light-dark schedule (a half-hour light, then half hour of darkness), over 80 percent of the rats' paradoxical sleep occurred during the dark segments. Even without cues of light and darkness, the animals continued to show a circadian distribution of paradoxical sleep. However, this pattern was altered by the removal of the adrenal glands.

Other researchers have begun to condition animals to alter the phase or even the period of their usual sleep cycle. In some instances animals have redistributed sleep epochs around the clock, as they learned to sleep for food rewards. Drs. M. B. Sterman and Dennis McGinty have trained cats to produce bursts of EEG activity resembling the spindles of sleep. In studies of sleep, Dr. Sterman had noticed that some cats emitted spindle-like bursts during waking at times when they learned to inhibit a behavior. The EEG pattern seemed to relate to activity within the sensorimotor area. Cats were placed in special cages where they could move freely while being recorded through wire electrodes. A filter device activated a relay whenever the cat emitted the sensorimotor rhythm: thus, the cat's brain waves directly controlled a switch that presented the animal with a reward of milk. In this feedback situation, the cats began to produce the rhythm often, standing very still, at attention. The spindle EEG patterns seemed to occur when the cats suppressed certain aspects of motor activity.

The training was visible in their sleep. When compared with control cats, they exhibited more spindle activity. Their sleep might be said to have improved for it showed fewer bursts of phasic activity (the twitches that interrupt sleep), and they had longer epochs of deep quiescent sleep. Even two months later, they still showed less phasic muscle activity, and longer epochs of uninterrupted sleep than untrained control animals. The effect of the training had persisted.

Hopefully, human beings also will learn to use EEG feedback to prolong sleep cycles and reduce phasic muscle activity. Certainly, what we call "mental habits" have an impact upon behavior and also physiology, but untrained people usually have demonstrated greater skill in damaging than in improving their sleep. Dr. Sterman's experiments lend hope that the quality of sleep may be improved by voluntary manipulations of conscious states.

Certainly, there have been some indications that people can learn to alter their usual rhythm of REM sleep. Dr. Steven J. Ellman, of City College in New York, has trained subjects to shift the preponderance of their REM sleep from the latter part of the night, where it usually falls, to the early portion of the night. Dr. I. Lewin and coworkers, at Bar Ilan University in Israel, have conditioned volunteers to enter REM sleep when a tone was sounded during their sleep in order to avoid an electric shock. Those subjects who did learn to enter REM sleep promptly out of Stage II sleep were doing so at an unusual time—during the early hours of night.

Perhaps, people can be conditioned to control some of the phenomena currently defined as REM sleep, including eye movements, imagery, EEG configurations, and muscle relaxation. It will be interesting to know whether the autonomic and metabolic aspects that coincide with the REM rhythm also can be shifted voluntarily from their usual circadian phase.

Rhythms in Brain Biochemistry

Rhythms of activity and rest in animals have led many scientists to suspect that the circadian pendulum of our days and nights might be governed by slowly shifting chemistries in the brain. Decades ago, one popular theory supposed that daytime muscular activity built up an accumulation of natural biochemicals that produced changes in cortical cells. It was thought that such a hypnotoxin stupified and sedated the brain, causing sleep. This humoral theory of sleep evolved into more sophisticated formulations, inspiring experiments in which blood or spinal fluid was transferred from tired or sleeping animals into the brains of rested animals.

In a transfer experiment conducted by Dr. John Pappenheimer and his coworkers, at Harvard Medical School, dialyzed (protein-free) spinal fluid was taken from sleep-deprived goats and injected into the lateral ventricles of the brains of rats and cats. Injections from sleep-deprived goats made them lethargic, but cerebrospinal fluid from rested goats did not. The filtration of the spinal fluid excluded all large molecules, implying that a very small molecule was active, perhaps a kind of natural sedative.

One possible candidate, which appears to be produced in the brain, is 4-hydroxybutyrate. When injected into the spinal fluid of sheep, this substance has caused them to sleep for 36 hours. Dr. T. Hayashi has reported a rhythmic rise in the levels of this compound in chicken brains. It might be one biochemical route through which rhythmic changes in the nervous system enhance sleep.

Some of the other substances seem to have very broad effects upon the nervous system, affecting mood and behavior. Norepinephrine (also known as noradrenaline), serotonin, and dopamine are often referred to as biogenic amines; that is, biologically potent molecules of the amine family which are nitrogen compounds structurally related to ammonia and to the more complex molecules that form protein. Norepinephrine which is thought to be produced throughout the nervous system is related to adrenaline, (which is produced by the medulla of the adrenal glands, and suffuses into the blood in times of emergency with an immediate impact upon the release of energy). Although the enzyme that forms adrenaline has been found in the brain, it is far less prevalent in the nervous system than norepinephrine. Norepinephrine is ubiquitous, acting like a messenger in the tiny synapse between one nerve cell and another. Nerve cells fire a chemical molecule from their many branches to cross the synaptic gaps to other cell receivers, thus interacting with receptor nerve endings. After its brief errand the chemical transmitter quickly vanishes.

There are about ten billion neurons in the central nervous system of an adult human being. Each cell has about a hundred receiving synapses and a hundred projections that terminate on other cells. This astronomical network is the key to all we experience as life, our survival functions, and our feelings. It is not known whether neurotransmitters merely grease the path for a bio-electric code of impulses, or whether they modulate the frequency and extent of nerve firing. However they perform their roles, they have a pervasive effect throughout the brain and in the peripheral nervous system.

Serotonin has been considered a prime candidate for a physiological hypnotic. Its precursor, tryptophan, has been used with sedative effects by a number of researchers, but notably in studies by Dr. Harold Williams at the University of Oklahoma Medical School. Serotonin (5-HT, or 5-hydroxytryptamine), can be produced out of 5-hydroxytryptophane, and is generously distributed in animal and plant tissue, found in bananas, pineapples, in nuts, and insect venoms. Under the right conditions, it may contract smooth muscle, cause constriction of blood vessels, dilation of skeletal muscles, and complex changes in blood pressure. It may facilitate the release of adrenaline thereby influencing energy in the body, and perhaps it also acts as a transmitter. The precise action of serotonin is still a matter of conjecture.

Peripheral levels of serotonin in blood have been shown to rise and fall in a circadian rhythm in man. In 1960, Dr. Franz Halberg and his associates collected blood and urine every four hours from patients in a Minnesota State Hospital. When they later compared results on normal and mentally retarded people, they found that the mentally retarded individuals had lower levels of serotonin, but both groups showed a circadian rhythm with highest blood levels around 5 a.m.

One would not expect blood and brain levels of serotonin to show peaks at the same time. In the 1950's, the Minnesota team had found a drop in serotonin levels at the time just prior to a mouse's time of peak activity. In man, an end product of 5-HT metabolism (5-hydroxyindoleacetic acid, 5-HIAA) rises to peak levels in urine between 6 to 9 a.m., just around the time that a person would be awakening and getting out of bed.

Dr. W. B. Quay, at the University of California in Berkeley, found peak levels of serotonin in the rat brain and pineal gland around midday. Levels were several times higher at noon, when the animal normally would be sleeping, than at midnight.

In the early 1960's, Dr. Werner Koella, at the Worcester Foundation of Experimental Biology, showed that serotonin could generate sleep in cats without passing through the blood-brain barrier. It could activate serotonin-sensitive cells in a region of the brain stem called the area postrema, a frontier between body and brain. Since then, a number of scientists have explored the importance of serotonin in sleep and have injected PCPA (parachlorphenylalanine) into animals to block serotonin. (Among those who have collected data on PCPA are Drs. Michel Jouvet, Werner P. Koella, William C. Dement, Elliot Weitzman, Allan Rechtschaffen, Vernon Pegram, and Thomas Crowley.)

A single injection of PCPA is sufficient to cause insomnia in a cat for five days. It also reduces the sleep of rats and monkeys. A depletion of serotonin in brainstem regions was one striking effect of this drug. Another was an effect similar to phenomena seen after REM deprivation. Dr. Dement observed his cats very closely after injections of PCPA. They were not merely restless and sleepless, they acted as if hallucinating. Meanwhile, as EEGs indicated, their visual systems were inundated by waves of PGO spikes. The cats twitched, became hypersexual, and stalked other cats with a strange, abnormal aggressiveness. When the drug wore off, or whenever an animal received a serotonin-precursor, enabling his brain to reassemble a supply of serotonin, the insomnia would recede and the animal would subside into sleep. Subsequent experiments combining PCPA and a monoamine oxidase inhibitor, however, restored the EEGs of sleep, but biochemical assay revealed no change in the brain level of monoamines that might explain a return of slow-wave sleep and suppression of REM sleep.

It has seemed unlikely that serotonin, or any single biochemical substance, would play the role of a sleep transmitter. Drs. Arnold Mandell, Charles E. Spooner, and Don Brunet have demonstrated that a number of amino acids produce soporific effects in chicks. When tryptophan, tyrosine, glutamic acid, histidine, aspartic acid, leucine, glycine, and others were injected, soporific behavior followed. In a survey of the literature, these researchers found that adrenaline and norepinephrine had been known to generate lethargy and sleep in animals and man. Thus, a wide range of amino acids and a variety of metabolic transformations may modulate the excitability of the central nervous system, making it impossible that the rise or fall of any single substance dictates the complex state we know as sleep.

Recently, Dr. Ernest Hartmann, at Tufts University School of Medicine, along with Drs. Thomas Bridwell and Joseph Schildkraut, have injected rats with alpha-methyl-*para*-tyrosine, a substance that effectively decreases brain norepinephrine without affecting other amines. This manipulation increased the amount of paradoxical sleep. L-Dopa, which increases norepinephrine, had an opposite effect and decreased sleep somewhat. Yet, L-Dopa had increased REM time in depressed human patients when administered by Dr. Vincent Zarcone and his coworkers at the Stanford University Medical Center. If much attention has centered upon the possible roles of the adrenergic compound, norepinephrine, and around the indole, serotonin, still another major transmitter substance seems to play some role in the complexion of sleep. Drs. Edward F. Domino and Marek Stawiski, at the University of Michigan, have reduced levels of acetylcholine through the blocking drug, Hemicholinium-3 (HC-3), which acts upon the subcortical levels in the brain without affecting that outer layer, the neocortex, sometimes described as the "thinking brain." After such a substance, distinct alterations took place in the sleep cycles of cats effectively increasing slow-wave sleep, and decreasing paradoxical sleep, as well as alertness.

A few of these important brain biochemicals have now been measured in brain tissue at short intervals around the clock, and several researchers have now reported circadian rhythms in their levels in the brain.

Within the middle of the brain is an elaborate structure known as the caudate nucleus—a relay center that acts as a kind of storehouse for many substances needed in brain function. In 1968-69, Drs. Alexander Friedman and Charles A. Walker were working at the Loyola University School of Medicine in Chicago. They measured levels of three compounds in the caudate nucleus and midbrain. One was histamine, a substance involved in allergy and related phenomena. The role of histamine in the brain is unclear: Drs. Friedman and Walker suggest that it may promote wakefulness. Dr. M. Monnier had demonstrated that histamine injected into the veins or brain ventricle can produce wakefulness in cats. Would norepinephrine and serotonin show rhythms similar to those observed in body temperature and adrenal hormones?

The Loyola researchers took samples of caudate and midbrain tissue at intervals around the clock from animals previously adapted to a standard light-dark schedule. They found that when the animals were awake, norepinephrine was at its peak and serotonin at its nadir. The opposite was true when the animals were asleep. These findings tended to support the hypothesis that serotonin is involved in sleep production. The fact that peak histamine and norepinephrine levels coincide suggested that these substances might both play a role in promoting wakefulness.

Studies by Drs. Richard Wurtman, at MIT, and Donald Reis, at the Cornell University Medical School in New York, have shown rhythmic changes in serotonin and norepinephrine in the pineal. Norepinephrine synthesis appears to be stimulated by light, but the serotonin rhythm persists even in blinded animals.

Recently, Dr. Reis and his associates have mapped rhythmic changes in norepinephrine levels and serotonin levels in 25 regions of the brain. Their subjects were cats, maintained on a strict regimen of 12-hours light and 12-hours darkness. Like the regional rhythm of norepinephrine, the serotonin rhythm differed from region to region. In the olfactory tubercle, for instance, it showed a 12-hour rhythm reaching a nadir at the time when the circadian peak was reached in the caudate nucleus, red nucleus, and inferior colliculus. A biphasic rhythm in the visual cortex showed a nadir in serotonin at the time of the circadian peak in the globus pallidus.

Most regions showed a cycle for either norepinephrine or serotonin, but not both. In those few places where both amines fluctuated, the rhythms were out of phase. Such a reciprocal rhythm was visible in the hypothalamus, in areas controlling body temperature regulation. The authors were struck by the regional specificity in the fluctuations of both amines, which suggested that the regional cycles are independently regulated, possibly reflecting fluctuations in activity, cell discharge, or metabolism in the amine terminals which are located in the brain stem. The complex distribution and timing of the monoamine cycles in the brain would make it difficult to connect a specific brain region with a rhythmic physiological function and a cycle of biogenic amine levels.

If some sleep researchers have focused upon serotonin, norepinephrine, and a few other substances that show circadian rhythms in brain and glandular tissue, other scientists have looked for clues to the sleep cycle in the activity and metabolism of neural cells. Dr. Holger Hydén and his associates in Stockholm have seen a metabolic coupling between neural cells in the brain stem and the glial cells surrounding them. The role of glia is not understood, although some scientists suspect that these cells help to nourish the neurons. Succinic oxidase is an enzyme important in energy metabolism. During sleep, this enzyme is exceedingly active in neurons and not so active in glia in the brain stem. During waking, however, the cells seem to exchange energy-producing roles: then the enzyme shows most activity in the glia. The rhythmic activity of the enzyme did not appear, however, during barbiturate anesthesia.

In addition to rhythms caused by enzyme activity and levels of biogenic amines; neurons, themselves, may alternate roles. Some brain cells have exhibited interesting rhythms of responsiveness. Drs. Madge and Arnold Scheibel, of UCLA, have recorded from cells in many regions of the brains of kittens. In one study of neurons in the reticular activating system of the brain stem, they continuously recorded impulses from eight units over eight-to-ten hours. Six of these cells had pronounced cycles. For two-to-four hours they were highly responsive to sensory stimulation. Then, for one-to-two hours they became unresponsive. During the unresponsive portion of the cycle, the cells seemed to have shifted their responsiveness to internally generated stimuli instead of external stimuli. The finding may begin to illuminate the way the brain processes information. It suggests that brain cells may have alternating functions, allowing them to respond to externally generated stimulation and then to shut off, as it were, and stop attending to the noises of the world, while they turn to respond to the internal signals of the body and brain itself.

The ubiquity of ultradian and circadian rhythms in the already sampled portions of the nervous system and tissue from the rest of the body will probably inspire a new strategy in the search for the "trigger" of sleep. It is no longer possible to assume that the discovery of a circadian rhythm in any important substance means that the fluctuation promotes sleep. Probably, Drs. M. B. Sterman and D. J. McGinty suggest, a time map of brain chemistry rhythms will run an analogous course to the neurophysiology of sleep.

Independent researchers, focussing on loci in the brain as different and separate as forebrain and brain stem, may someday put their results together to see how the entire system is organized. Electrical or chemical stimulation causes sleep of varying qualities and duration, depending upon which of the many regions of the forebrain, limbic system, brain stem, and even spinal cord are stimulated, and when they are stimulated. Perhaps this intricate and large sleep system responds to the changing balance of neurochemistry, along all of its pathways. The organization of these rhythmic changes must be dynamic and complex. Along with the alternations of activity and sleep are intermeshed a continuum of vital functions, outpourings of adrenal cortical steroids at certain portions of the cycle, of growth hormone during other states of sleep, each waxing and waning in tempo with the activity of the individual. The cycles of energy metabolism, cell division, and tissue repair must be integrated with important rhythms of psychological function including the sleep and dreaming that make it possible for such an organism as man to absorb and cope with all his recent experiences, and probably to store his memories so that he can retrieve them. In short, these many functions must be smoothly intermeshed with man's activity cycle if he is to continue adapting to his environment.

Time is the school in which we learn
Time is the fire in which we burn.

Delmore Schwartz

The force that through the green fuse drives
the flower
Drives my green age; that blasts the roots of trees
Is my destroyer.

Dylan Thomas

Time held me green and dying
Though I sang in my chains like the sea.

Dylan Thomas

Chapter III. The Development of Rhythms: Youth and Age

When a newborn infant first arrives home he is exasperatingly unpredictable. His hungers and needs, his moments of mirth or squalling, and his hours of sleep are out of step with the rhythm of family life. Only the love of parents for their own children and the protectiveness of the old for the young bend the household to seemingly erratic demands until the time, around the sixteenth week, when the baby begins to conform to the patterns of nightlong sleep and daytime waking that adults take for granted and even consider "natural." As exhausted parents understand, a new baby cannot be coaxed to be hungry at a certain hour or forced to do his sleeping at night and his crying by day. His wetting and excretion are "irregular" as well, and he shifts toward "regularity" at a pace dictated by the speed of brain development. If he happens to develop the habit of nightly sleep and daytime eating earlier than another child, it is not owing to any particular cleverness or exceeding love on the part of his parents.

As each baby, after his own fashion, develops activity and eating, sleep and waking habits consonant with those of his family, he becomes the embodiment of an important riddle. Does he exhibit a 24-hour rhythm because he has inherited biological "clocks" that are set for roughly this cycle, or might he learn any rhythm of sleep and waking if only he started early enough?

In attempting to answer this question and discover what rhythms exist in the developing infant, researchers have begun to discover that the rhythms of sleep and certain physiological functions deviate from those of the adult for quite some time into childhood. These findings may someday give pediatricians a way of knowing whether a child is developing physiologically at a rate that will allow him to perform with his peers, and whether, indeed, his internal systems are all in proper phase with one another. It may be possible to anticipate developmental desynchrony before a youngster suffers from the consequences of having one bodily system—such as sex hormone production or skeletal growth—far outrun the rest of himself.

Newborn Infants and Sleep

Much of the recent physiological recording of infants took its inspiration from the work of physiologists,

Nathaniel Kleitman and T. G. Engelmann, at the University of Chicago. More than 30 years ago they began recording observations of newborn infants in a systematic manner around the clock. In attempting to decipher the rhythms of the newborn child, they studied neonates who were raised on a self-demand schedule, in which it was the cry of the baby that dictated feeding and diapering, not some fixed clock hour. Contrary to the popular belief that newborns sleep 21-23 hours, Kleitman and Engelmann soon saw that most newborn infants were not sleeping away the day. On the average they spent about eight hours awake, but much of their wakefulness occurred at night when normally nobody would be looking. As these pioneer researchers reported, babies spontaneously alternated between naps and waking in cycles of about 50-60 minutes.

More recently, Dr. Arthur C. Parmelee, at the University of California in Los Angeles, has used refined EEG and other recording instruments to observe large numbers of infants during their first days of life outside the womb. He found that newborn infants exhibited signs of individuality at birth. Some babies slept nearly 21 hours a day at first, but others spent only 10-16 hours asleep, in what may be a clue to lifelong differences in sleep needs.

Sleep, itself, changes from infancy to adulthood. Infants spend half of their sleeping time in an active state resembling the adult's REM sleep. This is a time when the baby may sometimes appear to be awake, for he kicks, making small twitching movements of his fingers, and he may suck, smile, and grimace. A baby born ten weeks prematurely will be in this active phase of sleep all his sleeping time, but at term (40 weeks from conception) he will spend only half of his sleep time in this active state. Slowly, this active state diminishes, until by the time the child is two years old, he spends only about 20-25 per cent of his sleep in this REM phase.

At birth, infant sleep shows roughly a 40-50 minute cycle, according to the EEG records of Dr. Parmelee and his associates. The infant spends about 21 minutes in active, REM-like sleep, then drifts into deep, quiet slow-wave sleep for 18 minutes, then into another activated phase. Every three to four hours the baby awakens. Judging from very sensitive recordings of premature babies, Dr. Parmelee and his coworkers concluded that infants must begin alternating between an active, REM-like state and a quiescent slow-wave phase long before they are born. Recently, Dr. E. J. Quilligan, at the University of Southern California School of Medicine, recorded periods of rapid-eye-movements in animals in utero, coinciding with a drop in pulse rate and blood pressure. Long before birth, as instruments and mothers both have testified, infants demonstrate their own REM cycles of physiological change and activity.

Fetal Activity Rhythms

Even in the womb, an infant periodically kicks and moves and then seems to be quiet for a time. This fluctuation of activity in the fetus does not seem to be random. Dr. M. B. Sterman, at the Sepulveda V.A. Hospital and UCLA, had performed many studies of sleep in cats and was sensitive to all sleep events at a time when his own wife was pregnant. He accidentally noticed that the fetus would kick and move a good deal when she was in REM sleep, and he wondered if fetal activity followed a rhythm like that of the REM cycle. Eight pregnant women volunteered to sleep in the hospital laboratory and to wear several disc electrodes on their abdomens over the fetus, as well as electrodes designed to transmit their own brain waves, muscle tone, and eye movements. These women were recorded collectively for over 30 nights. Samples of their unborn infants' activities were obtained beginning at 22 weeks gestation until birth.

It appears that most mothers are unaware how very active the fetus is from early months, until birth, for the mothers would report kicks and movements, but hardly seemed to notice the considerable activity revealed by recordings. When Dr. Sterman and his associates analyzed their records, they found that there were two fetal rhythms. One was an intrinsic activity rhythm of about 30-50 minutes; the other was an apparently imposed 80-100 minute rhythm related to the mother's REM cycle. In later stages of pregnancy, the women occasionally had to interrupt their sleep to urinate so that their sleep cycles were broken. However, fetal activity rhythms remained stable even when the mother did not show her usual unbroken sleep cycles. The stable REM cycles in the fetus, despite interrupted sleep in the mother, suggested that some manifestations of the REM cycle were being transmitted during waking.

After birth, the infants were immediately studied by Dr. Parmelee. The 90-minute cycle disappeared. At birth and at two weeks, the infants showed a 40-minute cycle of respiration, EEGs, body movement, and eye movement. The REM cycle in infants runs about 40-47 minutes, begins in the womb, and continues after birth with waves of intensive sucking, kicking, and grimacing, followed by quiescence. This cycle lengthens to about 90 minutes when the baby is around 8-months old, and persists throughout life. Presumably the 90-minute fetal cycle is provoked by biochemical cycles in the mother during gestation, and only reappears months later when the baby's own nervous system is more mature. The infant also shows a roughly 180-minute cycle of hunger and feeding. The relationship between these rhythms during gestation and after birth is a matter of conjecture at present.

Dr. Howard Roffwarg and his associates, at the Montefiore Hospital in New York, have speculated that the activated REM-like state may provide the growing nervous system of the fetus with stimulation and instigation necessary in maturation. Miraculously, at birth, a child knows how to cry and suckle. Perhaps, he has had prenatal practice in survival behaviors, such as kicking and thumb sucking, during REM states. Surely, as intrauterine photography has shown us, the unborn have some experience in the limbo of the womb, sucking, kicking, and making muscular-neural preparations for survival after birth.

Although most parents are more aware of how long their infant sleeps than the changing quality of his sleep, they would probably not observe the slow lengthening of the sleep cycle. At three months it is only a couple of minutes longer than the roughly 40-minute cycle at birth, but the proportion of quiet sleep has increased. By the time he is eight months old, an infant spends twice as long in quiescent sleep as in active REM-like sleep, and his cycle begins to approximate the 90-minute cycle of adults. This increasing predominance of sleep characterized by large, slow waves on the EEG may indicate brain maturation. Innumerable experiments in neurophysiology have indicated that this particular sleep state is highly dependent upon the activity of the cerebral cortex. The UCLA team of Dr. Parmelee has conjectured that infant sleep may reflect the quality of the baby's wakeful maturation. Quiescent slow-wave sleep may develop at a rate parallel with quiet, attentive waking behavior, while the active REM state appears to represent the activity in more primitive brain regions and may parallel the restless and inattentive waking state. If these conjectures hold, the EEG of sleep may be a useful means of detecting or confirming developmental problems in young children. These studies may explain why the dominant 90-minute cycle, the rhythm of adult sleep, takes so long to emerge in the infant.

Mother and Child Compatibility

The study of time structure in infants also raises questions about the synchronization that may exist between mother and child, and the role this may play in all the interchanges from nursing and tendering to disciplining and teaching. Is there a physiological synchrony in the harmony of a good mother-child relationship? For instance, compatibility in parent-child relationships might be influenced by the compatibility of the activity rhythms. A mother who needed nine hours of sleep and who was a slow riser might have difficulty coping with a baby who slept only eight hours a night and awakened like a jack-in-the-box, becoming most alert and demanding at a time when she was discoordinated and sleepy. To this writer's knowledge, no controlled studies have been conducted to explore the possibility of physiological synchronization as a basic element in mother-infant relationships. A hint that there may be such synchronization comes from Dr. Boyd Lester at the University of Oklahoma Medical School. He has observed instances of sleep-cycle synchrony in mothers and infants as they were recorded simultaneously throughout the night in the medical school sleep laboratory. Some pairs of mothers and babies seemed to slip into REM sleep simultaneously—a synchrony that may have begun before birth, when fetal activity cycles were coincident with the mothers' REM cycles. Indeed, the 90-minute cycle in the unborn baby might be related to a metabolic ultradian rhythm in the mother.

Dr. Lester found that the mother-infant synchrony of sleep cycles was occasionally disrupted when the mother had come to the laboratory after emotional upset, such as a fight with her husband. On that night and subsequent nights, she and her child would not exhibit synchronous sleep rhythms. This is not a surprising observation given the known impact of novelty and stress upon the sleep rhythm.

One of the most salient responses of infants and children to joy and to stress is a kind of rhythmic, repetitive action. From the head-banging and rocking of young babies to the spoon-banging and marching of children, there is ample evidence that rhythm is rewarding in early youth. The security of repetition may be one aspect of this proclivity: children will often repeat a nursery rhyme in perfect meter, but using nonsense syllables. Why do youngsters love to beat rhythms against their cribs or on their high chairs? They will delight in marching around a room to music, or to rhythmic nonsense, forerunners of poetry. Perhaps, rhythm is actually one of the earliest and most inherent pleasures. One New York psychiatrist, Dr. Lewis Greene, has speculated that the unborn infant, floating in the darkness of the uterine waters, experiences the vibrations of his mother's heartbeat and the pulsing of blood vessels and abdominal muscles, and that these rhythms compose his environment. Judging from the pulse rates of animal fetuses, the unborn do respond to sounds. Then, at birth, there are no longer the familiar beats, but new noises, movements, tactile changes that may seem chaotic and unpredictable. Perhaps, the rocking and patting a mother gives her newborn child helps him bridge the transition from a world of simple beats into the chaos of living. As she holds the baby, their respiration may approximate a synchrony. She may, indeed, partly recreate the stimuli of breathing and pulse that were the baby's most significant rhythmic environment before birth.

Development of Circadian Rhythms in Sleep and Physiology

The newborn infant may seem to sleep more than he actually does because his wakeful periods are not always noticeable. At ten months an infant may sleep only three hours less than he did at birth, but now his waking hours cluster in daytime when his parents can observe him. He also eats and wets by day. Although newborn babies may seem to disregard day and night, infants show that they are diurnal creatures at birth. If fed according to their own demand, they will be more insistant about daytime feedings than at night when they will forget an occasional feeding.

Analyses of the sleep-waking rhythms of infants fed on demand reveal a circadian rhythm, but not a precise 24-hour rhythm. Studies by Drs. Kleitman and Engelmann indicated that newborn babies tended to wake and sleep in a roughly 25-hour period, an occurrence which has been observed in adults during isolation. On the other hand, Dr. Theodore Hellbrügge, of the Universitat München in Munich, Germany, also has collected observations on infants on self-demand feeding schedules. When analyzed by Dr. Franz Halberg, they suggested a shorter than 24-hour period.

The circadian rhythm, in phase with societal day and night, appears in the sleep-waking behavior of infants around 16-20 weeks. However, not all of the baby's physiological functions show rhythms in phase with those of adults. Indeed, one of the interesting observations of Dr. Hellbrügge is that the different physiological systems develop adult circadian cycles at different ages.

The Development of Physiological Rhythms in Infants

Dr. Theodore Hellbrügge has studied many infants on self-demand feeding. A large number of them showed a pronounced 24-hour rhythm and notable day-night difference in vigilance and electrical skin resistance during their first week of life. The conductivity of the skin may be related to changes in the chemistry and moisture of the skin that result from nervous stimulation. Thus, skin conductivity has been interpreted as an indicator of autonomic activity. Vigilance and skin resistance showed night-day differences just after birth. However, potassium and sodium levels in urine also may reflect aspects of nervous and corticoadrenal activity, and they did not show day-night differences. Indeed, as Dr. Hellbrügge found, the newborn infant had a pattern of sleep and waking, of eye activity, of body temperature, pulse, and excretion levels of potassium, calcium, and sodium that were not synchronized with the 24-hour period of the world around him.

Most of the physiological functions and behaviors studied showed an overall rhythm that was ultradian, shorter than the 24-hour day. By the time the infant was 14-20 weeks old, however, the pulse rate, temperature, urine excretion, and such electrolytes as potassium did begin to show a clear 24-hour rhythm. Hellbrügge's longitudinal studies point out that some functions seem to adapt rapidly, while others do not develop circadian rhythmicity until a child is several years old. The staggered acquisition of circadian rhythms does not suggest that we are governed by a single central mechanism or "clock" in the brain. Dr. Hellbrügge and his colleagues have wondered why different body structures developed 24-hour periodicity at different rates. When systems are implicitly coupled in their development—as are pulse, temperature, and certain urine excretion rhythms—does this imply that the heart and kidney systems are bound together in their rate of development?

A Timetable

A number of pediatric researchers and physiologists have contributed to the timetable of physiological rhythms in infancy (among them Drs. Theodore Hellbrügge, Nathaniel Kleitman, Arthur Parmelee, and Robert T. Franks).

In testing 300 infants for various bodily functions, Drs. Hellbrügge and J. Rutenfranz found only one rhythm that they could identify as circadian in the first weeks of life. This was electric skin resistance. Resistance was high in the morning and low at night.

In the second and third weeks of life the rate of urine flow exhibited a rhythm for which mothers are always grateful. Urine flow was greater by day than by night. Yet, this was a rhythm that did not seem to depend on the baby's habits of fluid intake, for the infant was drinking very nearly as much at night as by day.

An adult's temperature varies a degree or two in the course of a 24-hour period. This pronounced circadian rhythm of body temperature does not appear early: between five and nine months the higher daytime temperatures begin to become easily detectable. It is about this time, as Drs. Hellbrügge and Rutenfranz have shown, that one can see a circadian periodicity in the infant's level of blood sugar, in the constituents of the urine, and in the urine flow. Between four and twenty weeks, the infant is beginning to give signs of a circadian rhythm in his heart rate. Only much later, as a child of a-year-and-a-half or almost two, does he show a strong circadian rhythm in the excretion of creatinine and chloride. Much earlier he develops a rhythmic excretion of ions (charged elements) that are presumably active in nerve-cell firing—phosphate, sodium, and potassium. Some of these developmental patterns have been seen in animal studies. For instance, the blood sugar rhythm and the inferred rhythm of glycogen in the liver have been studied in chicks and rats. Neither of these species shows a circadian rhythm in these functions on the first days of life.

Hormones

The blood of an adult human being undergoes vast chemical changes each 24 hours. There is an approximately 70 percent increase and decrease in the blood levels of certain adrenal hormones from morning to night. Adrenal cortical hormones (17-hydroxycorticosteroids) have been interpreted as one index of adrenal activity. These hormones play a role in reactions to emergencies and prolonged stress and appear to modulate some nervous activity in the body as well as the use of energy. According to a recent study, the levels of cortisol in the blood are highest in a newborn baby when he is crying. This study was performed by Drs. T. Anders, E. Sachar, J. Kream, H. Roffwarg, and L. Hellman, at Montefiore Hospital in the Bronx, New York. In the adult, concentrations of 17-OHCS in blood and in urine fluctuate in a circadian rhythm, dropping at night during sleep and rising to highest levels in the morning in people on schedules of nocturnal rest and diurnal activity. No such secretion pattern was observed in the newborn infants studied at Montefiore. Cortisol was found to rise in newborn infants in response to arousal and stress, but showed no circadian rhythm or correlation with sleep schedule. Lack of a circadian rhythm in the newborn infant is not surprising, since the levels in adults may be related to their schedule of sleep and activity and also to light.

Another hormone, whose release is activated by the pituitary gland, is known as growth hormone. Adults show high levels of this hormone in the early part of the night's sleep, concurrent with their deep, slow-wave sleep. This relationship was not seen in infants studied at Montefiore. After about five weeks, growth hormone was higher during REM sleep. Moreover, during a quiet waking state, the infants showed base levels of this hormone that were roughly equivalent to those of adults, but until about ten weeks, the levels were higher during sleep than waking.

Although adrenal corticosteroids and growth hormone both depend upon activity within the pituitary gland, their circadian rhythmicity appears to be governed differently, judging from sleep reversal studies in adults. An inversion of sleep and waking produces an immediate adjustment of Stage IV sleep and growth hormone to the new schedule, while REM sleep and adrenal cortical steroids may be out of phase for over two weeks. Both of these hormones seem to take considerable time to develop their adult rhythms.

Dr. Roger Martin du Pan in Geneva has studied 27 infants from the time they were five days old until they were three months old. Even though the three-week-old infants began to sleep a bit more at night than during the day, adrenal hormones did not immediately reflect this shift. The circadian rhythm of 17-OHCS and potassium in urine occurred about a week later than the shift in sleep. This steroid rhythm was tracked by three-hourly urine samples, which showed that even the three-month-old infant had not approached the circadian rhythm of the adult. Adults examined by Dr. Richard Doe, at the University of Minnesota Medical School, excreted their least amounts between midnight and 3 a.m., and maximum concentrations between 6 a.m. and noon. Babies show a biphasic rhythm with one minimum between 3 a.m. and 6 a.m., and another between noon and 3 p.m.

In recent studies of young children, Dr. Robert Franks, of the University of Texas in San Antonio, found that blood levels of the 17-OHCS did not show a consistent adult circadian rhythm in children under two years. However, children ages three-to-thirteen years did show rhythmic change comparable to adults, suggesting that the development of the adrenal tides may occur somewhere around age three.

The rate at which an individual child develops adrenal rhythms and the peculiarities of his pattern may be illuminated by animal researches in laboratories, such as those of Drs. Seymour Levine, at Stanford University, and Robert Ader, at the University of Rochester. They have shown that rats have malleable hormonal systems, and that the rate of growth of adrenal responses can be influenced by manipulations of the mother before delivery or by stimulation and handling after birth. The maturation of the 24-hour adrenal rhythm and activity rhythm can be accelerated prenatally. Moreover, some of the adrenal responses of the infant rat are further altered by manipulations at critical periods.

Circadian Rhythms: Nature or Nurture?

We will never rear infants in constant darkness, or on a 30-hour day, or 72-hour day. But the rhythms of animals have been studied under constant conditions, and under unusual day-night schedules for generation after generation. The activity rhythm is particularly central and also easily measured. Moreover, other rhythms of the body are synchronized by the routine of waking and sleep with some exceptions. Conceivably, mammals could bequeath circadian rhythms to their young by rhythms of maternal activity. This rhythm might be transmitted to the young mammals through successive generations despite constant or bizarre outside conditions. Birds, on the other hand, can be removed from all contact with parental rhythms. Eggs can be hatched in incubators. Dr. Jürgen Aschoff, of the Max-Planck Institut für Verhaltensphysiologie in Erling-Andechs, Germany, has found that there is a circadian rhythm of activity in newly-hatched chicks. Lizards also can be incubated. Dr. K. Hoffman, also working at Max-Planck Institute, has incubated lizard eggs in a constant environment, and even the newly-hatched lizards spontaneously exhibited a circadian activity rhythm.

Dr. Aschoff raised several generations of mice under constant temperature and light. Their mean free-running period and daily activity rhythms remained around 25 hours, even to the last generation.

Dr. Edgar Folk, Jr., at the University of Iowa, raised a litter of rats under constant laboratory conditions with a foster mother who had a different rhythm from the natural mother. The foster mother and natural mothers were exchanged and placed with the litter at random times each day. Over the 24 hours the young pups nursed from both mothers about equally. These little animals showed an activity period of 24.5-25 hours, suggesting that they had inherited a rhythm of "almost a day" and that this was not imprinted by nursing.

Dr. Robert Ader, at the University of Rochester Medical School, has begun a long series of studies in which he has attempted to see whether rats could be trained to adopt vastly different rhythms than the 24-hour rhythms with which we are familiar.

Initially, a newborn rat shows no daily rhythm in his blood levels of the adrenal steroid, corticosterone. Adrenal secretion becomes rhythmic only after the animal is 21-25 days old. The influence of environment is considerable, for animals that are shocked and handled right after birth mature earlier. They show an adrenal hormone rhythm at sixteen days: five to ten days earlier than animals not handled. Apparently early experience speeds maturation and speeds the appearance of a hormonal rhythm.

This same accelerated maturation is seen with rats stimulated prenatally. These are the offspring of mothers that were handled daily throughout gestation. Even though these pups were then reared by control mothers that had not been handled, they exhibited an accelerated maturation of the circadian rhythm of adrenal activity and, indeed, an accelerated development of the 24-hour activity rhythm.

The prenatal environment appears to influence the rhythmicity of the young, yet it has been difficult to demonstrate that the activity rhythm of the mother is transmitted to the young. At the outset, it was believed that a laboratory schedule offering light from 7 a.m. to 7 p.m. and darkness for the remaining 12 hours would influence the mother's rhythm of activity and sleep, thus influencing the unborn. In order to investigate this possibility, the Rochester team began one study by reversing the light-dark schedule just as the rats gave birth. Perhaps the altered schedule might retard the development of a circadian hormonal rhythm in the newborn. This rhythm might not appear until they were more than 21-days old. Presumably, comparable neonates whose mothers' lighting schedules were left untouched would develop their adult rhythms faster.

Unfortunately, the experiment could not be as uncomplicated as changing the schedule of light and darkness might imply, for there is also the pronounced mothering rhythm to be taken into account, the rhythm of nursing, cleansing, and touching. This rhythm, which has been observed by keepers of rat colonies, can be perceived easily in the young. The pups, while nursing, show a whitish crescent shape on the stomach during the day, indicating milk-filled bellies, but this crescent is not visible at night. At a specific time of day, the mother regularly leaves the nest, the nest cools, and so do the young pups. Their stomachs are emptied at this time, giving an indication of their regular nursing periods and excretion periods.

Dr. Ader and his associates at Rochester have documented this mothering rhythm by building a cage mounted on springs. This cage had a barrier across the middle. When the mother jumped to the side where her litter stayed, the spring weight automatically recorded the event on a time sheet. When she jumped to the other side, avoiding her litter, there would be no record.

When living on a schedule of 12-hours light and 12-hours dark, the mother rat spends more time with her litter during the hours of light. Perhaps this is because she nurses more when she is inactive, or it may have to do with rhythms of pressure from milk against the nipples. In any event, it is during darkness that she spends relatively little time with her offspring.

What happens to the maternal rhythm if the mother lives in constant light during pregnancy and subsequent lactation? Animals reared on 12 hours of dark and 12 hours of light were submitted to constant light for the entire three weeks of gestation beginning with conception. Even in constant light these mothers retained their circadian rhythmicity and showed it in tending their young.

A generation was raised on constant light. These rats were impregnated and went through gestation in constant light. Yet, after the birth of their young, they, too, showed a circadian rhythm of mothering.

Eradicating light cues was not enough—not even for two generations. The researchers now tried to confuse the animal. They would take a rat on a regular twelve-hour-light and twelve-hour-dark schedule, and within the first four days after delivery, when she was nursing her young, they would invert the lighting schedule. During the next several days, the mother would simply shift her mothering rhythm, continuing to show a circadian periodicity.

Since 24 hours is the usual nurturing cycle, it seemed natural to wonder whether a twelve, a six, or an eight-hour day were equally feasible. Ader's group reared rats on a twelve-hour day (six hours of light and six hours of darkness). In this group of rats mothering activity was curious. The mother seemed to be guided by light and spent less time with her litter during the dark period. Initially, she spent as much of the light period mothering them during the first six hours of light as during the second. Eventually, she reverted to a 24-hour rhythm with a primary peak of mothering behavior during a six-hour period of light and a somewhat diminished peak during the following period of light.

In a subsequent study of activity rhythms, ten of these animals raised on a 12-hour day were shifted to a 16-hour day (eight hours of light followed by eight hours of darkness). They did not develop a 16-hour rhythm. They reverted to a 24-hour rhythm. Many of them seemed disorganized, became inactive, and indicated that there was something unusual about a 16-hour day. Further studies of this regimen may reveal what problems are caused by a 16-hour day. Mothers reared in constant light will be switched to a 16-hour day at parturition. If successive generations of rats cannot adapt to certain schedules of light and darkness, it will suggest that there may be some built-in tendency to follow certain rhythms and not others unless it is possible to demonstrate that our rhythms are entrained by geophysical fluctuations.

Until now at least, the manipulations of successive generations of animals have suggested that there is something special about the circadian rhythm. Rodents appear to adjust to some variants of the 24-hour rhythm of light and dark but not to others. Perhaps, from a long inheritance of evolution on a turning earth of night and day, the nervous systems of living beings have acquired a potential oscillation that is close to 24 hours, but not exactly an oscillation like that of rotating earth and moon. Such a rhythm may be entrained by light and social customs with a period that is latent at birth.

No doubt the question of nature versus nurture will inspire many more studies. At the other end of life, too, an understanding of the relation between internal and external synchronization of physiological rhythms may provide an early detection method for slow impairments of aging.

Age and Sleep Rhythms

At one end of the spectrum, infants sleep and wake in a rhythm that deviates from the 24-hour day of industrial society. At the other end, old people tend to return toward this polyphasic way of alternating between naps and waking. To what extent do these reflect the permissiveness of the outside world toward babies and older people, absolving them from schedules, and to what extent does this depend upon the state of the nervous system? Surely, sleep studies indicate that the electroencephalograph tracings of the young are clearly distinguished from those of elderly people. The deep slumber, usually characterized by large, slow changes in the EEG (the delta waves that signify changes of 300 microvolts over periods of a second or two) constitute one attribute in which the developmental and aging processes are exceedingly distinct. Behaviorally it is obvious. Youngsters between three and seven typically spend the first hours of the night in this state of profound oblivion. Often, indeed, young children are picked up by their parents and carted home after a dinner party, put to bed, and give the entire transaction no more wakeful cooperation than a grunt or two. Set on his feet in this sleep, a two- three-year old may collapse or behave like a sleepwalker. An eight-year old spends over three hours in this state while an adolescent will spend two and a half. This is the phase of sleep that is not observed in elderly people, who usually complain that they sleep too lightly. Indeed, one of the trends with age seems to be a decrease in the EEG amplitude, as found by Dr. Edwin Kahn and his associates at Mt. Sinai Hospital in New York.

Because volunteers in their twenties were so abundantly available for a decade of laboratory sleep studies, this age group has been the best documented of all. After first descending into sleep the person in his twenties spends about ten minutes in the lightest stage, descends into a slightly heavier stage, then deeper sleep—delta or stage IV sleep at the bottom of the well, and up again through the lighter stages to rapid-eye-movement dreaming for about ten minutes. Then, in 90-minute cycles (plus or minus about ten minutes), he will repeat the cycle with less heavy sleep at the bottom, and more REM dreaming, in periods that lengthen slightly toward morning. By the time a person enters his thirties, the very deep, slow-wave sleep is beginning to diminish.

As every older person knows, the subjective sign of this change is that he no longer sleeps so deeply. The unperturbed and imperturbable sleep of children has vanished by middle age. On the other hand, the proportion of REM in the sleep cycle remains the same throughout life unless a person has suffered brain damage from cardiovascular disease or other complications. The continuance of this rhythm in normal old people was reported by Dr. Irwin Feinberg and his associates, now at Langley Porter Neuropsychiatric Institute in San Francisco.

Some changes in sleep rhythms are an indication of illness within the maturing and the aging brain, but others, as Nathaniel Kleitman has noted, may only mean that the restrictions of society have loosened their hold on the individual because he no longer needs to meet daily schedules of job and family. A retired person is free to nap if he wishes. Many clinicians realize that the elderly patients who complain about insomnia are napping all day. Moreover, the older person's naps may begin to be spaced like the polycyclic sleep pattern of the infant.

Do Circadian Rhythms Break Down in Old Age?

Insofar as time structure may be crucial to health, one might expect it to break down with age. Dr. Norberto Montalbetti has studied the levels of 17-hydroxycorticosteroids in the urine of very old people. Working in a hospital near Milan, he took samples around the clock—every four hours. At first, it appeared that there was no circadian variation, but further studies on the same people revealed that their blood concentrations of 17-hydroxycorticosteroids followed a normal circadian rhythm. There had been a very slight (not statistically significant) shift so that the peak concentrations occurred earlier than in normal people, perhaps due to sleep disturbance. Older people with abnormal urinary steroid rhythms were probably showing undetected illnesses of the urinary tract. Other researchers in Florence, Italy, suggested that the decline of gonadal hormones could influence the overall timing of an individual as he grew older. Still, the old people studied by Montalbetti did not show glandular impairment. The only possible difference from young subjects was a modest change in the circadian phase of the peak concentrations of the hormone in the blood of the old people.

Dr. Dorothy Krieger and her associates at Mt. Sinai Hospital in New York, have measured blood levels of 17-OHCS in a comparable group of old people in a nursing home. No abnormal rhythms were detected.

Dissociation of Rhythms and Age

If several physiological rhythms begin to shift in their phase relationship with one another, will the change in the body's time structure begin to manifest itself in illness? Is there, indeed, any more dissociation of rhythms in older people than in the young? Probably it will be a long time before sufficient data has been gathered to give more than tentative answers to such questions.

One preliminary study has compared circadian rhythms in a group of young and a group of older men. Dr. Harold A. Cahn, at Utica College in Syracuse, and Drs. Edgar Folk, Jr., and Paul E. Huston, at the University of Iowa, undertook the difficult job of making measurements at short intervals around the clock. It is very hard to do this with animals, often impossible with people. This study is unusual in that it specified hourly collection of samples over a 33-hour period to measure four physiological indicators. Since the peaks of body temperature, urine flow, heart rate, and potassium excretion usually occur within the same four-hour span, it would be possible to see discordance if one or more of these peaks moved out of the usual phase. Three groups of volunteers were studied, young men in their twenties and thirties; a group of healthy older men in their late fifties and sixties; and a group of men from the psychiatric clinic, depressed people in their late forties to late sixties.

Every volunteer was studied individually in a controlled experimental chamber where light was scheduled, food and water measured, and all external disturbances screened out. Each person spent an initial eight hours adjusting to the chamber, during which hourly readings were made. He received light meals, each equal to the others in calorie content. Every hour he was asked to lie down in darkness for half an hour in order to eliminate any possible influence from a rhythm of muscular activity.

Of eleven young men, all but two showed perfect internal synchrony, and two showed a slight dissociation of one rhythm. There were six elderly men, and three of them exhibited dissociation. That is, there was no neat parallel between the peaks and troughs of their four measured functions. Peak urine flow, for instance, might occur during the afternoon, yet the peak potassium excretion would be way out of phase, occurring at night when urine flow was low. There were five depressed patients, and four showed dissociation, indeed, a great degree of desynchronization.

Most of the daily potassium excretion occurs at the time of maximum urine flow. Since potassium may influence the function of the nervous system, the movement of peak excretion out of phase with other functions might be symptomatic of illness. The authors wondered whether this discordance might be one indicator of an early change in timing that could lead to depression. Although there have been a number of studies of adrenal steroid levels in depressed patients, few studies have involved the round-the-clock sampling of four or five functions that might indicate whether there was desynchrony of phase relationships among normal rhythms. Thus, this study may be a first indication of a tendency toward dissociated rhythms in older people, particularly among depressed older people.

A dissociation of physiological rhythms might emanate from many origins—organic illnesses, lack of social pressures and schedules, irregular habits, virus, fever, shock, emotional stress. All of these may contribute to the increasing load of mental and physical disease that a person suffers as he grows older.

It is not yet known whether rhythms and social schedule are important in treating geriatric disease. Some day an understanding of circadian systems may become an important part of medicine for the aged and may offer insight into the extreme sensitivity of older people to slight changes in schedule, small doses of drugs, and moderate alterations in light, temperature, or noise level.

It is possible that internal discordance evolves when a person's activity and rest schedule is not strong enough to force him to follow an unvarying routine of sleep and wakefulness. Do infants and the aged both testify to the fact that circadian rhythms are synchronized by our schedules? Circadian rhythm studies are just beginning to look useful in the diagnosis and treatment of developmental problems in babies and problems of aging. By studies of internal phase relationships in elderly people, it may be possible to forecast oncoming symptoms such as depression, from the phase shift of a few internal functions, and to learn what importance social scheduling may have in the well-being of the person. At both ends of life, moreover, it may be possible to learn some of the implications of the "different drumbeat" each of us embodies. Time structure may be one of the profiles that should be obtained early. Even at birth some infants sleep only ten hours a day while others sleep as much as twenty hours, and differences in sleep needs are apparent early in life. Studies of evolving circadian rhythms in young animals and babies are likely to reveal other individual differences in "timing" that may be useful clues to lifelong attributes. These idiosyncrasies in "beat" may indeed presage subtle differences of temperament and aptitude that are not yet articulated or defined in the language of time.

That period of twenty-four hours, formed by the regular revolution of our earth, in which all its inhabitants partake, is particularly distinguished in the physical economy of man. . . . It is, as it were, the unity of our natural chronology.

C. W. Hufeland, THE ART OF PROLONGING LIFE

London, 1797

Chapter IV. Within the Compass of a Day

Technological society tends to treat individuals as if one were like another, and as if a man at a given hour were the same at any other hour. This attitude, while conveniently flexible, may be based on an unrealistic image of human beings. In the detailed fabric of our existence, our days and nights, our times of trouble or confusion, our hours of strength and clarity may be biased by a complicated intermeshing of biological time cycles.

The unit of a day is only one of many periods that finds expression in our physiology and behavior, but it is our most important social-time unit. The extent to which we change within this social period of 24 hours is too considerable to ignore, and an understanding of these changes may help to forestall some of man's illnesses and to aid in treating others. Techniques that offer a quantitative picture of our changes over time are also likely to influence our own uses of ourselves as beings with time structure who can self-consciously develop habits for greater comfort and health. Only a rare few people have an untutored grasp of their own timing: some need no watch to know the time, and others sleep and awaken to the beat of some internal clock. Despite the large role of biological cycles in our existence, they influence most of us without our awareness. We use phrases of acknowledgement, vague references to evening fatigue, or a "best" and "worst" time of day. Some people take afternoon naps during a regular daily "lull" in attention, and others are aware of recurrent hunger. We usually urinate soon after rising in the morning, but never ask why it is then, rather than at night that the need is most obvious.

As the hours go round, we perform a complex series of cycles internally, a progression of changes that alters our performance, our senses, the way we metabolize food, the symptoms of our illnesses, our vulnerability to stress or disease, and even some of the subtle displays of vitality and idiosyncrasy we fondly know as personality. In a brief space it is possible to offer only a glimpse of the many changes that take place within us each day.

Body Temperature

In the course of a day there is one familiar signpost of our changing physical state: our internal temperature systematically rises and falls over the course of a day, as does our skin temperature. The temperature varies about a degree or two each twenty-four hours, with almost clocklike regularity. A person's favorite hours of the day are likely to coincide with the high-point; this usually occurs in the afternoon and evening for a person who is active by day and sleeps by night.

One difference between people who sleep very well and those who sleep poorly, between people who love to bound out of bed in the morning and those who drag around for an hour or two before they come to life, may be the shape of this temperature cycle and its metabolic concomitants. One of the first evidences of this individual difference arose in a study by Lawrence Monroe, at the University of Chicago. He reported that a good sleeper's temperature tended to rise before he awakened in the morning and reached "normal" about the time he got up. But the poor sleeper's temperature did not rise so early and was still rising long after the person was up and awake. It neither dropped so low, nor did it rise before he awakened. Although these results must be considered tentative, they may offer first clues to physiological propensities in so-called morning people and night people.

The high-temperature period is temporally associated with the kinds of muscular tensions we display when we are awake. The low-temperature period is associated with the relaxation of the musculature during sleep. Many researchers have attempted to show a cause and effect relationship between muscular activity and the temperature cycle by putting volunteers in bed for several days and taking their temperature around the clock. As Dr. Nathaniel Kleitman reported in his book, *Sleep and Wakefulness*, the rhythm persisted despite inactivity, showing that the temperature rhythm neither depends upon muscular activity nor food intake. Its origin is not clear: it may be the intermeshing of several metabolic processes. Whatever the source, it is hard to alter the 24-hour period of the temperature cycle in a normal adult.

A famous expedition was led by Drs. Mary Lobban and Peter Lewis, who took students and colleagues to Spitsbergen, Norway. Here, in the summer, the difference between day and night is almost imperceptible. In 1953, they studied subjects who lived a 22-hour day. In 1955, the subjects lived a 21-hour or 27-hour day. Although some functions seemed to adapt to the short or long day, other functions continued to follow a 24-hour day.

The stubbornness of the roughly 24-hour temperature cycle does not mean that it cannot be changed. However, to date, normal, healthy people have not exhibited significantly different temperature rhythms, even when they have lived long, erratic days and remained in isolation as long as six months. The temperature rhythm may be controlled via the hypothalamus, a region of the brain that is known to regulate temperature, and where malfunction can result in inexplicable fevers. While the rhythm may adapt to periods shorter or longer than 24 hours, the range of adaptation may be narrow. No healthy person, for example, has yet produced a body temperature with a rhythm of 7 or 30 hours.

At about the time of day when our body temperature is normally reaching its peak, various other functions also are changing. The speed of pulse, pulse pressure, pulse-wave velocity, circulating blood volume—all have been shown to have a circadian rhythm. Since this rhythm persists in people who are kept in bed and who take meals at regular intervals around the clock, it is not dependent on exercise or eating. Even when meals are equally spaced, the pulse rate seems to be high at about the time the temperature is highest, and to drop during the night along with blood pressure in a slow decline very similar to that of the temperature rhythm.

Circadian rhythm of the peak body temperature and peak excretion of adrenal steroids (17-OHCS) remain in phase even when people live in isolation from time cues on activity-sleep cycles of 25 or 26 hours. Drs. Alain Reinberg, Jean Ghata and associates, in Paris have analyzed the physiological records of a man and a woman who independently spent over three months in a cave; the peak excretion of urinary steroids (17-OHCS) continued to precede the peak rectal temperature even when the circadian rhythm was extended from 24 hours to almost 25 hours.

The Urinary Electrolytes

Many of us do not notice our rhythms unless they go awry. When awakened out of a deep sleep in order to urinate, we realize this is an unusual time for the need. Why is it that we sleep through the night without urinating, although we drink as much at dinner as we do with other meals of the day? Since ancient times, people have been aware of a circadian rhythm in the volume of urine flow. In 1890, a German investigator named Lahr experimented on himself by staying in bed and taking fluids around the clock. He reported that his urine flow was still rhythmic, declining to a nadir during the eight hours of sleep. Similar studies have been performed on night nurses, and people sleeping with indwelling catheters.

Dr. J. N. Mills and his associates, at the University of Manchester in England, have been particularly interested in the great changes that one can observe in kidney function around the clock. They studied fasting people and people eating meals spaced evenly around the twenty-four hours. They have kept people in bed and studied people who slept at unusual hours.

The kidneys are huge bean-shaped organs about four-and-a-half inches long, two-inches wide, and an inch-and-two-thirds thick on either side of our spinal column, stretching up from the lumbar, or bottom region of the spine, into the central thoracic cavity. They are detoxifying organs, a kind of filtering and secreting system. One measure of kidney function is its rate of filtering metabolites and unwanted wastes from the blood and expelling them from the body in urine. What we eat and drink is metabolized, converted to energy, into tissue or placed in storage. The by-products are constituents of ammonia, which is transformed into urea and uric acid. We excrete less urea at night than by day.

The rate of urine excretion may be the consequence of a number of different rhythms. One may be the hormone known as ADH, or antidiuretic hormone, which comes from the posterior pituitary gland. Dr. E. Szcepanska and his associates, in Warsaw, have shown that there is a pronounced circadian rhythm in concentrations of ADH in the blood of normal people. The same rhythm accompanies changes in diuresis and specific density of the urine. Various hormones might govern the flow or retention of water within the body, but what we excrete is more than water. The urine contains a group of ions—potassium, chloride, phosphate, calcium. These electrolytes, as they are called, influence transactions in the nervous system and influence events in the membranes of cells. Researchers have been interested in the balance of excreted electrolytes as a possible clue to the shifting background that influences central nervous system transactions. Drs. J. N. Mills and S. W. Stanbury, at the University of Manchester in England, have found that phosphate has its lowest excretion rate just after a person wakes up.

Studies of the rhythms of sodium and potassium excretion have led to conflicting results, an inconsistency that may result from the fact that subjects were not all on standardized diets and schedules of eating and sleeping. In general, however, it does appear that both sodium and potassium are excreted mostly around midday and afternoon.

The circadian rhythm of potassium excretion in urine is particularly easy to detect and study because of its high amplitude. As Dr. Alain Reinberg and his coworkers in Paris have shown, it may be used as a reference rhythm along with the rhythm of body temperature. According to Dr. Theodore Hellbrüge of the Universitat München in Germany, this potassium excretion rhythm appears between the fourth and twentieth week of life in infants. The peak and trough both shift in phase when a person changes his schedule of sleep and activity; and the rhythm is altered in people with adrenal insufficiency, as shown by Drs. Halberg and Reinberg. Even when people are isolated from time cues during experimental sojourns in caves, the potassium excretion exhibits a persistent circadian rhythm extended over the longer day-night period of the individual. Healthy subjects who retire at 11 p.m. and rise at 7 a.m. show their peak potassium excretion at midday, between 10:30 a.m. and 2:30 p.m., as Drs. Alain Reinberg and Jean Ghata have shown, but peak concentration in red blood cells occurs later, around 7-8 p.m. These peaks and troughs may describe subtle but pervasive changes in the tone of the nervous system, and also the rhythmic activity of the adrenals.

Magnesium and calcium are influential electrolytes. Dr. Vincent Fiorica and his coworkers, at the FAA Civil Aeromedical Institute in Oklahoma City, looked for dietary or other external factors that might determine the calcium and magnesium rhythms. Their eight volunteers were sampled around the clock, while living on a controlled diet. The two electrolytes followed the rhythm of their activity and rest, but unlike the peak excretion of potassium and sodium which occurs during waking, the peak excretion of calcium and magnesium occurs during the sleeping hours.

There are individual differences in electrolyte rhythms, as one might expect. However, it is hard to compare individuals unless they are acclimatized to the identical schedule of sleep and meals. Moreover, one can only obtain an individual's profile by looking at several cycles simultaneously. Do his minimum blood levels of 17-OHCS coincide with minimum body temperature, pulse, urine flow, and potassium excretion?

Creatinine: A Physiological Constant?

For many years, researchers who wanted to see whether the levels of various substances fell within the "normal" range, would use creatinine as the basis of comparison. This is a urinary by-product of muscular activity. Creatinine was considered a constant, and its level in the urine was taken as a reference standard until the 1950's. Dr. Per Vestergaard, now at Rockland State Hospital in Orangeburg, New York, decided to take round-the-clock urine samples from normal individuals. He found that creatinine, like almost all physiological products we can measure, is not constant but fluctuates in a circadian rhythm.

"The Urinary EEG"

The kidney normally clears the metabolic byproducts of our many systems of tissue growth, protein utilization, and hormonal output, and it also clears the biochemical wastes that indirectly result from our emotions. Since one can sample urine without hurting a person or limiting his freedom, urine analysis has become a means of indirectly observing activity in the nervous system. Many investigators interested in the physiological aftermath of stress, but unwilling to disturb their volunteers by taking blood, have examined urine samples, sometimes at four-hour intervals, and mostly just once a day. The urine contains traces of the activity from a complicated group of glands, which includes the pituitary, the thyroid, the gonads, and the adrenal medulla and cortex. There are more than 40 substances secreted by the adrenals alone. Some are considered hormones, others precursors to hormones or metabolites showing hormonal activity, and they play diverse roles in regulating body functions.

In moments of peril, the brain signals the pituitary gland to release a well-known hormone, most frequently referred to by its initials, ACTH (Adreno-Cortico-Tropic-Hormone). This messenger hormone travels to the adrenals through the bloodstream and stimulates the outer layer of the adrenals—the cortex, to release corticoadrenal hormones such as cortisol, cortisone, and corticosterone into the blood, thus mobilizing body tissue and indirectly releasing sources of energy to muscle groups and to nerves around the body by releasing extra blood sugar. Some hormones, among them aldosterone, regulate the retention of water and salt in the body. A group of steroid hormones known as 17-ketosteroids includes the male hormones testosterone and androsterone. These are secreted in the testes as well as the adrenals and, of course, influence sexual function. But the control of the secretion and release of these hormones also lies in the nervous system, and it is for this reason that scientists sometimes seek indirect traces of emotional reactions in the hormone products of the urine.

For over a decade, researchers have suspected metabolic flaws as a cause of schizophrenia and have expected variability in the output of the adrenal glands among schizophrenics. Dr. Per Vestergaard has studied urine samples taken every two hours from both normal people and schizophrenics. He began to see that normal people excreted these steroids in as variable a manner as the patients. Some individuals showed a consistent output regardless of changes in the climate, whether they were getting married the next day, doing their income tax, or had played a stiff game of tennis. Factors that supposedly influence the hormonal output did not seem to change their consistency. Yet, other people showed tremendous variations in situations that would not have been called stressful. The measurement of a single hormone in the urine tells very little about emotional lability or physical activity.

Many urine studies have been inspired by a classic study done about 25 years ago by Dr. Gregory Pincus, whose research has helped to develop contraceptive pharmacology. Dr. Pincus studied the excretion of 17-ketosteroids in seven young medical students, from whom urine was collected at regular intervals, night and day, for several days. The ketosteroid excretion was high in the morning and low at night. Today, with better controls and methods, such studies of daily rhythms in urinary constituents could become useful for medical diagnoses and psychological research.

Most of us lead lives so busy that we do not notice we feel differently about the same things at different hours of the day. We may be unaware of the changing chemistry of our urine, yet it indicates a history of changing emotions and glandular responses to the day's events. These are biochemical traces of feelings superimposed upon the cycle of changes that occur in the body we inhabit. The body is, indeed, a habitation in constant flux. We may not sense the tracks of change, yet we know people simply do not behave the same way at night as in the morning. And why should they? They are totally different from bone to blood.

Blood

Whole blood is composed of many varieties of cells that we divide into white and red. Their relative numbers, the populations of certain kinds of white cells relative to red cells, may give an index of liver or adrenal function and of health or infection. By separating blood samples into their components and counting different kinds of cells under the microscope, it has been possible for researchers to estimate the proportions of different cells at different times of day in a normal person. The numbers of blood cells can indirectly indicate what is happening in the cortex of the adrenal glands. Before it became possible to specifically measure the adrenal hormones, patients suspected of adrenal disease were given blood-cell tests. Doctors counted the white cells known as the eosinophils after injections of cortisone. When the white cell count was high, it was found blood cortisol and cortisone would be low, and vice versa.

In the early 1950's, Dr. Franz Halberg and his associates began to demonstrate that the eosinophils had a marked circadian fluctuation in blood serum. In hundreds of experiments, mice and rats were maintained on uninterrupted schedules of 12 hours of light alternating with 12 hours of darkness in laboratory conditions kept as constant as possible. These standardized conditions, a necessity in time-series studies, became the prototype for many laboratories. So long as the rodent lived on a rigid schedule of 12-hours light and 12-hours darkness, the adrenal hormones would drop to their lowest level in the blood at the beginning of the light span, the beginning of the rodent's time of rest. Hormone levels would reach their peak at the end of the illuminated rest span, the onset of darkness, just before the rat began his nocturnal activity. Under controlled conditions, the rhythm of the adrenal hormone cycle, as seen from blood counts, or later from direct assay, was so regular that it could be used as a pointer. By knowing an animal's phase on his cycle, it was possible to know how he would respond to various kinds of external stimulation, to stress, or cortisone. Because light and dark are the primary synchronizers of adrenal circadian rhythms in rodents, a shift in the light-dark schedule would be followed by a shift in the adrenal or blood-cell rhythm. A shift of a few hours might require four days of readjustment, while an inversion of lighting would entail nine days or more before the cycle came back into its usual phase with the external lighting regimen.

The relationship between the adrenal cortical hormone rhythm and populations of white blood cells was explored extensively by Dr. Halberg. Blinded mice showed rhythms that deviated from the precise 24-hour cycle, running between 23 and 25 hours. The eosinophils, lymphocytes, and leukocytes all showed a pronounced circadian rhythm along with other white cells. However, the circadian rhythm became almost invisible in animals whose adrenals had been removed. Moreover, under conditions of constant light, the rhythm appeared to vanish.

Drs. John Pauly and Lawrence E. Scheving, at the Louisiana State University Medical Center in New Orleans, have continued and refined the exploration of rhythms in blood-cell populations along with other characteristics of the blood. Rats kept in 12-hours light and 12-hours darkness showed a rhythm of neutrophils, eosinophils, and lymphocytes, almost in perfect synchrony with peaks in the early hours of inactivity and light. Under continuous darkness this synchrony persisted, but when the animals were kept in constant light the circadian rhythms of these blood cells became dissociated.

Not all of the white cells have the same phase. In a study of pregnant women it was found that the total count of leukocytes, of segmented neutrophils, and stab neutrophils rose to a maximum during the day and fell to a minimum in the early hours of the morning, while the eosinophils and lymphocytes showed the reverse trend. Monocytes rose during the evening and fell during the early morning hours. In 1962, the New York Academy of Sciences published a volume, *Rhythmic Functions in the Living System*, edited by William Wolf, in which several authors cite the studies of many blood rhythms known up until that time.

Many properties of the blood change in a circadian manner. The speed of blood coagulation, for instance, shows a circadian fluctuation.

Drs. Scheving and Pauly, along with Dr. Tien-hu Tsai, have recently examined fluctuations in a variety of blood proteins. Blood that has been separated into its components, or fractionated, contains a number of proteins. These include mucoproteins, albumen, and the several globulins. Gamma globulin is the fraction of blood serum containing most of the immune antibodies to viruses, bacteria, and other foreign proteins. In recent studies by the Louisiana team, it has become evident that levels of mucoproteins change by as much as 41 percent and gamma globulin by as much as 28 percent each day. In rats on a schedule of 12-hours light and 12-hours darkness, the rhythm of gamma globulin showed a peak during the last hours of darkness and a nadir during the first hours of darkness. The animals seemed to have highest levels of gamma globulin during the last six hours of their circadian activity period—a rhythm that should be of interest to immunologists.

The Liver

Aristotle thought that the liver was the seat of the emotions. The liver's indirect effect upon behavior is so dramatic that one can see how he might have conceived this idea. Although most of our cells synthesize glycogen, a basic substance we use for energy, it is only the liver that has the capacity to store it and to regulate glucose or blood sugar around the clock, even when we are asleep. We must have glycogen at all times, and when a person is starving for food it is taken from the liver. Glycogen is the food of the brain; it is particularly essential at night when glucose from food is not available for energy.

Even under conditions of thirst and starvation glycogen rhythms in the livers of animals and birds have been seen to endure for quite some time. Dr. Halberg and his associates have elegantly displayed the circadian rhythm of glycogen deposit in mice. Even when the animals were deprived of food the rhythm persisted. Its persistence, even to the death of the animal, suggests that time structure must be essential to the creature's hepatic system.

In the human being, the glycogen curve appears to begin its circadian descent around late afternoon. By the early hours of the morning (3 a.m.-6 a.m.), the liver has used up much of its glycogen. This rhythm is relevant in the treatment of diabetics and in understanding responses to poisons or drugs.

The liver, in its rhythmic way, detoxifies the metabolities of many foods and drugs. Its role in the metabolism of carbohydrates and proteins is quick to affect the brain. When a person has cirrhosis, his liver can no longer handle nitrogen properly. Along with a multitude of other changes, nitrogen accumulates and may begin to form ammonia. If ammonia goes through the blood and interferes with energy metabolism in the brain, the behavioral effects may be very striking and may resemble temporary psychosis. In moderate form, all of us have seen the indirect effects produced by changes in liver activity.

Many people show irritability when blood sugar becomes low. The economy of our metabolism appears in hunger periods, which occur at about four-to-six-hour intervals. Although we blame this hunger on the capacity of our gastrointestinal system, it may not actually originate there. Hunger, as obesity studies have shown, may have very little to do with the stomach and a great deal to do with the brain. The content of the blood is one of the signals to the brain whether or not we need to lower temperature or to eat more. Because the body depends on the liver as a reservoir of energy-producing glycogen, it may be a rhythm of the liver within our economy of activity that influences our tendency to want to eat.

There is a possible relationship between energy metabolism and the behavior of certain psychotics. People who have suffered schizophrenia or manic-depression chronically are not usually sick all of the time, but they have sick spells at intervals—sometimes only two months a year. Often a telling signpost that a normal period is ending is an increasing anxiety and restlessness. A close observation of the way some of these people eat is revealing. They start by abandoning their normal diet of meat and vegetables. At the beginning of a psychotic crisis, they may gobble up doughnuts, bread, ice cream, potato chips, acting as if they are ravenous for carbohydrates and sugar. Indeed, they act as though they are starved of blood sugar in a manner known as hypoglycemia.

To a moderate degree, normal people sometimes suffer from the mental symptoms of a very mild hypoglycemia. It is a syndrome that often goes undetected. One of the symtoms is that the person is constantly reaching for something sweet, and that he becomes irritable, depressed, or anxious. This may explain food cravings seen during depression and exhibited by some women at certain points of the menstrual cycle, according to a recent report by Drs. S. L. Smith and C. Sauder. Hypoglycemia sometimes may be the root of problems in seriously neurotic people who can find no origins for their ill moods. Since there are pronounced circadian rhythms in blood sugar levels and the rate at which the blood will dispose of glucose, a blood sample taken at only one hour of the day may not yield much information.

Energy and ATP

The liver exerts a profound, if indirect, effect upon emotions and behavior and also regulates hormone levels by breaking down adrenal hormones. Energy in the form the body uses is hard to measure, but animal studies have revealed that there is a rhythm in the production and breakdown of the body's most fundamental energy unit. This is ATP, adenosine triphosphate.

Dr. Colin Pittendrigh and his associates have measured variation in the breakdown of ATP in the livers of hamsters. Within each cell there are, of course, many components. The mitochondria have become known as the power houses of cells. When a cell requires energy, its water breaks up the bonds of the ATP molecule in a process known as hydrolysis. During the course of hydrolysis, one of the phosphates is stripped away forming ADP (adenosine-di-phosphate) and releasing energy.

In studying the hydrolysis of ATP in hamsters, the Princeton team raised their animals in an environment with 12 hours of light and 12 hours of darkness. Since they are nocturnal, the animals were most active during darkness. Then, at intervals around the clock, livers were excised and put into a centrifuge, a device that spins substances at such high speeds that the components separate by weight. The mitochondria were separated from surrounding tissue, and ATP and ADP measured. It was discovered that ATP was being transformed to release energy 25 percent more rapidly during the night, when the animals were active, than it did by day.

A relationship between the breakdown of ATP and behavior has been demonstrated by Drs. E. D. Luby, J. L. Grisell, C. E. Frohman, H. Lees, B. D. Cohen, and J. S. Gottlieb at the Lafayette Clinic in Detroit. This group has studied people during prolonged sleep deprivation. After a waking vigil of about one hundred hours, people began to hallucinate and show slightly psychotic behavior. They would see fire bursting from the walls and suspect their friends of conspiring to kill them. They also showed changes in motor performance and in energy transfer systems.

After about four days of sleep deprivation, blood samples indicated that sources of ATP and ADP were beginning to run down in sleep-deprived subjects, and that was when behavior changes began to occur. At first, energy production seemed to increase by an emergency process of synthesizing substances that are not usually utilized for energy, but after the fourth day this emergency system also seemed to be depleted. Once the individuals went back to sleep, the energy systems returned to normal and so did their behavior. This study suggested that sleep must be central in the body's rhythm of energy production and metabolic functioning.

Nutrition

People who want to lose weight are often advised to eat their heaviest meal in the morning. Although this advice began with a hunch, there is now some evidence that there are rhythms in the formation of proteins in the body. These might tell us the most efficient time at which to eat carbohydrates, proteins, and other food elements of our diet. Recent findings suggest that breakfast may, indeed, be a good candidate for the biggest meal of the day.

Amino acids are molecules that make up the building blocks of proteins. Since the mid-1960's, a number of biologists and endocrinologists have found a 24-hourly rhythm in the levels of amino acids in the blood of young chicks, mice, and human adults. In 1967-1968, a number of young endocrinologists and biologists were separately working on rhythms of amino acids in the blood of human adults. The daily variation of one of these, tryptophan, had been traced in human adults and mice.

Tyrosine is another one of many amino acids we utilize to make food proteins and hormones. Tyrosine happens to be essential in producing melanin, a substance that determines skin color. It is also a component of thyroxine, the thyroid hormone that helps to control body metabolism. It is a precursor for brain transmitters, epinephrine and norepinephrine. Tyrosine may be used in the process of transforming fats or proteins into sugars for body energy.

Tyrosine transaminase is an enzyme that transforms tyrosine into a precursor of sugar. As might be expected, it is found in the liver. By 1966, it was well-known that the production of glycogen followed a circadian rhythm in liver tissue. This suggested that an enzyme that manufactures sugar would also show a roughly twenty-four hour rhythm. Drs. Richard Wurtman and Julius Axelrod, of the National Institute of Mental Health in Bethesda, tracked this rhythm in the livers of rats and attempted to find out how the periodicity was controlled.

The nocturnal rat was kept on a schedule of light from 8 a.m. to 8 p.m., and dark for the remaining 12 hours. The animals ate very infrequently and rested most of the daylight hours becoming active at night. During the first half of the day the liver enzyme showed little activity, but as the day wore on it slowly increased, reaching a peak about two hours after the onset of darkness, then dropping off throughout the night in the rat's active period.

The activity of the liver enzyme is now known to be enhanced by adrenal corticosteroids. Was the rhythm driven by the adrenals? The investigators removed the adrenals from some of the rats, but the cycle of enzyme activity continued. The cyclic rise and fall of liver tyrosine transaminase continued even after the removal of the pituitary gland.

Recently, Drs. Ira Black and Julius Axelrod have found that the neurotransmitter, norepinephrine, could regulate the activity of this tyrosine degrading enzyme by decreasing its activity in the liver. The rhythm persisted in constant light or darkness yet shifted its phase when the light schedule was reversed, suggesting that the cycle was driven by a "clock" mechanism within, although it could be synchronized by light. Because the rhythm was suppressed at peak levels by transection of the spinal cord and chemical interruption of nerve impulses, it appeared to be caused by some central brain mechanism.

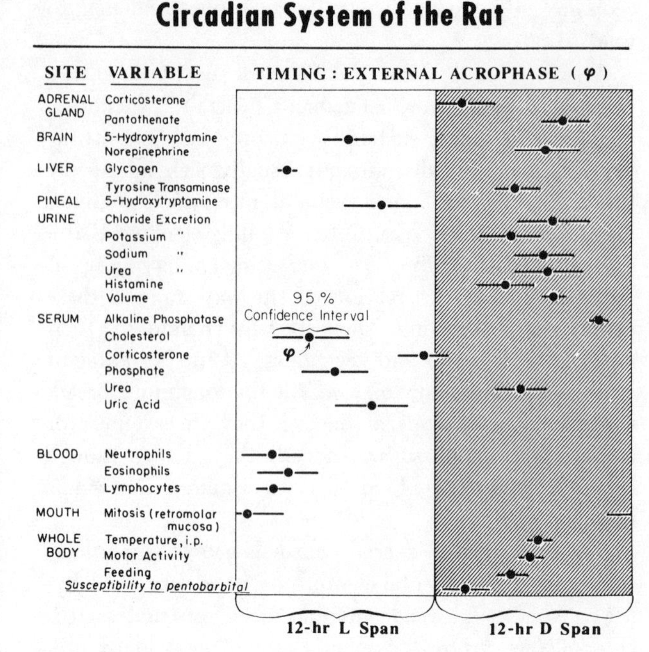

Acrophase, (represented by a dot) the peak of the circadian cycle, is shown here with reference to the light-and-dark cycle of the rat. The original data came from the Chronobiology Laboratory at the University of Minnesota, except for data on liver tyrosine transaminase (J. Axelrod), blood leukocytes (J. Pauly and L. Scheving), brain 5-hydroxytryptamine, and susceptibility to pentobarbital (L. Scheving), and on urinary volume and histamine (C. Wilson).

The local, biochemical mechanism of the rhythm was clarified by an ingenious series of experiments, in which Drs. Black and Axelrod increased tissue levels of norepinephrine and suppressed the enzyme cycle at low levels. When they blocked norepinephrine by injecting animals with the drug, alpha-methyl-*para*-tyrosine, the enzyme activity in the liver increased. When they injected norepinephrine into animals, the enzyme rhythm almost vanished. Indirect increases of norepinephrine were caused by the drugs that inhibit the enzyme monoamine oxidase, which ordinarily breaks down norepinephrine. After monoamine oxidase inhibitors, the tyrosine transaminase rhythm was suppressed.

Only in test-tube observation did it become quite apparent that both tyrosine transaminase and norepinephrines competed for the same cofactor in the liver—pyridoxal phosphate. This vitamin is necessary to the enzyme action. Since norepinephrine appears to suppress the tyrosine transaminase rhythm by competing with the enzyme for the cofactor, it seems possible that the enzyme rhythm might be caused by variation in the ratio of norepinephrine to pyridoxal. This work suggests how body biochemistry may be influenced by the brain.

In human beings, blood levels of amino acids are used for diagnosing certain metabolic diseases. Tyrosine concentrations, for instance, may allow a doctor to distinguish between people with normal and hyperthyroid conditions. Clinicians taking tyrosine levels at 9 a.m. have reported higher concentrations in people with overactive thyroids than among healthy individuals, but these observations were made before it was widely known that tyrosine levels rise and fall in the blood plasma in a daily rhythm.

Dr. Wurtman and his associates at MIT first observed the rhythm of human tyrosine levels in a study of six healthy students. Their blood was sampled and analyzed while the young men stayed on a special diet. During the twenty-four hours of assay they had a diet containing a normal amount of protein. They were kept in bed for the seven hours between midnight and 7 a.m., and blood samples were taken at three-to-five-hour intervals.

The blood analysis showed that tyrosine concentrations reach their high point at about 10 a.m. and the lowest point around 2 to 4 a.m. Exercise only enhanced the rhythm. Two subjects were required to walk a treadmill from 4 to 4:45 p.m.; their tyrosine levels went up 8 percent. Although the amount of tyrosine in the blood seems to depend upon the amount of protein one eats, its rhythmic fluctuation does not. When eight men were put on a very low protein diet for two weeks, their levels of plasma tyrosine became extremely low. Nonetheless, the rhythmic rise and fall seemed to persist. The greatest rise in amino acid levels occurred when the men were asleep. Activity, the direct effects of muscular exercise, and the rhythms of activity and rest did not seem to be the source of the rhythm. Moreover, if rat studies can suggest what happens with human beings, the tyrosine rhythm of man does not depend upon the adrenal glands.

Tyrosine is not the only amino acid that shows a rhythmic change in blood concentration, as a study of 23 young men soon showed. Staying at the MIT Clinic, they were divided into three groups. Each group received a different diet: one an ordinary three meals a day, a "house diet" containing about 1.5 grams of protein per kilo of body weight. A second group received a formula diet of egg protein, while the third was placed on a very low protein diet. During the next 24 hours, blood samples were taken at three-hour intervals except during six hours left undisturbed for sleep. The blood was analyzed for levels of 16 amino acids. Some variation stemmed from the diet, but most of the amino acids appear to follow a rhythm like that of tyrosine, fluctuating even more over the 24 hours than did glycine, alanine, or glutamic acid, which comprise about 80 percent of the free amino acid pool in the blood.

Recent amino acid studies clearly imply that the "normal" plasma level depends upon the time of day at which blood is sampled. All of the plasma amino acids appeared to follow a twenty-four hour rhythm in volunteers who were living a 24-hour day. The rhythm, itself, continues whether or not a person happens to eat a high protein diet, but it may be influenced by adrenal steroids. Dr. Wurtman and his associates were able to shift the peak of plasma tyrosine rhythm by an injection of a synthetic hormone, dexamethasone. Cortisone also can depress the plasma tyrosine concentration, and, in addition, insulin may influence the plasma rhythms of amino acids.

The conception of a "normal range" of an amino acid in the blood now must incorporate an individual's biological time of day and his diet. Normal tyrosine levels might be only half as high at 2 a.m. as they would be at 10 a.m.

If concentrations of amino acids in the blood limit the amounts of these essential substances in the organs, they may indirectly limit the rate at which muscles, brain, and other organs can synthesize needed proteins. Thus, the availability of amino acids may determine the fluctuations in such organs as the adrenal and its production of hormones.

The Timing of Protein Utilization

While the MIT team was clocking levels of amino acids in the blood of student volunteers, another group of medical researchers was making a similar study at the U.S. Army Medical Research Institute of Infectious Diseases at Frederick, Maryland. Drs. Ralph D. Feigin, Albert S. Klainer, and William R. Beisel measured whole blood and serum amino acids levels in six normal men in an attempt to see whether the rhythms were altered by heavy protein diets, by extreme exercise, or by a shift in the wakefulness and sleep cycle. For five days, the military volunteers stayed in the metabolic ward of the hospital, where diet, sleep, and activity were carefully controlled.

One amino acid, methionine, showed a 100 percent increase in the evening. Isoleucine also increased an enormous amount. All of the 18 amino acids measured (except for citrulline) showed considerable change between morning and night, exhibiting a circadian rhythm. During another phase of the study, three men consumed very heavy protein test meals at different times of day. Blood samples were taken before the test meal, and again repeatedly throughout the next six hours.

There were many variations in diet. Two men received approximately twice their usual dietary protein, while another two fasted for a day. Five-hundred grams of liver eaten at 8 a.m. led to a very swift rise of amino-acid levels. Yet, the same protein meal eaten at 8 p.m. did not elevate blood levels of amino acids above the fasting value generally found at that time. In fact, there was a decrease. So, indeed, an evening protein meal did not prevent the usual nightly decrease in concentration of blood amino acids. Manipulations of diet showed that it didn't matter whether a person reduced his protein intake to 10 percent of the normal level or increased it to twice the normal level. The blood rhythm of amino acids in each subject was about the same. In fact, a single day of starvation had no detectable effect on the amino acid concentration nor on the periodicity. Despite day-to-day differences in amplitude, the period and phase of the rhythm remained constant for all the men studied.

Sleep Reversal

In a further experiment, six volunteers were brought into the Army hospital for controlled studies of the timing of sleep, activity, and meals. They were studied for three days on their normal schedule; then were made to sleep by day and be active at night. This inversion was continued for about ten days, after which they were returned to their normal schedule and studied for the following week. Every day at four-hour intervals, blood was sampled, and during the phase shift blood was obtained around the clock at two-hour intervals. Rhythms of body temperature, excretion of sodium and potassium, and changes in amino acid periodicity were also recorded.

Normally, when they slept at night, the peak concentrations of amino acids in the blood occurred somewhere between noon and 8 p.m. However, after the shift in the sleep-wakefulness cycle, their periodicity reversed and the peak concentrations came between 4 a.m. and 8 a.m. When they were returned to their original usual schedule, the amino acid rhythm also returned. The amino acid rhythm seemed to be related to the rhythm of sleep and waking, but not the temperature cycle. The blood amino acid rhythm adapted rapidly to the phase shift, even in men whose body temperatures and urinary rhythms resisted the inverted schedule for many days.

Three of four men did not reverse their temperature rhythms until the sixth or seventh day after the phase shift. Two men promptly showed a reversal in their rhythm of sodium excretion, but a third only reversed after ten days, and one never did show adaptation but excreted large quantities of sodium in his urine during his sleep period. Two subjects seemed to lose the periodicity of sodium excretion. Still, within 72 hours after resuming their usual schedule, all the volunteers showed their usual cycles of urinary electrolyte excretion.

Many manipulations were used to attempt to decipher the basis of the amino acid rhythm in the blood. It was more dependent upon the time of protein intake than the amount ingested, for even two weeks on a very low-protein diet did not change the diurnal rhythm of plasma tyrosine. Because the rhythm did not follow the body temperature rhythm during a 180-degree change in sleep-waking schedules, it would seem to have a different relationship to metabolic rate than does body temperature. Exercise increases both metabolic rate and rectal temperature, but it produced no detectable change in the blood amino acid levels.

Because the blood amino acid levels so rapidly adapted to a new schedule, even when body temperature and urinary rhythms lagged, this rhythm would seem to be more easily synchronized by external factors than is body temperature or urine rhythms. The circadian amino acid rhythm did not seem to depend upon rhythms of hemoglobin or any other blood elements that normally reach a peak in the early morning. Insulin levels rise after protein meals, and insulin may enhance the transfer of amino acids across the membranes of cells. If the rising levels of insulin after protein meals facilitate the use of these amino acids in the body, insulin is also likely to affect their rhythm. Since growth hormone and thyroid hormones also play important roles in regulating the way we utilize proteins, these may also interact with the rhythm of amino acids.

Food and Stress

The efficient use of food also depends on the stresses we experience. In 1965, Dr. N. S. Scrimshaw, of the Department of Nutrition and Food Science at MIT, and his associates, studied effects of sleep loss and sleep reversal on 19 students. The diet was divided into equal portions, consumed at 8 a.m., noon, 5 p.m., and 10 p.m., while the students kept their normal daytime cycle; then during reversal of day and night, the same four-hour spread of meals took place at night.

One way of measuring protein utilization is to see how much nitrogen is excreted in urine. Sleep loss seemed to be accompanied by an initial retention of nitrogen, as well as of sodium and water. The first 24 hours without sleep was associated with a drop in body temperature, pulse rate, and increased retention of nitrogen. On the second day of sleeplessness, however, there was an increase in nitrogen excretion, which would have corresponded to a 12 percent increase in the body's requirement for protein. After the second day of sleep loss, the subjects required even more protein.

Effects of sleep reversal do not show up immediately. If there was no observable effect on nitrogen excretion during the first full day of reversal, after seven days the nitrogen loss amounted to about 6 percent of the average daily protein requirement.

Although some students hardly reacted, other students were extremely reactive and showed as much as a 20 percent increase in nitrogen excretion during the sleep deprivation in the five days after their 180 degree phase shift. They would have had to compensate by adding protein to their diets. This study does suggest that the efficiency with which a person utilizes food is related to the time of sleeping and waking and the regularity of the schedule. Protein is not utilized evenly around the clock, and some people need to compensate more than others for interruptions of their regular schedules.

Circadian Rhythms in Taste, Smell, and Hearing

Some evidence suggests that it may be physiologically efficient to eat a big breakfast. Yet, people throughout the world seem to prefer a large midday or evening meal as the social and culinary event of the day. This may have something to do with blood levels of hormones that create a rhythm of taste and smell acuity each day. It may be no accident that dinner smells so good, or children's play seems a little louder at night when people cannot tolerate as much noise as in the morning, and that lights may feel irritatingly bright; sensitivity does not remain the same at all hours of the day.

Indeed, our sensory keenness probably fluctuates in a circadian rhythm, judging from the work of a group of endocrinologists led by Dr. Robert I. Henkin, at the National Institutes of Health in Bethesda. The work was originally oriented around the peculiarities of patients suffering from insufficient output of adrenal cortical hormones, either because of Addison's Disease or pituitary problems. Such people often suffer from extreme and continual fatigue. Sometimes they seem to crave salt. This is interesting in the light of earlier work by Dr. Curt P. Richter, at Johns Hopkins University in Baltimore. Richter showed that rats whose adrenals had been removed would avoid a solution of salt that normal rats would drink. The adrenalectomized animals did not dislike salt, as he discovered, but presumably they had been rendered so sensitive to a tiny amount of salt in water that normal intensities were too strong for them. This suggested a curious link between lack of adrenal hormones and taste sensitivity.

In 1962, Dr. Henkin and his associates compared taste acuity in thirteen normal volunteers, two patients with anterior pituitary insufficiency, and seven patients with Addison's Disease. First thing in the morning each person rinsed out his mouth. Then drops of distilled water were placed on the tongue. The water began to contain tiny amounts of salty, bitter, sweet, or sour substances (sodium chloride, potassium chloride, sodium bicarbonate, sucrose, urea, hydrochloric acid). How many increments of the taste substance had to be added to the water before the person detected that it was different from plain distilled water?

The patients with adrenal cortical insufficiency were first tested without any drug treatment and subsequently while on medication. Without medication, these patients had much more sensitive taste than normal people. Their detection was at least 150 times as sensitive as that of the normal person. Later, however, after hormone treatment to make up for their own insufficiency, these patients were no longer any more sensitive than normal volunteers. Adrenal hormones appeared to be affecting taste sensitivity.

Two general types of adrenal steroid hormones were administered in the study, one a carbohydrate active steroid like cortisone or prednisolone, the other a sodium-potassium active steroid, deoxycorticosterone—DOCA. DOCA did not alter taste sensitivity, but treatment with carbohydrate-active steroids reduced taste detection to normal. The investigators inevitably wondered whether the carbohydrate active steroids had a similar influence on other sensory modalities.

They tested olfaction in much the same way they had tested taste, using the bottles that had contained the taste substances. The patients could detect smells even more acutely than they tasted; for instance, salt has little or no smell to most people, but it contains enough chlorine gas to be detected by patients with adrenal insufficiency. Again, treatment with DOCA did not diminish their detection, but after treatment with carbohydrate active steroids their smell sensitivity for all vapors gradually diminished to normal. Patients with adrenal insufficiency provide a marked contrast with those patients whose adrenal disease causes overproduction of carbohydrate active steroids. Cushing's Disease is one example, and patients with this kind of illness could not taste or smell even highly concentrated substances.

Visual and Auditory Perception

Normal subjects and patients with Cushing's or Addison's Disease were exposed to a flash of light while wearing electrodes on the skull that would transmit changes of polarity from the brain to the electroencephalograph machine to be amplified in the form we know as brain waves. By repeating the light stimulus and repeatedly recording the brain's EEG response, the research team was able to average the EEG responses and gain an estimate of the speed and intensity of the brain's response to the light flashes. Addisonian patients were recorded in this manner without drugs, then after treatment with DOCA, and after receiving a carbohydrate active hormone. There was no ambiguity about the effect of the hormone. They showed a far more rapid response to the light flash after receiving the carbohydrate active hormone. Thus, these steroids seemed to be increasing the speed of the brain's response, which suggested that they must be directly affecting nerve tissue. The curious role of these hormones in sensory perception was elucidated further by tests of hearing.

Not only does the person with Addison's Disease detect tastes and smells more sensitively than a normal person, but he can detect sounds inaudible to the normal person. An "average" person can hear low tones of about 50 cps, a rumble, and tones as high as 15,000 cps, just lower than a dog whistle. As a person grows older this frequency range shrinks, and most older people cannot hear much over 10,000 cps. People with Addison's Disease don't follow this rule. They detect sounds softer by 20 decibels than a normal person of the same age, particularly in the middle range of frequencies that include most voice and music sounds. The older patients, moreover, showed an expanded sensitivity to high frequency sounds, hearing much higher frequencies than normal people of the same age. In short, the person with low adrenal cortical steroid output seems to be unduly sensitive to all sensory stimuli, and Addisonian patients are easily disturbed by noises and are uncomfortable in a normally noisy room. Yet, their range of hearing and detection of soft sounds diminished to normal after carbohydrate active steroids.

Sensory Detection Versus Integration

When asked to rate the loudness of different sounds or repeat lists of words spoken through a distortion filter, Addisonian patients did surprisingly badly. Even though they had been given practice lists of words, and could hear soft sound below the range of normal people, they could not reliably repeat word lists presented or recognize filtered speech. Strangely, they could not judge loudness as well as normal people, nor discriminate as well as normals between a steady and warbling tone. Although deficiency of adrenal hormones rendered them more acute in detecting sounds, they lacked an auditory sense of direction, proving very inaccurate in localizing sounds. Again, their entire performance returned to normal after they were treated with carbohydrate active steroids. It appeared that these steroids influence the nervous system in a manner affecting the integration of sensory stimuli.

In subsequent studies of cats, the NIH team soon detected adrenal steroids in brain tissue and in the spinal cord in concentrations that might permit an impact upon nervous activity. When they adrenalectomized cats, the levels of these steroids fell in the brain and spinal cord as well as other body tissue.

The speed of nerve conduction was next studied in cats, in normal people, and in patients with adrenal insufficiency. The researchers gave a slight shock to the ULNAR nerve at the elbow and timed the response of muscle contraction in the hand. Again, the outcome was unexpected. Animals that had been adrenalectomized and untreated patients with adrenal insufficiency reacted much faster than their normal counterparts. Treatment with DOCA did not slow the response, but after the carbohydrate active steroids, their responses were that of the normal counterpart.

Further cat studies corroborated these findings. Axonal conduction was faster in adrenalectomized cats than intact creatures, but the curious thing was that conduction across the synapses was slower. The synapse is the tiny gap between the contacts of one nerve cell and another. In studying nerve transmission, the researchers implanted microelectrodes on both sides of the synapse. Now they began to see the role of adrenal hormones in the speed of nerve transmission.

In a normal cat, the nervous impluse was conducted at the speed suitable for the message to be integrated. When the cat's adrenals were removed, the axon conducted more rapidly than normal and the synapse more slowly than normal, so that signals were no longer processed at speeds normal for cell-to-cell communication. This curious effect of adrenal hormones on the speed of axonal and synaptic transmission might explain why Addisonian patients were overly sensitive to taste, smell, and other sensory stimulation, yet had difficulty judging, discriminating, and integrating this sensory information.

Since carbohydrate active steroids influence the excitability of the nervous system, one might expect daily fluctuations in sensory acuity and performance in normal people as a consequence of the circadian rhythm of adrenal cortical steroids. The NIH team wondered whether these changes would be sizeable enough to detect, and they set up a round-the-clock study in the metabolic ward of the Clinical Center at the National Institutes of Health in Bethesda.

Circadian Rhythms in Sensory Keenness

During a period of 36 hours, tests were made every four hours of taste, smell, hearing, and the velocity of nerve conduction in normal volunteers. The results are of interest to a variety of clinicians, psychologists, and pharmacologists.

A normal person's sensory acuity appears to reach its maximum around 3 a.m. (if he goes to sleep regularly around 11 p.m.). Ordinarily, people do not listen to music or eat snacks at this time of lowest steroid levels. After 3 a.m. (depending upon the person's sleep routine) there is a sudden drop in acuity as steroid levels begin to rise. During the day, as steroid levels decline, acuity increases. Around 5 to 7 p.m., people reach a high point of sensory acuity, beginning to resemble that of Addisonian patients. At the end of the day rather than the beginning, they can do things that require keen detection of taste, smell, or hearing, but not requiring fine discrimination. It is interesting to know that there are also detectable circadian rhythms in the brain-wave patterns we use as crude indicators of the background state of consciousness.

EEG Rhythms

Several years ago, Drs. Gilbert Frank, of the Hospital of the University of Washington, Seattle, and Franz Halberg, of the University of Minnesota, and their coworkers sampled blood cortisol and EEGs around the clock for 30 hours. Their 19 subjects were maintained on an identical schedule of rest, of waking, and meals for a week beforehand and continued a normal routine of waking activity. Three were the scientists, themselves, who recorded their own EEGs, and took their own blood samples every three hours.

This study revealed clear evidence of a circadian rhythm in the human EEG output, judged by frequency and amplitude, even in the scientists who were too busy to sleep. A number of round-the-clock studies of brain-wave patterns have been accomplished with permanently implanted animals. Month-long studies of implanted monkeys have been conducted at UCLA by a team of scientists, preparatory to launching an experimental monkey into orbit. Dr. T. Hosizaki and other members of this group have published autospectral densities derived from their recordings from many parts of the brain. Drs. Thomas Crowley, at the University of Colorado in Denver, Daniel Kripke, of the Albert Einstein School of Medicine, Franz Halberg, and G. Vernon Pegram, at Holloman Air Force Base, have analyzed continuous EEG recordings in rhesus monkeys. They have reported a circadian rhythm modulating the frequency and amplitude of the EEGs. In addition to records from animals, there are a few 24-hour records from neurosurgery patients in whom deeply implanted electrodes have been necessary for diagnosis. In the absence of biotelemetry, however, the picture of circadian rhythmicity in the brain waves of human beings remains largely restricted to the hospital setting. In 1965, Dr. D. Ivanov and his coworkers in the Soviet Union reported their analysis of EEGs that had been recorded by biotelemetry from freely moving volunteers who were living out their usual schedules of work and recreation. In the translated summary of this work, it appeared that the EEGs in the alpha range (9-13 cycles per second) were faster around 5 p.m. than in the morning, although by absolute count, there were more alpha waves during sleep than during the waking portion of the day. The analysis of bioelectric intensity did, indeed, imply that there is a circadian modulation of the frequency and amplitude of the EEG rhythms that we interpret as "background states" of consciousness.

Future research will quickly tell how much these circadian modulations of the EEG, or cycles of steroid levels, influence the behavior of human beings. Most social groups gather in evening to converse, eat, play music, gathering at a time of relative steroid deficiency and sensory heightening. Whether a person lives in a primitive or highly industrialized society, his skill demanding tasks are likely to occur during hours of higher steroid levels, when integrative faculties may be at their peak. Indeed, there is already a growing literature that indicates how performance is correlated with our physiological cycles within each day.

Performance and Fitness

Although circadian rhythms in performance and acuity are issues of great economic interest, influencing the safety provisions of airlines and the performance of pilots and air traffic controllers, there is at present no display of the phase of a man's different physiological functions, performance, and stress resistance around the clock, indicating what to expect of an individual at any given hour. A spate of recent studies gives some indication of the way such a map will look when it is finally possible to correlate biotelemetry measurements of a person's physiological rhythms with his performance at various tasks.

Even before systematic, round-the-clock studies, it was suspected that body temperature and performance were related. Today, with fine temperature readings that indicate ultradian rhythms (30-minute cycles in cats) as well as the circadian rhythms reaching a daily high point and nadir in late sleep, some researchers suspect that many brain functions may be modulated by the underlying metabolic rhythms that cause these changes in body temperature. Individuals, moreover, will vary somewhat in their temperature rhythms, some showing a faster rise in the morning, others a steeper fall at night. These are sophisticated differences, for an individual's normal body temperature is not apt to change by more than 1.20 degrees each day.

This is a very small change to measure over the course of the 24 hours. It has been exceedingly difficult to develop performance tests refined enough to indicate changes of similar magnitude. This has been accomplished by an English scientific team at the National Medical Research Laboratories in Cambridge. Drs. Robert T. Wilkinson, Peter Colquhoun and their associates tested Navy recruits on vigilance and arithmetic, and other criteria after various schedules of sleep or sleep loss. Dr. Wilkinson had previously found that a highly motivated subject could compensate for sleep loss on short tests, lasting only an hour or so, but not, if he spent a full eight-ten-hour day in the laboratory doing various hour-long tasks. Vigilance, for instance, might be gauged by his ability to detect a faint signal in a noisy setting, listening to a train of beeps through earphones for the one that was unlike the others, perhaps a fraction of a second shorter or longer. A person's performance could be rated on the percentage of total signals detected, number of errors, and the time it took him to respond after the signal. In one series of studies, oral temperatures were taken hourly during waking and every two hours in sleep. Temperature rose an average .56 of a degree in the first three hours of waking, and remained stable changing only .4 degree during 60 percent of the day. When signal detection, response latencies, and calculation outputs (on a calculation test) were computed with regard to each subject's oral temperature, these aspects of performance changed with temperature, although decision performance and calculation errors did not. The most consistent improvement in performance efficiency occurred during the first three hours when temperature rose most. Peak performance coincided with the time of peak temperature, and low performance coincided with the interval of lowest temperature.

It has been thought that a metabolic rhythm may underlie the body temperature rhythm, and that this may act as a modulator on the central nervous system. Twenty-four-hour EEG recordings do suggest that a circadian rhythm modulates the frequency and the amplitude of the brain-wave pattern. Thus, a 24-hour record might show that a person's alpha rhythm was about 9.5 cycles per second at one hour, and 10 cps at another, with a change in amplitude of several microvolts. There appears to be a parallel circadian rhythm in the way an individual estimates short-time intervals (such as 10 seconds or 2 minutes). Such estimates might be influenced by slight changes in the frequency of the EEG background rhythm. For instance, Dr. Walter W. Surwillo, at the University of Louisville in Louisville, Kentucky, has reported that a subject would estimate a 10-second interval fast (9.74 seconds) when his alpha rhythm was fast, but would give a longer than 10-second estimate when his alpha rhythm was slower. Drs. Wilkinson, Donald Tepas, and many others have looked at EEG responses of the brain at different hours around the clock. These brain-wave responses are the immediate pattern emitted by the person's brain after he has seen or heard or felt a test signal. Dr. Ernst Pöppel has examined changes in EEG patterns in relation to circadian changes in time estimation, adding numbers, vigilance, subjective fatigue, body temperature, keenness of hearing, and potassium excretion.

Critical Flicker Fusion Frequency

One way of measuring acuity is to present a subject with a flashing light. If a light is turned on and off at a high rate, it begins to look like a steady beam and is said to fuse. The rate at which the light is flickering when it first appears to fuse is called the critical flicker frequency, often abbreviated CFF.

Drs. John F. Walsh and Henryk Misiak, at Fordham University in New York City, gave tests of this sort to students every three hours over a period of twelve hours. They saw that the students fluctuated between perceiving the light as flickering and perceiving it as fused. Since the CFF test is used in evaluating drugs and the sensitivity of the nervous system, it is well to know that an individual's reaction may fluctuate a good deal even during the daylight hours. The Fordham study suggested an inverse relationship between CFF and time of day. The highest values were obtained at 8 p.m. and 11 a.m., suggesting a possible relationship between acuity and the mid-points of the activity cycle.

If performance fluctuations are now described in local clock time, in the future they are likely to be described with reference to the physiological phase of the individual. As a person slowly and almost imperceptibly changes physiologically around the clock, so, too, does the complexion of his performance. A group of researchers led by Dr. K. E. Klein, at the Institute for Flight Medicine in Bad Godesberg, Germany, have begun to construct a phase map relating performance to biological change. They collected oral temperature, cardiac output levels of white blood cells, measured simple reaction time to stimuli heard and seen, and computed individual variance in reaction time. In addition to studying staff members at the flight institute, they also studied nine students for plasma protein, plasma aldolase, and adrenal carbohydrate steroids.

Another group of students was subjected to a series of very complex tests, each requiring coordination and continued performance for over ten minutes. They were given a physical exercise test on a bicycle that was attached to a voltimeter which not only measured the amount of work, but also allowed the researchers to evaluate and predict the maximum oxygen uptake. Subjects were tilted to 90 degrees on a tilt table for about 20 minutes and tested while holding their breath, then while hyperventilating. Blood pressure and pulse responses were recorded. Altitude tolerance was tested in a low-pressure chamber.

Mental performance reached its high point between 2 and 4 p.m. when reaction time and complex psychomotor coordination were also at their best. The poorest performance was observed at the dead point between 2 and 4 a.m. This is a familiar curve and the rhythm of physical fitness was expected to be similar.

Two things are very important in judging performance rhythms. One is the range of oscillation or amplitude over 24 hours. Dr. Klein's study indicates that this is especially significant in aerospace medicine. For example, the predicted maximum of oxygen consumption, the time of useful consciousness under conditions of too little oxygen, and the pulse pressure when a man was on a tilt table showed considerable variations over time. These were of notable magnitude, as were changes in plasma protein over the course of a day, plasma aldolase, blood eosinophils, and the plasma 17-OHCS. On the other hand, temperature ranges were small. Reaction time and psychomotor performance oscillated moderately (10 to 20 percent of the day's average). The resting cardiac output did not oscillate so much in the 24 hours, although a person on a tilt table showed a considerable range in blood pressure. Heart rate on the tilt table did not change much. This research may indicate the range of vulnerability of physiological systems during diverse activities and stresses as the individual changes around the clock. During the 24 hours, as Dr. Klein and his associates found, psychomotor performance showed variation of a magnitude comparable to the performance decrement caused by .09 percent blood alcohol.

Another important aspect of rhythmicity in human efficiency and tolerance for stress is the timing of peaks—the phase. When is a man most effective at psychomotor tasks? When is he most effective at visual detection? Careful thinking is apparent in the German flight institute study. Cycles of mental performance, tolerance for lying on a tilted table, and physical fitness show their peak, on the average, from about 1 to 7 p.m. The poorest responses occur between 2 and 6 a.m.

One variable under study showed a completely inverse cycle. This was the time of useful consciousness in a pilot with too little oxygen. He was almost 50 percent more resistant to anoxia at 3 a.m. than he was at 3 p.m. A healthy man is not equally efficient for all kinds of stresses at every hour. Oxygen consumption may be lower at 3 a.m., making altitude adjustment easier even if this is a nadir for other kinds of responses.

The integrity of the 24-hour rhythm in all of our chemical and neural systems is such that we are bound to find peculiarities in our abilities over the 24 hours. We are bound to find that there are some things we can do better at certain times because the levels of certain hormones are higher in our system at that point. Or, if we imagine ourselves to be like 24-hour clocks with millions of hands, some of these hands are in an advantageous direction at 3 a.m., others at 10 p.m. We need to know the particulars. Which functions can we perform at what hours?

Drs. K. E. Klein, H. M. Wegmann, and H. Bruner have published the charts of their many physiological and performance tests in *Aerospace Medicine*, in May 1968. Perhaps, in the future each person will have a phase map showing how one organic system behaves relative to the others at each point in the day.

Time charts may advance our understanding of performance, especially when they are completed by accurate statistical analyses and amplified by further experiments. Today, psychologists perform very sensitive tests on people, trying to determine their thresholds of hearing or taste or to judge the speeds of reactions in their nervous systems by their responses. However, if there is no control for biological time of day, one study simply will not be comparable to another.

Individuality in Tempo and Rhythm

Individual variation in time structure appears to be quite as pronounced as is variation in athletic talent, height, and temperament. We do walk to different drumbeats. But each of us is also a different person within the 24-hour cycle. How we feel, how well we perform, our moods and alertness, our sensitivity to taste, smell, and visual objects, our enjoyment of food or pleasure in music, all are changing around the clock.

Our ability to do complicated, coordinated tasks with our hands, to think out intricate problems, discriminate, to withstand an environment with too little oxygen, or even to stand on our heads is not constant throughout the 24 hours. We are different persons at 10 a.m. than at 10 p.m.

Some physiological functions, urinary constituents, such as electrolytes, adrenal hormones, or metabolites of substances used in nervous transactions, might be used to indicate time of day within the body, like the hands of a clock. Their phase relations to each other are related to sensory acuity, reaction time, to pulse, body temperature, blood pressure, and pulse pressure. Dr. Pöppel has found that the longest reaction times in one of his studies and the longest estimated time intervals were nearly in phase with the minimum values of pulse, temperature, and urinary electrolytes. Although differences between the proverbial "owls" and "larks" may result in part from habit, they cannot be laughed away as temperamental idiosyncrasies. Animal studies suggest that there may be individual differences in timing that are inherited.

Dr. Franz Halberg and his associates, at the University of Minesota School of Medicine, have demonstrated with inbred mice that the circadian rhythm of adrenal hormones and body temperature differs from one strain of mouse to another. Genetic inheritance may also shape an individual's activity cycle, just as it appears to influence the amount of sleep an individual requires, a quota that is evident at birth.

The rhythms of activity and rest are extremely revealing in animals and reasonably easy to record simply by mounting a cage so that movements in the cage will make an inked mark on a slowly rotating drum of paper. Dr. Curt Richter has been astonished by the accuracy of what he calls a "24-hour clock" in rodents and small mammals. Even under uncontrolled conditions in which the laboratory animals were exposed to noise, handling, and the famous humid heat of Baltimore summer and damp cold of winter, many hundreds of records on hundreds of rats showed an unbelievably stable activity rhythm.

In addition to recording thousands of inbred rats, Richter trapped large fierce Norway rats in Baltimore alleys, and compared them with their domesticated and inbred cousins. Wild rats, even in cages, maintained an undeviating 24-hour day. Moreover, when blinded, their day became only slightly shortened. One Norway rat, who had grown up in the hurly-burley of the back alleys, not on a conditioning schedule of laboratory illumination, "lost" only seven minutes in seven months after blinding.

Richter's research led him into almost every area of physiology, and his studies of the activity rhythm of rodents caused him to build ingenious cages, enabling rats to climb, nest, dig and do many "natural" movements besides running in a mesh running-wheel. By observing individual rats closely and paying attention to the peculiarities of their activity records over many days, weeks, and sometimes months, Richter saw that they were idiosyncratic. They did not distribute their energies identically during their activity period. Some, like "lark" types, showed their most intense burst of energy early after rising; while others seemed to reach their peak toward the end of the activity period, like the "owls."

As scientific studies continue to promulgate information about these rhythmic changes, cultivated human beings will be able to foresee their own sensitivities and schedule their activities more wisely. In medicine, the charting of human-time structure should permit a change of vision comparable, perhaps, to the change brought about by high-powered microscopes as they revealed the structure of animal tissues. The changing physiology of an individual, during the social unit of 24 hours, gives a slight clue to the circadian rhythms that are beginning to be discovered in drug reactions, in vulnerability to stress, toxins, and pain, as well as to infection and allergy.

Human Circadian System

SITE	VARIABLE	N of SUBJECTS
BRAIN	EEG, Total	16
	" Delta (< 1-3.5 Hz)	16
	" Theta (4-7 Hz)	16
	" Alpha (7.5-12 Hz)	16
	" Beta (13-30 Hz)	16
	" Mental State	
EPIDERMIS	Mitosis	193
URINE	Volume, Rate of Excretion	1
	Potassium, "	5
	Sodium, "	1
	Hydroxycorticosteroid, "	4
	Tetrahydrocorticosterone	8
	Tetrahydrocortisol, "	8
	17-Ketosteroid, "	4
	Epinephrine, "	1
	Norepinephrine, "	1
	Aldosterone, "	4
	Magnesium, "	8
	Phosphate, "	10
	pH	10
	Sodium/Potassium	10
BLOOD	Polymorphonuclears	15
	Lymphocytes	15
	Monocytes	15
	Eosinophils	11
	Hematocrit	4
	Sedimentation Rate	4
	Ca++	4
	Na+	4
	pCO2	4
	Viscosity, Shear Rate	4
	Screen Filtration Pressure	4
ERYTHROCYTE	K+	4
PLASMA or SERUM	17 OHCS	13
	Testosterone	4
	5-Hydroxytryptamine	5
	Protein	4
	Protein-bound Carbohydrate	4
	Hexosamine	4
	Sialic Acid	4
	Na+	4
	Ca++	4
WHOLE BODY	Temperature (oral)	11
	Physical Vigor	10
	Weight	10
	Heart Rate	10
	Blood Pressure – systolic	10
	" " – diastolic	10
	Expiratory Peak Flow	10
	Respiratory Rate	10

TIMING: EXTERNAL ACROPHASE (φ)

φ

95 % Confidence Interval

24 HR = ACTIVITY SPAN + REST SPAN

Analyses by the Chronobiology Laboratory, University of Minnesota.

Not chaos-like, together crushed and bruised,
But, as the world harmoniously confused:
Where order in variety we see,
And where, though all things differ, all agree.

Alexander Pope

So in one place the blood stops, in another it passes sluggishly, in another more quickly. The progress of the blood through the body proving irregular, all kinds of irregularities occur.

Hippocrates

Chapter V. Circadian Rhythms in Cell Mitosis and Illness

When people with peptic ulcers forget to eat or drink something to coat their stomachs they can expect to feel pains at about the same time each day. Most ulcer patients have not realized that the very regularity of the pain is a sign in their favor. Quite a few diseases show circadian rhythms in their symptoms and shrewd physicians have realized that these rhythms might be useful.

Long ago, Dr. M. Arborelius, in Hälmstad, Sweden, noticed that patients with ulcers or cancer of the stomach complained of hunger-like pains, which were indeed suppressed by eating a meal. However, there was a striking difference among patients. Some patients seemed capricious and irregular in the time of their complaints. Dr. Arborelius realized that clinicians must be overlooking useful information in the patients' feelings.

After studying some 200 patients (86 had cancer, the others, benign ulcers), Arborelius found that none of the ulcer patients but almost all of the cancer patients showed irregular rhythms of sodium chloride excretion. Irregularity was, perhaps, a discriminator for cancer. A doctor who relied on X-ray diagnoses would not have been as accurate in separating these patients as a clinician who merely listened to their complaints of pain and made a note of the hour. After a week or two it would have been clear that the ulcer patients had pain at regular intervals, while the cancer patients were unpredictable.

Roughly three decades elapsed between Arborelius' observations on the rhythmic symptoms of ulcer patients and a study using blood cortisol rhythms in ulcer patients as a possible clue to the origins of the illness. Recently, Drs. B. Tarquini, R. Orzalesi, and M. Della Corte, at the University of Florence in Italy, compared eight patients suffering from peptic ulcers with fifteen normal control volunteers. All lived on the same schedule of meals and activity for a week before the study so that, as nearly as possible, they would be in phase with one another. Then, for 24 hours, at regular intervals, blood samples were drawn and analyzed for the adrenal hormone, cortisol.

The ulcer patients differed from the normal people in both the amplitude and phase of their cortisol rhythms. The researchers felt that this corroborated suggestions that adrenal function may play a role in the genesis of peptic ulcers. In the early 1950's, Dr. Hans Selye had proposed that peptic ulcers might be the result of stress as expressed through adrenal hormones. But early attempts to uncover differences between the ulcer patients and normal people by blood or urine steroid levels produced nothing. The Italian team had inferred that changes in the adrenal cortical output might be too subtle to detect, excepting by a study of circadian fluctuations. So it proved. The circadian cortisol rhythms did seem to differentiate between normal people and ulcer patients.

Studies of nightlong sleep have indicated that the levels of adreno-corticosteroids rise in the blood toward morning in bursts that appear to be almost in phase with rapid-eye-movement sleep. The study of hormonal secretion, by measurements of blood levels throughout the night, has revealed some problems of interpretation that apply to all studies of fluctuating blood levels. Cortisol appears to enter the blood in sizeable bursts that show high levels and then quickly subside. The recent work of Dr. Howard Roffwarg and a team of medical researchers at Montefiore Hospital suggests that either a large number of subjects, or sampling at roughly 15-minute intervals, may be necessary to obtain values that would give a fair estimate of the individual's circadian rhythm.

If adrenocortical "puffs" into the blood occur in a rhythm like that of REM sleep, nightlong recordings indicate that there may be a similar periodicity of abnormal gastric secretion in ulcer patients. Drs. R. H. Armstrong, Anthony Kales, and their associates at the University of California in Los Angeles, have recorded nightlong brain-wave patterns in normal persons and duodenal ulcer patients who slept with stomach tubes. The ulcer patients secreted abnormal amounts of gastric acid at night periodically and preponderantly during that period of heightened autonomic activity, REM sleep. The normal persons who also slept with stomach probes showed no such periodic secretion.

As 15-30 minute increments of gastric secretion were collected and analyzed it was clear that the ulcer patients secreted far more at night than the controls, with an output of gastric acid that was 3-20 times the amount found in the normal control subjects. Although the normal comparison subjects seemed not to secrete more during any particular phase of sleep, the ulcer patients did their abnormal secretion preponderantly in phase with the REM sleep cycle. Certain other nighttime symptoms such as migraine headaches also have been found to occur in an ultradian rhythm very nearly in phase with that of REM sleep.

Periodicities of symptoms may provide clues to underlying mechanisms in some illnesses. Circadian symptom rhythms have been recorded in allergies, kidney disease, epilepsy, tuberculosis, depression, and other emotional illnesses, as well as in diabetes, cardiac illness, and many glandular diseases. Although this chapter offers a mixed assortment of examples in which rhythmometry may help in the diagnosis, understanding, or treatment of an illness, it only emphasizes the integrity of time structure within a healthy individual. Today, rhythmometry studies are beginning to become useful in diagnosing and treating cancer and certain endocrine illnesses. If few researchers volunteered for the rigors of time-series studies in the past, the future is likely to encourage an increasing number. Many of these will be due to the pioneering work of Dr. Franz Halberg (whose early studies are briefly summarized in *NIMH Program Reports, Number 3*, January 1969). The biological time structure of illness has been called chronopathology. This includes displacements of normal behavior to abnormal phases of the sleep-waking cycle such as eating and alertness at night with sleepiness by day, or physiological rhythms with an amplitude of change that is abnormal, creating fluctuations that are invisible in healthy persons.

Arterial Disease

There are many diseases in which temporal changes offer useful data for diagnosis. Some of these are respiratory or obvious glandular illnesses. Others involve the circulatory system.

Recently, for instance, Dr. V. Bartoli and his coworkers in Florence, Italy, have found that it is possible to detect a circadian rhythm in the flow of blood in calves of the legs of people with peripheral arterial disease, but not in normal people. In many instances, people have failed to find a circadian rhythm in a physiological function only to discover later that the rhythm existed, but that the change was of low amplitude. Sometimes, healthy volunteers may have too "flat" a rhythm to detect.

In healthy subjects limb circulation appears to be fully independent of the "biological clock" and to behave differently compared to subjects with peripheral arterial disease, in whom 24-hour periodicity can be demonstrated. Patients with intermittent spasm or obstruction showed a circadian rhythm with a peak in early evening and a nocturnal trough with a nadir between midnight and 4 a.m. Perhaps, a rhythmical reduction of limb-blood flow is one of the mechanisms underlying the nocturnal pain of subjects with limb ischemia, a peripheral arterial disease in which there is a decrease in circulation. Here, the pronounced rhythmicity of blood flow in peripheral artery disease may begin to explain some of its recurrent symptoms.

Endocrine illnesses are often described as if they were the pure results of too much or too little of a hormone that might suppress or cause overproduction in another gland. A new set of endocrine relationships has begun to appear as a result of time-series studies. These are the phase relations among rhythms of specific glandular activity, especially the adrenal and its related kidney functions. In collaboration with Dr. Franz Halberg, Dr. Frederic C. Bartter and Catherine S. Delea at National Heart and Lung Institute in Bethesda, studied the rhythms of a number of urinary products and blood constituents in groups of normal subjects. This gave a picture of the relationships among certain adrenal steroids, as well as their relation to the numbers of lymphocytes and other blood cells and the urinary output of sodium and potassium. Patients with certain endocrine illnesses, such as aldosteronism, showed urinary aldosterone rhythms out of phase with those of normal subjects. Aldosterone is a steroid that causes the body to retain sodium and excrete potassium; overproduction of it is sometimes found in patients with hypertension.

The NIH laboratory has studied adrenal hormones and kidney function in hypertensive patients. As phase maps of body rhythms evolve, they may help to explain why certain symptoms may occur at particular intervals. In addition, they are beginning to provide data relevant to practicing physicians. Bartter and Delea found some patients who had normal blood pressure at 8 a.m., but high blood pressure at 6 p.m. (the hour when normal blood pressure begins to reach its daily peak). This means that any insurance examiner who judged a person from a single blood pressure reading around 10 a.m. would be deluding himself. The person might have hypertension, yet appear normal.

Rhythms of Cell Division

For many decades, it has been observed that the cells of the body do not divide at an even rate, the same at all times of day. Nor is the rate the same in all seasons of the year. In 1917, a Dutch physiologist observing some newborn kittens saw that the cells in the cornea of a two-day old kitten divided most rapidly around 10:30 p.m. and least at 10:30 a.m. In 1939, Dr. Z. Cooper observed that cells from the human prepuce skin divided in a circadian rhythm. In the 1950's, Dr. Franz Halberg and his associates began tracking circadian rhythms of cell division in the ear skin and mouth tissues of mice and hamsters under controlled conditions. They were able to do a close mapping of the mitotic rhythms relative to light and dark schedules, relative to adrenal rhythms, and manipulations such as adrenalectomy. The methods evolved in the Minnesota laboratory are slowly being adopted elsewhere, and data have accumulated that may begin to illuminate physiological processes, for example, by showing the phase relationships of peak cell division in the tissues of a single organ.

The most pronounced circadian rhythms occur in cells known as renewal populations, surface tissues such as skin, ears, mouth, tongue, eyes, and hair, that are in constant contact with the outer world. Eroded and damaged each day, they must be replaced at a rapid rate. In adult human beings, for instance, skin cells divide mostly between midnight and 4 a.m. when a person normally is asleep, as demonstrated by Dr. Lawrence E. Scheving at the University of Louisiana in New Orleans. The rate at which new cells are produced roughly equals the rate of loss. It can be observed by watching mitosis or by counting radioactively-labelled cells. A radioactive substance, known to be used in cell replication, is injected into tissue: the amount of radioactive substance incorporated by cells is taken as a measure of mitosis.

Drs. Lawrence E. Scheving and J. E. Pauly have used radioautographs of tissue to track circadian rhythms in the division of corneal cells from rats. Animals were kept under standardized and controlled conditions and samples were taken at intervals around the clock, using protocols like those evolved by Dr. Halberg and his associates. Cells and cell nuclei were more heavily radioactive at certain hours than at others. The peak in radioactive labelling occurred around noon, the midpoint of the animals' rest period. Earlier studies of tissue regeneration and of circadian mitotic rhythms had suggested that cell mitosis might be linked to the adrenal cortical hormone rhythm. However, the corneal mitosis rhythm was observed in animals without adrenals or pituitaries, suggesting that it did not depend upon the rhythmic output of those glands. Moreover, the rhythm persisted in constant darkness. When animals were kept in constant light for long periods, however, the mitosis rhythm disappeared. In addition, after two weeks of light there were significant retinal changes, such as the disappearance of the rods.

The influence of light had been demonstrated a decade earlier by Drs. C. P. Barnum and Franz Halberg in studies of circadian rhythms in the blood cells known as eosinophils in mice. The circadian rhythm was persistent in animals who lived in constant darkness for 13 days: but after 9 days in constant light the rhythm vanished. The rhythm reappeared if the animals were immediately put on their former schedule of alternating light and darkness every 12 hours.

Biological and physiological rhythms seem to be calibrated to rhythms in the natural environment, and the biological potency of light is such that no creature can be considered a closed system. Thus, the control of light, darkness, temperature, noise and other major attributes of the environment are necessary for studying physiological time structure, as Dr. Halberg and his coworkers have shown. It is in the development of methods and instruments for such study that the Minnesota laboratory has contributed uniquely.

During a monumental decade of research, Dr. Halberg's laboratory attempted to accumulate enough data for a phase map that would show how cell division was organized into organic function. At first the laboratory concentrated on the adrenals, whose rhythmic outpourings of hormones regulate carbohydrate, protein, and electrolyte metabolism, influencing sensory acuity, sexual behavior, and many other functions. Glands with such pervasive influence and rhythmic output might also guide rhythms of tissue regeneration, along with rhythmic outpourings from the pituitary, that small gland at the base of the brain whose secretions regulate activity in thyroid, in the adrenals, gonads, and growing skeleton, controlling salt metabolism, and water retention in the body.

Dr. Halberg and his associates then began to focus upon the rhythmic processes of cell division within tissue. They used the liver because it is a large, influential organ and regenerates swiftly after surgery. Thus, they could use tissue both from young animals in a state of rapid growth and from adult animals after surgery. The glycogen level of the liver had long been known to vary in a 24-hour rhythm. By injecting a radioactive phosphate, a substance used in cell division, it was possible to measure the relative uptake of radioactive substance in tissue fractions taken at different hours around the clock. In the course of innumerable studies, it was possible to assemble an indication of rhythmic changes in different components of liver tissue, from the very tiny microsomes to the heavy nuclei. The laboratory then began to investigate mitosis rhythms in the fundamental nucleic acids, DNA and RNA. DNA—deoxyribonucleic acid—is believed to transmit hereditary characteristics from one generation to another and provides a template for each cell in the body. RNA—ribonucleic acid—is found in the nucleus and in cytoplasm around the nucleus. Among other functions, it appears to regulate the rate at which each kind of protein is produced.

One might expect that RNA and DNA would be most active at times when many cells were preparing to divide. Moreover, one would not expect to find that DNA and RNA were simultaneously at their maximum synthesis, but perhaps would be staggered in their peak activity.

To determine the cycles of DNA and RNA synthesis, large groups of animals on rigid environmental schedules were considered as one individual. Sizeable sub-groups were killed and their tissue analyzed at short intervals around the clock. One early study involved 140 inbred young mice. Groups of 20 animals were injected with radioactive phosphorus and killed two hours later, each group at a different hour around the clock. The experimenters began at 8 a.m. one day and finished their last group at 8 a.m. the next morning.

They did, indeed, find that DNA and RNA exhibited a 24-hour rhythm of relative specific activity and labelled phosphorus uptake. DNA and RNA did not reach their peaks simultaneously, but out of phase. Judging by phosphorus uptake, the peak DNA activity occurred as RNA activity was dropping to its nadir. The Minnesota laboratory was the first to determine a circadian rhythm in DNA and RNA activity. Since then cytochemistry and histophotography have added complementary information about mitosis in nucleic acids. Biologists, using the photographic technique, may inject a precursor such as radioactive thymidine, which will be used by DNA in forming a new molecule. Radioactive thymidine will label only those cells that happen to be doubling their DNA at the time of injection. The tissue's radioactive emissions will darken the film emulsion, and the density of dark spots in the photograph will indicate how many DNA molecules took up the radioactive substance in the process of duplication. The photograph of tissue taken at one time of day easily can be compared with the photograph of tissue taken at other hours, giving a time scan of DNA synthesis.

Using a number of techniques, the Minnesota laboratory began to draw up a map of the phases of DNA and RNA synthesis, in relation to the phase of mitosis in certain cells. These, in turn, could be related to rhythmic increases and decreases in the levels of substances such as fats and sugars in the liver. After many experiments using thousands of animals, and many replications, Dr. Halberg's laboratory was beginning to evolve a time-map of liver function. The several recorded rhythms were intermeshed as though they had been choreographed.

Rhythms of Mitosis and Metabolism in the Liver

During the time of maximum cell mitosis, certain metabolic functions such as the deposition of fats (phospholipids), and the formation of RNA appeared to slow down. These metabolic activities almost stopped at the time of the peak mitosis, although only a relatively few cells were actually dividing. RNA and phospholipid labeling reached their peak about six-eight hours before the peak formation of DNA. Maximal RNA activity preceded the peak glycogen content of the liver by about eight hours, preceding maximal cell mitosis by about sixteen hours. Thus, each of these biological functions showed a circadian rhythm, and there was an orderly sequence of rhythmicity that could be specified in terms of phase differences.

A cell-activity cycle in the immature liver was specified by these studies. For about eight hours there is a stage when most of the dividing cells are completing their final step, telophase. During this eight-hour interval there is a peak time for the incorporation of radioactive phosphorous into RNA and phospholipid. Next comes another eight-hour stage that culminates in a peak incorporation of the tagged phosphorus into DNA, thus indicating peak DNA synthesis. At this point, the 24-hour round is completed by another 8-hour stage which includes the peak of cell mitosis. In the first four hours of this span, glycogen levels reach their daily high point.

Thus, the various stages of cell metabolism and mitosis in the liver follow a reproducible sequence that can be timed by reference to a lighting cycle and reliably detected.

What one sees in this mapping of rhythms is one of the biological mechanisms that permits smoothness and continuity in the functioning of an organism through an overlap of one process with another. Populations of cells do not all begin DNA synthesis at once; mitoses are not begun in unison. Instead, there is a trend provided by rhythmic change in which numbers of cells are behaving the same way at roughly the same time. Gradual rhythmic shifts allow the body to change without suffering the great disadvantage of discontinuity.

As the Minnesota team extended its analyses to the cells of the kidney, adrenal cortex, skin, pancreas, small intestine, and to the pituitary, hypothalamus, and other portions of the brain, they inevitably found circadian rhythmicity. The same phase relationships that had appeared between DNA synthesis and mitotic rhythms in liver cells, however, were not necessarily to be found in cells of other organs. There were fixed sequences, phase relationships, among cellular rhythms within an organ, but all organs did not show the same pattern. In some tissues, such as liver or skin, one would see that maximum cell division took place during the animal's sleep, while in other tissues, the adrenal gland, for example, peak cell division occurred during the period of high motor activity.

As they asked how a particular rhythm was governed, the Minnesota team would try to shift or eradicate the rhythm experimentally. The rhythm of liver glycogen levels were of particular interest. Although each cell in the body uses glycogen, only the liver can perform the function of long-term storage that is essential for regulating blood sugar. If liver glycogen did not reach its normal levels at the right time, as in some liver diseases, the whole system and notably the brain would be deprived of energy during times of daily demand.

Glycogen levels are related to food intake, of course, and many people have thought that the liver's rhythm of glycogen concentration must be related to feeding schedules. However, Dr. Halberg and his associates demonstrated that the circadian glycogen rhythm persisted in the livers of mice that had been completely deprived of all food and liquids. The peak glycogen level in the rodent liver occurred at the end of the span of darkness and activity, and glycogen content declined during rest, even when the animals were hungry and thirsty. Liver glycogen was one of the first physiological rhythms to be studied, beginning with the Swedish investigator,

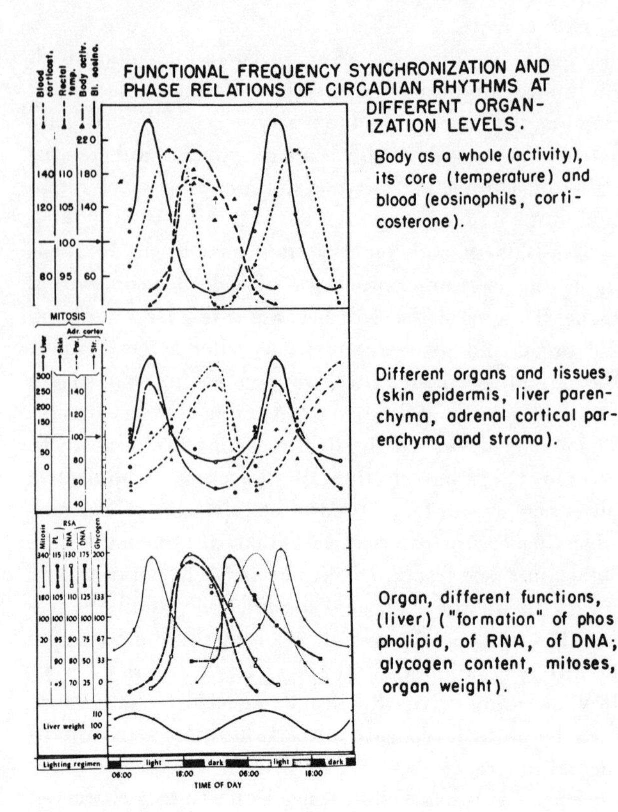

This figure from Franz Halberg shows internal timing of several physiological rhythms in rodents on a standardized routine of light and darkness.

Dr. E. Forsgren, who saw that liver glycogen fluctuations offered a clue to the nature and treatment of diabetes.

In the normal person the carbohydrates, sugars, and proteins of food are transformed into usable form. The body continually breaks down these substances adding a phosphate to form adenosine triphosphate, or ATP, the energy-bearing unit within each cell. Two pancreatic hormones, insulin, and glucagon, are also important in this process in balance with a hormone from the pituitary gland. Most of us have known someone with diabetes mellitus, someone with vague symptoms such as excessive thirst, abnormal weight variation, lack of energy, and faintness. The diabetic person can suffer from hunger even when he is well fed, for he does not produce enough insulin at the right time and cannot break down into usable form the sugars ingested in a normal diet.

Even when a person is resting his brain consumes 25 percent of the oxygen used by his entire body. Unlike the rest of the body, the brain has little in the way of carbohydrate reserves and depends upon constant nourishment from glucose in the blood. A decline in the glucose supply means a drop in the brain's main source of energy. Thus, a seemingly slight interference with blood glucose can have a pronounced effect upon mental function. This effect is often witnessed in people who have episodes of low blood sugar–hypoglycemia. Frequently, they have symptoms such as fatigue, sweaty hands, irritability, anxiety, inability to sustain attention, and lack of motivation.

In the morning after sleep and every time he goes without food the diabetic person suffers from this form of brain starvation. The hormonal balance required to metabolize and store sugars in useful form has been disrupted, and his brain may lack sugar at the same time he is excreting unused sugar in urine.

Liver glycogen rhythms have been studied by Dr. Arne Sollberger at Yale University. Human beings show a glycogen curve that begins to descend in late afternoon, falling throughout the night so that by morning the liver has used up most of its glycogen. At this time a diabetic patient without food might respond badly to insulin.

Clinical studies of diabetes were in full swing at the Karolinska Institute in Stockholm in the 1920's. There, Drs. Jacob Möllerstrom and Arne Sollberger concentrated on daily rhythms of blood sugar and rhythms of excretion of citric acid, pyruvic acid, acetone, ammonia, and sugar. Dr. Sollberger found abnormal rhythms of acidity in both blood and urine in patients. When the acid-base balance of the blood was disturbed patients would show excessive acidity in their blood, with peaks around 4 p.m. and 4 a.m. Blood acidity is taken as a sign of considerable pathology, and it is accompanied by a variety of symptoms, such as nausea and headache. It is possible to see how a disruption of intermeshed rhythms might result in a transient acid-base imbalance.

Diabetic patients lack the insulin needed to convert sugar into useful forms. Until recently, they were prescribed insulin on the assumption that the body needed the same amount at all times. However, in studies of healthy adults in Paris, Drs. Canivet, Lestradet, and Deschamps have detected a circadian rhythm in blood insulin and blood glucose. This suggests that therapy might be designed to restore artificially the insulin rhythm in the diabetic person. There are other reasons why rhythmic hormone replacement might be efficacious.

Insulin affects other hormones. Recently, Dr. M. Serio at the University of Florence in Italy, studied a group of diabetic patients during and before insulin treatment. During insulin treatment they no longer showed their usual circadian rhythm of cortisol levels. However, when the insulin treatment was stopped the adrenal rhythm reappeared. Dr. Serio has observed that hypoglycemic drugs seem to cloak the cortisol rhythm and conjectured that they did so by affecting the hypothalamus.

Other studies in the laboratory of Dr. Halberg at Minnesota, suggest further reasons why the diabetic patient might particularly notice disturbances in the insulin rhythm at certain hours of the day.

The Pancreas and Insulin Rhythms

Insulin secretion occurs in the pancreas, which acts both as an exocrine gland (secreting substances through a duct), and as an endocrine gland (secreting hormones directly into the bloodstream). To manage these different functions it has many different kinds of cells. Acinar cells are part of the exocrine system, producing digestive enzymes, the biological demolition corps that breaks down fats and sugars into useful components. Alpha and beta cells are a part of the endocrine system. Alpha cells produce glucagon, which ultimately increases available blood sugar by transforming glycogen into glucose. Beta cells produce insulin, which in turn helps to regulate glycogen and, in effect, lowers blood sugar levels. These three kinds of cells are actively dividing at different times, operating as it were, in opposition.

Drs. Walter Runge, Franklin Pass, and Franz Halberg found that the mouse pancreas exhibits a distinct chain of peak phases in the mitotic rhythms of certain cells. That is to say the peak mitoses in beta cells precedes the peak of alpha cells, which reach their peak eight hours ahead of the acinar cells. Thus, there is a circadian rhythm of mitosis that may not cause, but which does parallel, the usual levels of insulin in the blood. In the delicate balance of carbohydrate metabolism, levels of insulin must be timed to match and counterbalance the action of a hormone from the pituitary gland. This means that therapy aimed at approximating the normal phase relations of endocrine rhythms in the body may need to reproduce the correct phase of rising and falling insulin levels in the body instead of merely increasing deficient levels.

Seasonal Variations

Just as some patients notice that their symptoms wax and wane each day quite a few patients with endocrine illness feel more intense symptoms during certain seasons of the year. Dutch and German scientists have recently shown that there is a seasonal difference in the amount of glycogen stored in liver cells. Not only did the amounts differ in May, January, and July, but the circadian phase was different, suggesting a response to the changing length of the day. Studies of rodents indicate that liver cells contain a minimum of glycogen around midnight in January, a minimum at 8 p.m. in July, and 10 p.m. in May. The amount of liver glycogen found in January is double the amount found in July. When first observed this seemed too startling to be believed, but a replication has produced the same result. It was not a technical error.

Dr. H. von Mayersbach, of Hannover, Germany, has been a pioneer in the exploration of cell rhythms. While tracking DNA and RNA rhythms in liver tissue, he found that laboratory animals showed subtle but pronounced changes in tissue structure with seasonal change. They also showed a sex difference. Sex differences in the DNA content of spleen, skeletal muscles, and brain will have some impact on histology laboratories where tissue is being examined for abnormality.

Dr. von Mayersbach and his associates encountered some seemingly inexplicable problems in staining tissue for microscopic study, since in one of their studies the microscopic count disagreed with the biochemical analysis. The discrepancy underlined a serious problem in medical laboratory work. If tissue is taken from the body at one hour of the day, the fixative, the chemical that is used to make it visible under the microscope, will not act in the same way that it acts if that tissue were taken from the animal at a different hour. In other words, tissue changes so considerably around the clock that histochemical substances will interact with the tissue unevenly, depending upon the biological time of day the sample was taken.

These findings may help to solve some of the traditional riddles posed by the patient with a subtle disease. The histology laboratory of the clinic must now cope with cyclical change. At one testing a patient may seem to have signs, but further blood or tissue samples taken at another time give a different picture of his probable cause of ill health.

Circadian Rhythms in Glucose Tolerance

In a glucose tolerance test a person receives an intravenous dose of glucose after a night's fast. Then he continues to fast while his blood is sampled over the four or five hours following the injection. The purpose is to see how fast the glucose is dispersed from the blood and presumably converted into other useful forms of energy. Most often, the sugar is given by mouth—with questionable results. In some studies, for instance, the subject who is "normal" in the morning can be "diabetic," in the afternoon by changing the hour of oral sugar ingestion.

Recently, Dr. Robert Abrams, an endocrinologist at New York Downstate Medical Center in Brooklyn, and his colleagues made a study of 16 healthy Navy men between 19 and 32 years old in the Clinical Investigation Center of the Naval Hospital in Oakland, California. Each subject stayed in the hospital for four days of study, being fed equal calorie meals every six hours. On the fourth day they received injections of glucose every six hours beginning at 6 a.m. Injections came just before their meals, so that they had fasted almost six hours before each shot. In order not to disturb sleep or hormones the infusions were made through a tiny catheter that was implanted in the arm. Glucose was infused slowly in one arm. Blood samples were taken from the other arm five minutes before the infusion and at short intervals during the subsequent hour. This gave a picture of the metabolism of the glucose in the first few minutes, the subsequent twenty-minute interval, and hour after injection. Urinary glucose, blood insulin, and plasma steroid concentrations were also measured.

The blood insulin responses to glucose were typical of normal people. There would be an abrupt increase in blood insulin, reaching a maximum about five minutes after the sugar infusion and declining so that in ten minutes it approached the preinjection level. By an hour later it was down to the normal level. There was not much variation in the insulin response after the four different glucose injections, suggesting that there is not much variation with time of day. It also suggests that the rate of glucose disposal is not directly tied to the blood insulin level as one might have expected. Each man showed a tremendous variation in blood levels of triglycerides—fatty acids—during the disposal of the injected glucose. The disposal appeared to be related to triglyceride values.

The researchers were surprised at the high fatty acid content of the blood, especially since the levels were measured after more than five hours of fasting. A careful look showed that the triglyceride levels were not related to meals, but appeared to rise and fall in a circadian rhythm with a high peak at about 6 a.m.,coincident with a high point in adrenal cortical hormones, but a low point at noon, much earlier than the adrenal nadir.

Dr. Abrams and his coworkers decided to explore the influence of adrenal hormones on glucose disposal. The rate of glucose disposal was clearly not an artifact of disordered sleep, nor fasting, nor the timing of injections, but it did seem to be related to adrenal hormone rhythms. Two volunteers were given injections of a drug that suppresses the pituitary hormone, ACTH, which normally would cause the adrenals to release hormones into the blood. After a shot of dexamethasone their blood showed very low levels of the adrenal cortical steroids, 17-OHCS, and the circadian rise and fall of this hormone was altered. So, too, was their rhythmic increase and decrease in the rate of disposal of glucose.

It has long been known that adrenal hormones, which can be measured as 17-OHCS, must influence the use of glucose. Other scientists have observed that an oral dose of the adrenal hormone, hydrocortisone, produces an excess of blood sugar some time later. The reaction may take from three to six hours. The Oakland team found a very sharp increase of adrenal cortisol in the blood about 6 a.m., and six hours later a very sharp drop in the glucose disposal rate, which fell to its trough about noon.

The lag times between the peaks of these various rhythms—the adrenal hormones, triglycerides, and glucose disposal rate—may be clues in attempting to find out how diabetes mellitus begins. Dr. Abrams sought to undertake a study of metabolic rhythms in pre-diabetic patients and also patients whose illness had begun late in life, appearing when they were in their thirties and fourties. This kind of diabetes, he postulated, might be related to erratic schedules of living, to the desynchronizing effects of stress or trauma, evolving into the elusive diabetic syndrome from slow aberration in the rhythmic utilization of carbohydrates. Ultimately, a metabolic illness, such as diabetes, involves the intermeshing of many biochemical gears. If one metabolic rhythm were out of normal phase, would it set off a concatenation of disorders?

Another disease, that also happens to be called diabetes, probably stems from another kind of origin—malfunction within regulatory centers of the brain. It often occurs in people who have had cardiovascular disease or strokes. Dr. Lawrence Kahana and his coworkers at Duke University have found a number of patients with endocrine symptoms due to brain damage or tumors. Some of these patients, with tumors in the region known as the optic chiasm or the third ventricle (a well of cerebrospinal fluid), often have diabetes insipidus. These diabetics do not show an abnormal urinary excretion of sugar, but they do show insufficient activity in the pituitary gland. In diabetes insipidus, one sees the effects of a flaw in the integration of the normally cooperative rhythms of brain and glandular functions.

The pituitary also secretes growth hormone, which is believed to stimulate bone growth and without which certain body cells deteriorate. Growth hormone also can prevent insulin from lowering blood sugar. Thus, it would seem to play some role in diabetes. In a study of squirrel monkeys, Dr. Robert Abrams found that the pituitary gland would secrete growth hormone whenever touched with an extract of brain tissue from the hypothalamus. Moreover, by successive hypothalamic lesions Abrams was able to localize the sites that produced exceptional blood sugar responses to insulin injection. Hypothalamic lesions led to an extraordinary drop in blood glucose and also an impaired growth hormone response. Previous studies have shown that damage in the region of the pituitary and hypothalamus interferes with growth in rats and produces sensitivity to insulin injections.

The current picture of diabetes appears to be more subtle and mysterious than ever, involving several glands, and perhaps mistiming within an archaic region of the brain. The etiology of the illness involves subtleties that could not have been measured a decade ago. Now some researchers conjecture that slight discordance among the balanced rhythms of metabolism could produce symptoms seen in the early stages of diabetes. Researchers and clinicians are also beginning to ask whether there are forms of cancer that are the consequences of altered time structure.

Cancer

The cancer cells that ravage parts of the body show a rampant growth that differs in tempo from the surrounding tissue. These rhythms of mitosis indicate an abnormal rate of multiplication that often seems to be apart from the circadian period of mitosis rhythms in normal surrounding tissue. In their extensive study of mitosis rhythms, Dr. Halberg and his associates found that normal cell populations showed their greatest reproductive activity during particular intervals in the 24 hours. Mitosis in the skin of hamsters and mice fell to its lowest point during the time of darkness when the animal was active, and rose to a peak during the illumination period when the animal was at rest. The circadian rhythm of peak DNA synthesis, RNA synthesis, and cell division, the experimentalists repeatedly demonstrated, could be at least transiently altered by shifts in the animal's light-dark regimen. As Drs. C. P. Barnum and Franz Halberg showed, a rodent left in constant light would begin to show some significant changes in the time structure of cell mitosis. After exposure to constant light the usual phases of peak DNA and peak RNA synthesis would be notably changed. Clearly, if such a basic change persisted it might affect the rhythmic reproduction of those cells and the effect of those cells in organic function. Thus, factors such as environmental light and an animal's activity rhythm might alter certain fundamental phase relations involved in the rhythmic reproduction of cells. Inevitably, one wonders about the bearing of such effects on cancer.

In collaboration with Dr. Mauricio Garcia-Sainz, at the Oncological Hospital, Mexico City, Dr. Halberg has analyzed cell mitoses in tissue from human beings and animals. One group of tissue samples had been taken at two-hour intervals around the clock from cancer patients before they went into X-ray treatment. A second set of samples was collected in the same manner from these patients after treatment.

By a strict count of dividing cells and objective method of data analysis it was clear that the cell proliferation was not circadian in its rhythm before X-ray treatment. Cell division rose and fell in a roughly 20-hour rhythm in some cases, or an 8-hour rhythm in others. This was markedly different from the 24-hour rhythm of surrounding tissue. It is no tautology to say that cancer cells are largely non-circadian in their mitosis rhythms. They might divide more rapidly than normal cells, yet still show the modulation of a 24-hour rhythm as they fluctuated in mitotic intensity. Lack of circadian modulation suggests some defect in the time integration mechanism of these cells, creating a form of temporal anarchy as well as histological abnormality. Thus, cancer cell activity falls outside of the overall temporal harmony of the body, which integrates function around the unit of a day.

Dr. Halberg's laboratory has conducted a series of studies in which cancer-producing compounds or tumor tissue were implanted in rodents. An early symptom of pathology in animals destined to have mammary tumors was arrhythmic mitosis in the skin of the animal's ear. Long before the tumor appeared the ear skin cells deviated from their former circadian rhythm of mitosis. Traces of abnormal cell behavior, thus, long preceded any gross physical signs, suggesting that human cancer might also reveal itself in the earliest stages by arrhythmic cell behavior at the surface of the body.

After X-ray treatment the picture changed. The mitoses in cancerous tissues were now more nearly circadian and there were fewer signs of fast mitotic cycles. The riotous proliferation of cancer cells might be described as the consequence of some flaw in the clocklike mechanism that usually creates temporal discipline in healthy tissue. It was the noncircadian mitosis that seemed to diminish after X-ray.

Recent analysis of biopsy tissue has suggested that not all kinds of cancer entertain the same mitosis rhythm. Halberg and Garcia-Sainz have found that cells from breast cancer do seem to show a circadian rhythm, whereas other cancers do not. Perhaps the location of the cancer or the influence of hormonal rhythms influences the mitotic cycle of the cancer cell.

The abnormal rhythms of cell mitoses are not the only irregularities of timing in cancer patients. One startling documentation obtained by the Minnesota laboratory is a group of rectal temperature charts on two cancer patients. The temperatures were taken every six hours over many days and were so erratic that they resemble no other patient records in the hospital. The irregular temperature patterns of patients living on a rigid hospital routine are a striking example of abnormal rhythms as a sign of ill health, and show that changes in the 24-hour rhythm may indicate serious illness. It has been conjectured that signs of phase or period change among easily measured body cycles such as temperature, could become early warning signs of pathology.

With timing as their focus. Drs. Halberg and Garcia-Sainz distinguished among the characteristic rhythms of particular human cancer cells. In deciphering the reproductive cycles of the cells they hoped to be able to clock the cycles of malignant cells. Then it might be possible to time X-ray therapy to hit specific cancers at their most vulnerable phase while avoiding side-effects on the patient.

In 1963 and 1964, Dr. Donald Pizzarello and his coworkers, at the Bowman-Gray Medical School in Winston-Salem, North Carolina, had found that the dose of X-irradiation that made rodents sick during the day would kill them at night. When laboratory mice were given does of X-irradiation at night during their active period they died, yet by daytime the identical doses merely made them sick.

A possible mechanism for the vulnerability to irradiation at a certain circadian phase has been reported by Drs. Y. G. Grigoryev, N. G. Darnskaya, and their coworkers at the 12th COSPAR meeting in Prague. They had irradiated mice at intervals around the clock and found what they interpreted as a pronounced vulnerability at two phases in the 24 hours. The deaths mainly resulted from damage to the tissues in which blood cells are formed. The Soviet team then studied spleen tissues after irradiation, which was administered at various times of day. Depending upon the hour of gamma irradiation, the spleen colonies were altered as much as 2.3 times. Ordinarily the peak production of bone marrow and spleen cells in mice coincides with the middle of the light period, while the minimum occurs during the dark period. The peak of mouse mortality from irradiation might be explained by the decrease in the total amount of blood-producing cells during that period. The second period of mortality observed might be attributed to a definite phase of the mitotic cycle in bone marrow. The double mortality peak observed by the Russian team suggests a biphasic rhythm of radiosensitivity over 24 hours, perhaps generated by a single

Circadian System of the Mouse

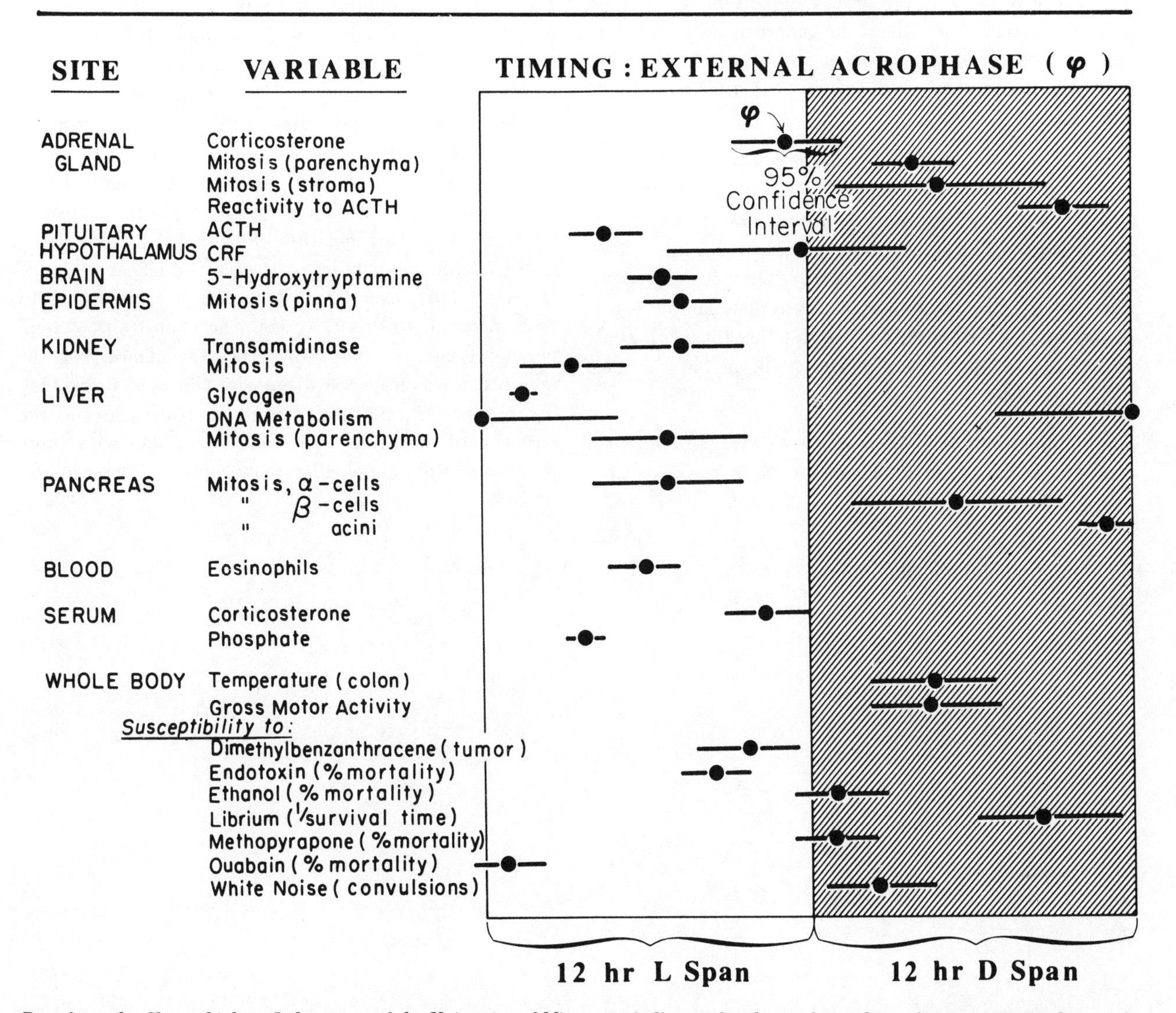

Data from the Chronobiology Laboratory of the University of Minnesota indicates the phase relationships of numerous circadian peaks of biological functions in the mouse. In addition this displays hours of maximum susceptibility to a tumor-inducing chemical, noise, and toxins.

mechanism, or by two cycles in separate blood forming tissues. Presumably the circadian rhythms underlying vulnerability to X-ray in mice exist in human beings as well.

A more potent treatment of tumors may evolve from studies of the effects of antimetabolites, such drugs as actinomycin-D, which slow down the process of protein synthesis. These drugs have been used to interfere with learning in studies of the memory process, and have been used by biologists to explore the basic mechanisms of RNA synthesis. If protein synthesis were programmed according to a circadian rhythm, it might best be blocked by interfering with DNA and RNA synthesis at a certain time of day. Dr. S. Cardoso, of the University of Firenze, Italy, suggests that appropriate timing of antimetabolites may inhibit DNA synthesis very efficiently.

Each day DNA synthesis increases and drops. There is a ten-to-one difference between synthesis at the peak and at the trough. Since tumors are generally out of phase with normal cells, it ought to be possible to hit tumors during their peak hours of cell division without inhibiting or damaging normal cells. Recently, Dr. Mauricio Garcia-Saniz of the Oncological Hospital in

Mexico City, has found that patients respond to X-ray treatment in a circadian rhythm. When precisely clocked, this differential response may mean that treatments can be calculated to attack the cancer at its vulnerable mitotic phase and yet minimize, by proper timing, the unpleasant and severe side effects experienced by the patient.

While research on cell rhythms may offer possibilities for the early diagnosis and timing of cancer treatment, other researches have suggested that cancer may be a disease to which a susceptibility is inherited. These studies fall in the realm of pure biology, and, for the most part, involve the rhythmicity of extremely simple organisms. Some of these studies suggest that there is a kind of "circadian clock" within each cell, perhaps in the DNA molecule.

If time structure is a heritable set of traits then there may be such a thing as a propensity or resistance to illnesses of "mis-timing." This may mean that it could be possible to predict propensities to illnesses such as diabetes mellitus, or depression, early in life. Moreover, if a person's speed of adjustment to phase shifts could be used as an indicator of temporal susceptibility, it might be possible to encourage vulnerable people to avoid irregular work-rest schedules, jobs involving rotating shifts, and to pay attention to their circadian rhythmicity in the manner that a person with tendency to obesity might give to diet. If phase-amplitude maps can be constructed for healthy individuals, a useful diagnostic profile may cut across the traditional categories of behavior and physiology. Responses to sleep loss, to sleep reversal, to crises, as well as certain traits of temperament and responses to tests of vigilance, may be relevant in predicting or diagnosing illnesses. Individuality in time structure may offer some early clues to the patterns of behavior that recur in people who show psychosomatic disease, allergy, or depression later in life.

The anguish suffered by men aware of the changes taking place within them and their inevitable end has caused sages and moralists to study the actual significance of change for individuals, societies, and the world.

Paul Fraisse, PSYCHOLOGY OF TIME

Chapter VI. Hours of Changing Susceptibility

The circadian tides that pervade our physiology have as a natural consequence daily rhythms in resistance or vulnerability to drugs, stress, allergy, pain, and infection. The timing of an event literally may tip the balance between health and illness, or survival and death. Inevitably these rhythms have ramifications throughout medicine, from the administration of drugs and vaccines, to the toxicity studies of pharmacology. Changes in susceptibility over the hours may be sizeable. In one experiment, for instance, a large dose of amphetamine killed only 6 percent of a group of rats when administered at the end of their activity cycle, but the same dose at the middle of the activity cycle killed 77.6 percent.

The practical implications of such variations in daily susceptibility have not been overlooked by such agencies as the U.S. Department of Agriculture, where Dr. William Sullivan and other scientists have conducted time-series studies that may enable farmers to time the spraying of pesticides so as to reduce both pests and the amount of toxins spread on the earth. The domestic fly, for example, is most vulnerable to pyrethrum around 4 p.m., while other insects have their peak sensitivity to other poison at different hours on their cycle.

The circadian rhythm of symptoms of illness or allergy may one day influence the strategy of medication, along with the rhythm of sensitivity to the drug. There are pronounced circadian rhythms in the responses of individuals to certain drugs. In some therapies hormones can be used to imitate the normal circadian rhythm, while other drugs may be used to deliberately alter the period of some rhythm, as tricyclic antidepressants are used to lengthen the sleep cycle. Sophisticated factors of timing may eventually enter into the strategy of drug use. Travelers who have crossed time zones and are phase shifting, may have an altered rhythm of response to their usual medications. Moreover, some drugs may cause an after-effect—a phase shift in the normal physiological rhythms of the treated individual.

The examples described in this chapter may indicate why the differential responsiveness of animals and man may bias the outcome in drug therapy, surgery, encounters with stress and infection—with ramifications throughout research. None of this differential response is really surprising. One might expect to find such changing reactions in a body where the changing phases of each organ make the individual physiologically different at one hour of day than another. Cells in some tissues are rapidly dividing, while elsewhere, organic activity may be reaching a relative slowdown. The specific reactivity of an organ may well depend upon the phase of its internal cycle.

The same rhythmicity seems to influence in vitro tissue. Living tissue, maintained in an artificial nutriment outside of the body, will persist in circadian rhythmicity as many studies have shown. Dr. Edgar Folk, Jr., and his coworkers at the University of Iowa, have shown that oxygen consumption in the isolated hamster adrenal gland rises and falls 60 percent above and below the daily mean each 24 hours; and a similar rhythm was seen in the secretion of corticosterone. Isolated hearts beat most rapidly at a particular interval in the 24 hours.

Drs. R. P. Spoor and D. B. Jackson at the University of South Dakota School of Medicine, have indicated what circadian rhythmicity may imply for many kinds of research. They found that isolated heart atria responded differently to test drugs, among them acetylcholine, depending upon the hour at which the heart tissue had been removed from the animal. Hearts were removed from rats at different hours of day and night and suspended in solution, after which the right atrium was isolated. Its contractions were measured after

Human Circadian System

Birth, Death, Morbidity, Susceptibility And Reactivity

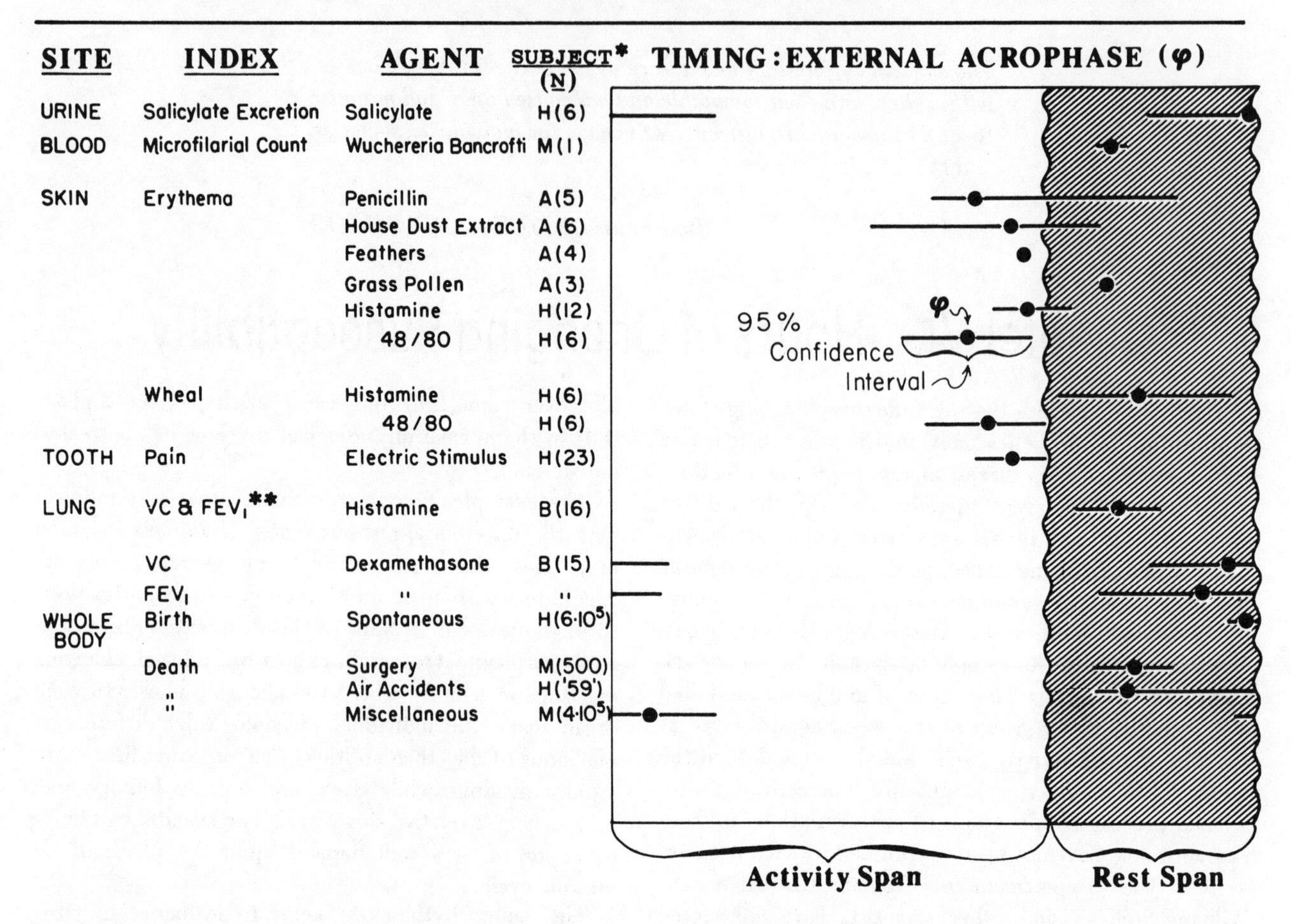

This summary of human susceptibility rhythms was prepared by the Chronobiology Laboratory at the University of Minnesota. *H= Healthy; M= Morbid; A= Allergic; B= Asthmatic, bronchitic, or emphysematous. ** VC= Vital Capacity; FEV= 1 -second forced expiratory volume. Acrophase (the peak represented by the dot), indicates the circadian phase of maximum response.

contact with acetylcholine. If isolated at 11 a.m., the rate of contraction decreased markedly. But if the heart had been isolated at 11 p.m., the rate of decrease showed much less response to acetylcholine. Thus, the possibility of a significant circadian change in responsiveness may apply to any fragment of living tissue, making time a useful factor in research.

Much of the present knowledge of hours of changing responsiveness began with the numerous rodent experiments of Dr. Franz Halberg and his associates, at the University of Minnesota Medical School. They showed that an animal made sick by a drug or bacteria at one hour of the light-dark regimen might die from the same dose if it were injected at another phase of the cycle.

Susceptibility to Bacteria and Infection

In 1954, the Minnesota laboratory began injecting inbred mice around the clock with the bacteria (Brucella) that cause undulant fever. A gauge of changing sensitivity was the number of animals in each injection group that died and the length of time before they succumbed. Standardized injections of this disease proved to be least injurious during daylight hours, whereas they were almost uniformly lethal during the animals' 12-hour period of darkness (and activity).

Subsequently, mice were exposed to an intestinal bacterium (E. Coli) that is very harmful to rodents. Groups of mice were injected with constant doses at four-hour intervals around the clock. Few mice died if injected in the middle of their activity cycle, but the same dose was lethal to those mice receiving it at the end of their rest span. By graphing the mortality rates against the hours of injection, Dr. Halberg drew a curve showing susceptibility rhythms.

The bias of timing upon survival or death is extremely dramatic, yet, perhaps, it is not surprising, considering rhythmic variations in possible immunity factors, such as gamma globulin. This is the blood factor that contains the most immune antibodies to bacteria and viruses. Drs. Lawrence E. Scheving and John Pauly recently found that gamma globulin levels in rats on a controlled lighting schedule reached a peak during the last six hours of darkness, and a nadir in the first hours of darkness. This bottom level of gamma globulin occurred just at the end of the animal's rest span, the time when mice had proven most vulnerable to E. Coli, and a time when other experimental mice had been far from their peak resistance to pneumonia.

Dr. Ralph Feigin and his coworkers at Washington University, St. Louis, have challenged mice with the same doses of pneumoccoci at different hours of day and night. The mice survived best if injected at 4 a.m., at the peak of their activity cycle. Dr. Feigin and his associates suggested that this might be a phase in which hormonal levels and other body chemistry were optimum for fighting infection.

Given the influence of circadian rhythmicity in vulnerability to pneumonia, time of day might also be expected to influence reaction to vaccination. Dr. Feigin and his associates had previously tracked the circadian rhythm of certain amino acids in human beings, finding lowest levels around 4 a.m. and peaks around 8 p.m. They then found they could upset the amino acid rhythm with injections of a vaccine for a virus disease known as Venezuelan equine encephalomyelitis. Men immunized at 8 a.m. showed less distrubance of amino acid concentrations than did men who received the vaccine at 8 p.m. Because this vaccine contained a live virus, the scientists inferred that time of day and the amino acid rhythm probably influence the body's response to infection.

Dr. M. Smolensky, at the University of Minnesota Medical School, Minneapolis, has examined the responses of mice to an anti-inflammatory drug at different hours. The response curve was revealing. Given at certain hours the drug caused severe loss in weight. This side effect was reduced by placing the animals on a new light-dark schedule, shifting their phase in the manner of east-west jet travelers. The study suggested that the timing of this drug might be critical in determining the severity of side effects. Such a factor could be especially relevant in medicating human infants for whom weight loss might be dangerous. Circadian patterns of man's susceptibility to inflammation have not been tracked in most instances, although there are a few illnesses that are legendary for their rhythmic symptoms.

In some diseases a parasite population makes a cyclical migration from one part of the body to another. The microfilaria known as Wuchereria bancrofti, tiny worms responsible for elephantiasis, infests millions of people in Asia and Africa. The adult worms live in the lymph vessels, but the larvae inhabit the bloodstream. By day they absolutely vanish, appearing by the millions each night. Dr. F. Hawking, of the National Institute for Medical Research, London, showed that in monkeys the microfilaria actually accumulate in the lungs during the day: their migrations can be influenced by oxygen.

The rhythmic shift of this parasite is evident enough to be measured in man, and a current study, coordinated by Dr. Franz Halberg, will examine the precise rhythms of the parasite in Ceylon and India. Relative to the circadian rhythms of temperature and urine constituents, do these parasites have the same phase relationships in each geographical region? Whether or not there are local differences, it will be useful to know host and parasite vulnerability rhythms in attempting to time medication for maximum effectiveness and minimum side effects.

Audiogenic Seizures and Epilepsy

Dr. Halberg and his coworkers have performed a series of studies on rodents and also surveys of some epileptic patients that may have some potential significance for human medicine. They began with studies of audiogenic seizures.

Minute-long bursts of loud noise have remarkable effects upon certain inbred mice. A noise such as an air blast comprising many high tones will cause rats and mice to dash around in an uncontrolled way, finally falling into convulsions and dying. Noise does not always produce this sequence. Sometimes the animals respond by crouching, as if offended, or by walking around, without running. Audiogenic seizures long have been a curiosity among psychologists and neurophysiologists. They have been measured after the administration of certain drugs in explorations of the ways in which drugs affect brain excitability. The fact that noise causes a sequence of activity that sometimes ends in death, and sometimes not, used to be considered a curious phenomenon.

In 1955, Dr. Halberg and his colleagues began time-series experiments subjecting groups of inbred mice to noise on a controlled environmental schedule. Each mouse was tested in a stimulator cage and exposed to a ringing of electric bells for a minute. The noise was about 104 decibels, roughly as loud as a nearby jet engine but with considerably less energy. The animals seemed impervious by day but highly susceptible at night. Convulsions and deaths were high during the activity span, and peak sensitivity seemed to coincide with the daily peak in body temperature.

"Time of day" within the nervous system is a potent concept. At any specific moment, while a creature is on a controlled and unvarying schedule, phase relations among internal systems are poised in a certain balance. But when the creature's schedule is shifted this internal environment goes through a period of transition, too.

When the Minnesota group replicated its study of audiogenic seizures, one group of mice was put on an inverted schedule in which the light-dark regimen had been shifted by 12 hours so that the span of darkness conformed to the comparison group's span of light. The peak susceptibility ordinarily occurred during the dark span, but the phase shift had increased the overall susceptibility to noise. The mice on the inverted schedule were not only highly susceptible to convulsion and death from noise during the dark span, they were even more susceptible during the illuminated daytime hours.

The mouse reaches its highest pitch of activity in the hours of darkness that correspond to early morning for man. Thus, this kind of convulsive periodicity seems to bear some resemblance to the morning type of periodic seizures seen in epileptics. The effect of a phase shift in increasing the hours of vulnerability to sound-induced convulsions suggests that a similar phenomenon might be seen among epileptics who travel by jet, and thereby undergo a phase shift that may increase their vulnerability to seizures at all hours of the day and night.

Halberg and his group tabulated the seizure hours of a large number of epileptics at a state hospital, and found seizures clustered toward morning. In one study, EEGs were recorded at 90-minute intervals around the clock along with blood samples. Brain-wave abnormalities were found at the time of day the epileptic patient usually suffered his seizure, even if he had no attack on that day. Moreover, when three of the patients were studied ten years later their EEGs still showed spikes and paroxysms at about the same time of day. This suggests that with a relatively rigid social routine the 24-hour distribution of abnormal brain activity is remarkably stable, a finding that has potential usefulness in the administration of anticonvulsive drugs as well as in understanding mechanisms underlying epilepsy.

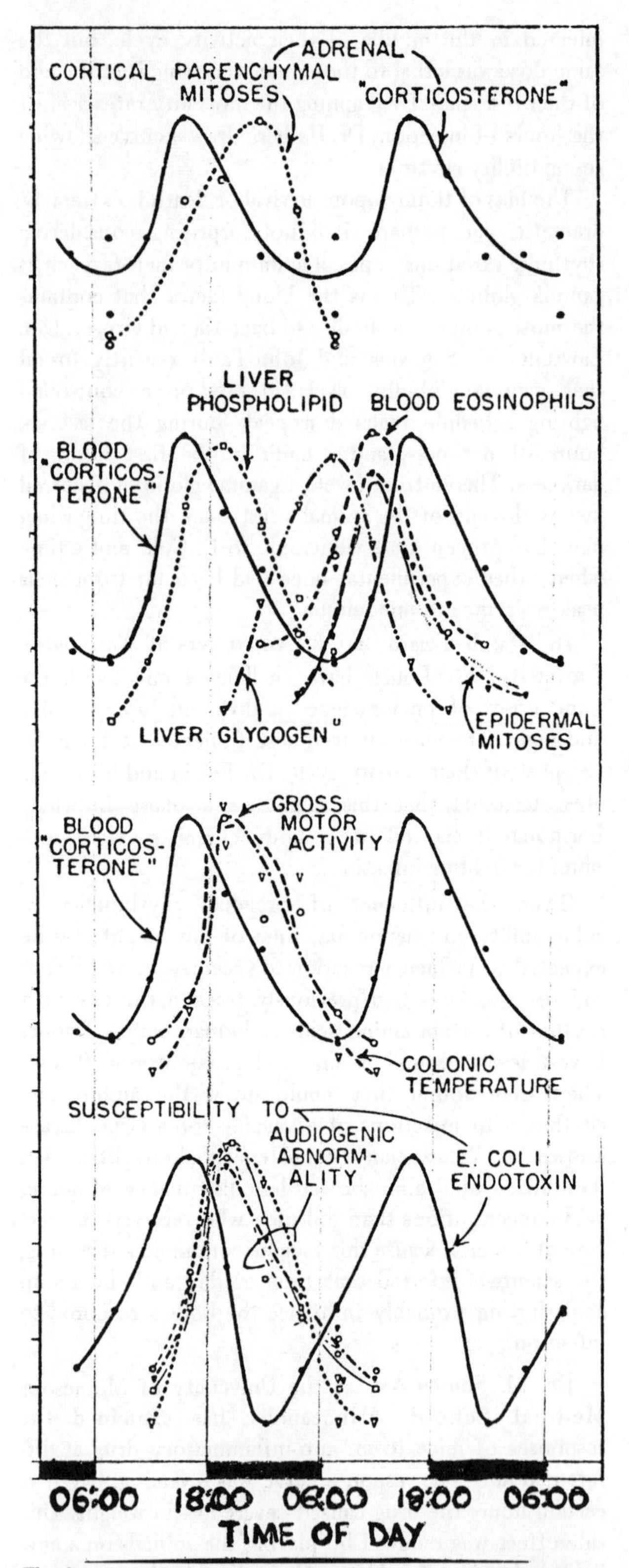

The figure, reprinted with permission from Franz Halberg, shows results of periodicity analysis of mice on a controlled light-and-dark routine. It is possible to see the phase relations of peaks of circadian physiological rhythms and of susceptibility to noise and bacteria with reference to the light-dark regimen.

Anesthetics, Alcohol, and Stimulants

Unlikely as it might seem, the rodent is not a bad test animal in evaluating drugs for man, since it more nearly resembles man in metabolism than many other animals. Using controlled lighting schedules, so that all the test animals were positioned at known phases of their cycle, these experiments all followed roughly the same procedure. Out of a large pool, sometimes containing hundreds of inbred animals, a subgroup was injected or tested at a specific hour and subsequent groups were tested at two or four-hour intervals around the clock for one or two days.

Mice injected with alcohol—a large dose akin to a quart of vodka for a medium-sized man—found it lethal at the end of their rest span (8 p.m.) and beginning of activity. Sixty percent died if given alcohol at their awakening time, but only 12 percent were killed at the onset of inactivity.

Responses to major surgery may sometimes stem from the biological time of day at which the individual was operated upon as well as the particular organs affected. One factor in response to surgery is the influence of the anesthetic, and these depressants may indeed have pronounced cycles of maximum impact.

The clinical dangers of an anesthetic, best known as halothane, prompted Dr. Halberg and his associates to conduct a time-series study of mice. Susceptibility to halothane was influenced by age, but a ten-minute exposure to a standard dose of the anesthetic might kill either 5 percent of the animals or 76 percent, depending upon biological time of day. Maximum sensitivity seemed to occur at the middle of the active time when the animal was least sensitive to other toxins.

Dr. Lawrence E. Scheving and his associate, Dr. John Pauly, have looked at effects of depressants and stimulants on the nervous system at points around the clock. Dr. Scheving's laboratory has shown that the amount of time a rat will sleep after an injection of the barbiturate, pentobarbital (Nembutal), depends upon when it is injected. At one time of day the hypnotic will cause 50 minutes of sedation, whereas the same dose causes a 90-minute sleep if given at another time of day. Mortality curves for high doses of barbiturates also follow this rhythm. An animal living on a light-dark schedule that approximates the hours of natural light, with 12 hours of light and 12 hours of darkness, will be most vulnerable to sodium pentobarbital at the early part of the dark cycle, between 9-10 p.m.

The harm done by nicotine or the stimulant, amphetamine, depends upon the time of day it is received. Drs. Scheving and Pauly kept rats on an artificial day-night schedule and injected potentially lethal amounts of amphetamine at two-hour intervals. Around 6 a.m. (the end of the activity period) mortality was low—6 percent; but at midnight (the time corresponding to the peak of activity) 77.6 percent died. This meant that there was a period of very high resistance to the drug and a period of great vulnerability. When human beings take amphetamines to enhance alertness and energy or to reduce appetite, are there corresponding time periods when the drug does harm?

When rodents were kept in continuous light (perhaps akin to continuous darkness for man), Dr. Scheving found that the colony rhythm flattened and the animals became more resistant to amphetamine. On the other hand, when rats were blinded the vulnerability rhythm to amphetamine shifted notably. Because light and darkness and the phase of the animals' schedule produce such dramatic changes in vulnerability, toxicity studies should specify the light-dark cycle of the animals tested and the hour at which the substance was used.

Dr. Scheving's laboratory has evaluated toxicity rhythms under a number of conditions and for a variety of substances, among them pentobarbital, amphetamines, strychnine, and nicotine. The effects of the drugs were clearly related to the animal's phase in the circadian cycle. Some drugs act upon the endocrine system: metyrapone, which acts on the pituitary reserve of ACTH, is most lethal at the beginning of the dark period of the activity cycle. So is acetylcholine, a neurotransmitter. Unlike acetylcholine, the anti-anxiety drug, chlordiazepoxide (Librium), was found to be most lethal to rats in the middle of their dark span, the middle of their activity.

These results could have very strong implications, since drugs affecting the nervous system are being bought in ever-increasing volume. Most of these drugs are not accompanied by literature that specifies timing. Some medications are not to be taken on an empty stomach, some are given ad hoc for pain or motion sickness, and hypnotics are usually prescribed before bedtime. Apart from the kind of practical advice that would suggest a person retire at once after taking a rapidly metabolized barbiturate such as hexobarbital, the doctor could not expect to find any statement about the hours of activity or rest when a drug should have its maximum effect.

In some instances, buckshot capsules are designed with coatings that release pellets of the drug continuously for as long as 18 hours as if the body were constant. A doctor has no way of calculating drug responsiveness in a patient so as to offer the minimum dose with optimum effectiveness. At present, indeed, it is impossible to specify at what time of day psychogenic drugs might be most helpful. A few intuitively gifted clinicians have experimented with pulsed doses of drugs.

Some years ago, Dr. Heinz Lehmann of Douglas Hospital in Montreal, noticed that some of his psychiatric patients were not sedated by the hypnotics they took at bedtime. He and his staff experimented with parceling out the doses, giving a fraction in mid-afternoon, a fraction in evening, and a final dose before bed. Under this regimen, patients who had been irrepressibly agitated were eased into sleep on less sedative than they had been taking before. With art, perhaps, drugs can be tailored to the beat of the body.

Today, in order to judge a person's neural time of day, it would be necessary to ask him to undergo a battery of tests. He might, for instance, estimate a two-minute interval by counting from 1-120. The speed of this count varies in a circadian rhythm, as demonstrated by a careful round-the-clock study made by Dr. Gwen Stephens at the University of California, Irvine, who recorded her own oral temperature, pulse, and estimated two-minute span, seven times a day for 36 days. Sometimes she counted slowly and sometimes she counted fast. Her most rapid counting and brief estimates correlated with her peak pulse rate in a pronounced circadian rhythm.

Rhythms in Pain Tolerance

In the 1970's, looking back on the evidences of circadian rhythms in the nervous system, it does not seem surprising that there might be circadian rhythms in tolerance for pain. In 1960, such a cycle was unexpected. Dr. Elliot P. Weitzman and his colleagues were trying to create a precise measure of pain tolerance in monkeys as a baseline for drug testing. Electrical probes were implanted in the ganglion of a nerve that branches from the muscles and skin of the face to the brain stem. This is the gasserian ganglion, located about an inch behind the ear. When given electrical stimulation in this ganglion people reported considerable pain, and monkeys have grimaced and pulled away, suggesting that they, too, find it painful. Using direct electrical stimulation, it was possible to control the voltage to make fine adjustments in the increase or decrease of sensation.

Monkeys, seated in a chair in a soundproof chamber, were trained to press a lever to lower the intensity of pain. Each lever press diminished the intensity by one step. There was a rapid succession of 25 increments between minimum and maximum intensity, so the monkeys had to press the bar rapidly to adjust the stimulation to a bearable level or turn it off altogether. Presumably, a small amount of stimulation was tolerable, for they never bothered to turn off the voltage completely.

In long sessions, lasting six to eight hours, each monkey would settle for a particular, presumably bearable level, indicating this by the rate at which he pressed the bar to maintain the shock at that particular intensity. Although the monkeys were individual in the level of shock they tolerated, each one was quite reliable. It was possible to repeat the procedure and watch the changes produced by drugs. When a monkey had taken morphine, he would tolerate a much higher current. When given barbiturates and tranquilizers, however, the monkeys seemed to fluctuate wildly, sometimes accepting a lot of stimulation and sometimes very little; a phenomenon replicated recently by Dr. Charles Stroebel at the Institute for Living in Hartford, Connecticut, in a study using constant light.

A normal monkey would bar press for over eight hours in order to maintain a constant level of stimulation. But if he were kept in the restraining chair for 24 hours, he would begin to exhibit some radical changes. After 16 hours in the chair, he would begin to tolerate extremely high intensities of shock; then he would move back to a lower level, slowly allowing the intensity to rise again until about 18 or 19 hours later, when he accepted almost the maximum voltage. However, when the clock came around to the time of day at which he began the endurance trial, the monkey again pressed the bar steadily and maintained the level he had originally found bearable.

The experimenters noticed that pain tolerance shifted according to time of day, and this may be true of sick persons. At night, a sick child becomes most cranky and restive, and in the darkness of a hospital patients grow plaintive and pathetic, requesting more pain killers and showing signs of suffering. Do they feel their symptoms more intensely then, or does night bring feelings of fear and loneliness? The monkey study suggested that there may be some rhythmic fluctuation in the tolerance of pain itself, a part of the creature's circadian rhythm, one that can appear in a caged animal living in constant light.

Neurologist Henry A. Shenkin at the Episcopal Hospital in Philadelphia, feels that pain may be related to the blood levels of adrenal hormones. Dr. Shenkin has conjectured that pain might influence the rhythm of plasma cortisol, and this could allow neurologists to distinguish between people whose pain derived from emotional sources and those who should be diagnosed for organic complaints.

Round-the-clock plasma cortisol levels were established for a number of people who suffered no pain and were not in the hospital. Dr. Shenkin hypothesized that pain would interfere with the normal circadian cortisol curve. He conducted a double-blind study in which patients were evaluated independently for pain while their plasma cortisol rhythms were charted. Among 66 patients, some had hernias, disc problems, cancer, and other obvious sources of pain. Another group of 45 patients reported pain for which there was no "objective" source. Thirty-eight were classified as having normal circadian rhythms of cortisol at normal levels in the blood. When compared, at 4 p.m. and 9 p.m., the persons with objective pain showed plasma cortisol levels that were about double those of the people with undiagnosed pain. People who feel miserable, but whose source of pain cannot be located, are particularly difficult to diagnose with accuracy. Dr. Shenkin believes that their normal circadian variation in plasma cortisol might be a diagnostic criterion for considering whether their pain might be due to emotional causes. In this instance there would have been an error of about 10 percent, for the diagnostician would have been misled in only 16 cases out of 111.

Responses to analgesic and sedative drugs may be influenced by rhythmic changes in the state of the nervous system. However, in examining fluctuations in toxicity, the activity of enzymes in the liver is likely to be crucial.

Ouabain: Kidney and Liver Enzymes

Ouabain is a substance somewhat like digitalis, a heart stimulant and diuretic that causes water to be released from tissue and excreted. Ouabain is quite poisonous, and when it is injected into mice in high doses death is usually swift—within ten minutes. When Dr. Halberg and his associates subjected mice to this toxin at four-hour intervals around the clock, it was clear that the poison was most lethal around 8 a.m. Those animals at the end of their day, so to speak, and about to rest, were most injured. Yet, many mice survived in the group that was injected at 8 p.m., at the end of the animals' rest hours.

A great portion of our response to drugs or toxins appears to be due to the activity or inactivity of enzymes, catalysts that destroy toxins in the liver and the kidney. Beginning with the work of Halberg's Minnesota laboratory, a number of these enzymes have begun to be time-charted. The homogenized tissue is placed in a biochemical solution that will be altered by the extent of enzyme activity. The side products of the biochemical reaction often can be gauged by various color tests. In actuality, tracking the daily rise and fall of enzyme activity entails the usual difficulties of time-series experiments. The rhythms are sensitive to changes in light-and-dark scheduling. In many instances they are influenced by diet, since a vitamin deficiency or protein deficiency can change the amount of available enzymes or smaller molecules known as coenzymes, often derived from vitamins.

Early experiments by Dr. Halberg's laboratory on the citric-acid cycle in adrenal tissue showed that coenzyme A, which is central to energy production throughout the body, follows a circadian rhythm. The peak level of this coenzyme occurred near the middle of the daily dark span in rodents, while adrenal corticosterone concentration reached its peak near the end of the daily light span.

Many biologists have looked at the effects of diets or stimulation upon the activity of a kidney enzyme, transamidinase. Perhaps some of the fluctuation they were seeing stemmed from its 24-hour rhythm. This enzyme, important to energy storage, was studied by Dr. Halberg's laboratory and proved to have a pronounced circadian rhythm. Its most intense activity came during the hours of light, when the rodents were resting, and it dropped to a trough during the dark hours of activity. On comparison, two strains of inbred mice differed in the phase of the rhythm: transamidinase activity rose and fell earlier with respect to the light-dark schedule in one strain, although the period of the rhythm was the same in both. This suggested there might be genetic orgins for differences in timing and thus in responsivity to drugs.

The manner in which enzyme rhythms might change our responses to drugs has begun to be unfolded by a number of laboratories. In 1967, for instance, Purdue researchers Drs. Frederick M. Radzialowski and William F. Bousquet described a study of circadian rhythms in a liver microsomal enzyme system of rats and mice that is known to break down steroid hormones and foreign compounds such as drugs. Thus, this enzyme system controls the intensity and duration of certain drug responses. Further studies of drug metabolism must be conducted with an understanding of the underlying rhythms of enzymes that act on the drugs in a circadian manner as this team suggested. Otherwise, fluctuations in drug response could be misinterpreted. This means, among other things, that it is essential to maintain the animal colonies used in drug response and toxicity studies under strictly controlled light-dark cycles. Other studies of liver enzyme rhythms include the already described rhythm of tyrosine transaminase, worked out by Drs. Ira Black and Julius Axelrod at the National Institute of Mental Health in Bethesda, those conducted by Dr. Richard Wurtman at MIT, and those of Dr. H. von Mayersbach at the University of Hannover in Germany.

Drs. H. von Mayersbach and R. P. Yap have demonstrated that there are enormous differences in the activity of esterase, an important liver enzyme that exhibits an unmistakable 24-hour rhythm. These researchers have shown that specific enzymes have different activity rhythms, all of them circadian, but reaching their peaks at different times. As with DNA, there are some sex differences: male and female livers show slightly different enzyme phases. Dr. C. Jerusalem and his associates have gone on to show that circadian rhythms in liver cells are not only biochemical but structural, and can be observed clearly under the microscope.

It was in the course of studying a muscle relaxant, succinylcholine, that one impressive enzyme rhythm was discovered. Dr. von Mayersbach and his associates at Hannover were interested in the mechanisms by which the body detoxifies succinylcholine. This relaxant had been used in place of curare during surgery, but had caused fatalities by stoppage of respiration. As the Hannover group injected rats with a dose of succinylcholine that was estimated to kill about half of the animals, it became clear that there was a biphasic mortality curve. If the same dose of the relaxant were injected toward noon or midnight, the mortality was about 10 percent, but if the drug were injected around 8 a.m. or 8 p.m., mortality rose to about 60 percent. The cause for this biphasic mortality was not at all obvious.

Esterase in the liver can act very much like another enzyme, cholinesterase. The team injected succinylcholine into the skin and then examined the activity of liver esterase. The activity of the liver enzyme was precisely the opposite of the mortality curve. It was most active at noon, and showed its nadir around 8 a.m. and 8 p.m. The esterase had a pronounced circadian rhythm, and this rhythm seemed to determine how rapidly the relaxant would be broken down and detoxified. The biphasic mortality curve appeared to be a consequence of the rhythm of this liver enzyme.

Rhythmic changes in vulnerability to toxic compounds may be predicted through studies of enzyme rhythms as well as by mortality rates. Such studies may predict where drugs may entail some risk of toxicity and how to time administration for least risk. In the treatment of many ailments, such as diabetes or Addison's Disease, hormones are administered as replacements for those the body failed to supply. In these instances a knowledge of the normal hormone rhythms may be some advantage. Indeed, Drs. Alain Reinberg and Jean Ghata have studied circadian rhythms in normal persons and patients with Addison's Disease, and have wondered if people could learn to time their own medications in a way that approximates a normal cycle by attending to their own feelings.

Self-Regulation of Adrenal Hormones

The adrenal hormone deficiency of Addison's Disease frequently leaves patients feeling fatigued. Drs. Reinberg and Ghata asked several Addisonian patients to apportion their daily drug dosage according to the one of three treatment schedules that made them feel best. One patient, a wall painter who had been a patient of Dr. Ghata for ten years, began to take his cortisol so that he had only a third at bedtime, and the rest upon rising. Without being instructed, he had approximated the concentrations that would be found in a normal person at these times of day and night, creating a normal circadian rhythm.

In collaboration with Dr. Franz Halberg, a longitudinal study was made of a girl with Addison's Disease. Hormone treatment that approximated the circadian rhythm of a normal person was most successful in eradicating the symptoms of fatigue. In Paris, since then, a number of patients have been placed on a self-regulation schedule. The efficacy of timing medication may be learned through a comparison of patients on non-adaptive schedules with those whose drug therapy approximates the normal circadian hormone rhythm.

It seems likely that in many endocrine diseases, abnormal rhythmicity plays some role. If carefully supervised patients were allowed to distribute their medication around the clock in the manner that gave them most relief, they might act in a manner like that of the Addisonian patient, who simulates normal time-structure.

Similarly, if patients with allergies were clocked in their use of hormones and anti-allergens, and were permitted to allocate medication according to the intensity of need, many of them would indicate that allergic symptoms also show a circadian rhythmicity.

Allergy

Asthma and related allergies are so common that they afflict about one out of ten Americans. Asthma causes about 9,000 deaths a year, and some 3,000,000 school children suffer from it. Difficulty in breathing comes from a swelling of the bronchial tubes and membranes, which can be caused by allergy to plant or animal proteins, drugs, pollutants, house-cleaning materials, or insecticides. Some forms of the disease seem to be nonallergic, and sometimes the precipitating mechanisms appear to be emotional. The asthmatic patient, particularly a child, who sometimes wakens from sleep gasping for air and feeling that he is about to suffocate, may rarely experience a symptom during the day.

In 1962, Dr. Alain Reinberg of the Centre National de la Recherche Scientifique in Paris, became interested in the fact that a dose of cortisone or cortisol diminished asthma symptoms. In attempting to see how the adrenal steroid pattern might be related to the symptoms, Dr. Reinberg studied eight asthma patients and four healthy subjects for abnormalities in their rhythm of excretion of adrenal corticoids. Both groups showed a decreased level of excretion at night and higher levels in the morning. During the study, asthma attacks occurred only when both potassium and adrenal corticoid excretion levels were low, in the early part of the night, a time when many people with asthma expect their most serious attacks.

There are other rhythmic effects, slight but detectable, that have been seen in asthmatic children. Dr. C. J. Falliers and his associates of the Children's Asthma Research Institute and Hospital in Denver, have studied children around the clock as they inhaled and then exhaled into a spirometer, in a test of vital lung capacity. In 1966, in collaboration with Dr. Halberg, nine normal and nine asthmatic boys (who were not taking medication) were given spirometer tests every four hours. Pulse, a urine sample, and rectal temperature were recorded at the same time. On most measures of circadian rhythms, the asthmatic children resembled their normal counterparts. However, the healthy children showed their peak phase of vital capacity between 1 p.m. and 4 p.m., while the average peak of the asthmatic child occurred around 2 p.m.

Children with severe asthma are relieved of their terrible symptoms by adrenal hormones, but these have the unfortunate side effect of stunting growth and delaying maturation. Drs. Falliers, Reinberg, and Halberg conducted a collaborative study, hoping that the side effects of hormones might be diminished by giving smaller doses at strategic times of day.

During a four-month study at CARIH, 28 youngsters were given prednisone, a synthetic derivative of cortisone. They received the drug on specific time schedules, and one measure of the drug effect was the time of the peak expiratory flow rate. The PEFR was gauged by having the child blow into the spirometer as hard and fast as possible, as if blowing out candles on a birthday cake. When asthmatic boys, not ordinarily taking hormones, were then given hormones and a spirometer test every two hours, it was clear that the hormones shifted the time of the PEFR. Moreover, the level of the child's output was changed only when he received the hormone before 8 a.m., not if it were given later.

During the study, youngsters who received prednisone at 1 a.m. showed their highest PEFR about two hours earlier than they had before. Those who received it at 7 a.m. were unchanged. Those who received it around 1 p.m. showed their crest PEFR about six hours later. Although these results did not indicate which timing was preferable for the child, the medical staff noticed that children who received their medication at 1 a.m. or 7 a.m. seemed to benefit more than the others. A subsequent study by Dr. Reinberg corroborated this observation, and it is hoped that further research will lead to a more precise way of choosing the best time of day to administer hormones so that they can be given in far smaller quantities.

Like asthma, other allergies seem to torture their victims at particular times of day. Dr. Alain Reinberg conjectured that this might be partly due to a rhythm in histamine response.

Histamine Skin Reaction

Histamine is one of the substances in skin that causes a response to insect bites and to burns. It is an amine, structurally related to ammonia, and among the causes of the flush in the skin during an allergic reaction.

Dr. Alain Reinberg and his coworkers at the Foundation Adolph de Rothschild in Paris, put six healthy adults on a standardized routine of sleep, waking, and eating for at least a week in the hospital before the testing procedure. Then they were injected with histamine under the skin. How big a reaction would they show in the reddened portion around the injection?

The intradermal injections were made at fixed hours, day and night. One thing was very clear: the largest histamine reaction and the biggest weal of the skin followed the 11 p.m. injection. The response of the skin would occur within 15 to 20 minutes after the injection, depending upon when it was given. Sensitivity of human skin to histamine or to a chemical that liberates histamine distinctly follows a circadian rhythm, with its peak at about the time of evening or night when adrenal corticosteroids are dropping to their lowest levels.

The effects of anti-histamine drugs were tested in the same way. The volunteers were injected with histamine, and then swallowed a constant dose of anti-histamine while maintaining a rigid hospital schedule. Some received the drug at 7 a.m., others at 7 p.m. Judging by the shrinking of the red welt, the morning dose had a far greater impact than the same dose at evening. The morning dose lasted about 17 hours, yet taken in the evening the anti-histamine effect lasted only a few hours.

Dr. Reinberg studied five patients who were extremely allergic to penicillin. Under the same carefully controlled conditions, they were given penicillin scratch tests, but their immediate skin reactions were different at different hours of the day. The greatest irritation and response of the skin to the allergen came at 11 p.m., the time that the excretion of 17-hydroxycorticosteroids was at its nadir. In several studies, it seemed likely that the rhythm of allergy might be related to the rhythm of the corticosteroids, and this might be a reference for timing medication.

In some instances, such as the control of fever, the duration of a drug effect may add to its efficacy. Aspirin is such a drug.

Salicylate

Drs. Reinberg and Halberg evaluated the rate at which the body gets rid of sodium salicylate, a compound similar to aspirin in structure and effect. Urine samples were taken from six volunteers who lived on a controlled schedule of eating and sleeping for over nine weeks. Each person received the dose at four different time points, and the tests were spaced a week apart. After receiving a gram of aspirin, the person's urine was collected at four-hour intervals for the next 48 hours. In order to ascertain precisely where the person stood on his biological cycle, the urine was analyzed not only for traces of the drug but for the adrenal hormone, 17-OHCS, and for potassium. By comparing the peak phases of 17-OHCS and potassium rhythms for several cycles, the researchers were sure that the person's rhythms were reproducible and stable. He could be compared with other people at the same biological time of day, with reference to these two functions.

The timing of salicylate evidently makes a difference in the durability of its effect. Taken at 7 a.m., salicylate seemed to linger in the body and was detected in the urine 22 hours later. If taken at 7 p.m., it stayed less long and the last point at which it could be detected was 17 hours later.

Whatever has been learned about the circadian distribution of drug responses under carefully controlled schedules must be considerably revised after any phase shift. Thus, a person traveling across time zones should not be surprised to find that he has a different response to a hormone or sedative that he ordinarily takes every day.

Phase Shifts and Drug Response

Dr. Franz Hallberg and his associates at the University of Minnesota made repeated studies of liver enzyme activity in animals. They manipulated the pronounced 24-hour rhythm of the rodents by a number of tactics. They found that starving the animals did not change the rhythm, nor did handling the animals or disturbance in the colony room. However, when mice were put on an inverted lighting schedule comparable to a 36-hour jet trip in daylight, it took two weeks before their enzyme rhythm readjusted to the new lighting schedule. After a major phase shift, the body's rhythms adapt, but different systems adapt at a different pace, some at once, others taking 9-12 days. This transitional phase, in which some systems have adapted but others are out of phase with the new activity and light cycle, may account for some of the discomfort of travelers and also some anomalous reactions to drugs and stress. The change in susceptibility is dramatically clear in animals.

Ordinarily, ouabain is most lethal to mice at the beginning of their rest span, following the active period of darkness. However, when the lighting schedule is reversed the time of susceptibility shifts. As Dr. Halberg and his associates found, there was a transition period that lasted for 17 days before the maximum vulnerability again coincided with the beginning of the light cycle and rest period.

Analogous experiments with noise showed that animals were generally more susceptible to audiogenic seizures after phase shifts. Now the animals became prone to convulsion and death at all hours of day and night. Even nine days later, the mice who had undergone inversion of their light-dark cycle showed greater nervous susceptibility to noise than siblings kept on an unchanging regimen of light and dark.

A complicated aftermath can follow phase shifts. An animal's maximum vulnerability to a given substance may change phase or may even extend over more of the 24 hours. Thus, in a long internal adjustment period, an individual could become more vulnerable to certain kinds of trauma, and less susceptible to others. Studies of drug responses to phase shifts have not been conducted with human beings, but it has been conjectured that man, too, would be altered in his susceptibility by a shift in phase, whether this came from rotating work shifts or travel. Rapid travel might, indeed, render a person more vulnerable to certain kinds of infections, nervous shocks, and toxins, while he became more resistant to others. In the instance of infection, a traveler might be forced to be active at times when his circadian immunity level, judged by gamma globulin levels, was particularly low. Similarly, he might encounter allergens at a time when he would normally have been in bed, and when his allergic responses are particularly intensified.

Timing is a new element in pharmacology. Today an LD50—a dose that kills half an experimental group of animals—cannot be a specific criterion for judging the toxicity of a drug, since the LD50 changes with time of day and is altered by schedule changes that create phase shifts.

Phase Shifts Caused by Drugs

As pharmacologists attempt to anticipate the responses of humans to new drugs, circadian rhythms become a complicating factor, for they can be altered by drugs. This may cause us to reexamine the effects of some staple drugs. Antibiotics are a case in point. Biologists have found that actinomycin-D can shift the circadian rhythms of synthesis within the DNA and RNA molecules, the genetic templates of every cell. This, in turn, may shift the circadian rhythm of certain nervous system activities.

Dr. Felix Strumwasser of the California Institute of Technology, has examined circadian firing rhythms in a giant nerve cell from a common tidal pool mollusk. The firing rhythms were shifted notably by injections of actinomycin-D. One way in which drugs may exert their influence on behavior and memory might be by shifting the timing of RNA synthesis, thus altering the time at which certain cell populations reach their peak of multiplication.

A number of researchers, sensitized to the impact of drugs on time-structure, have begun to report that barbiturates may shift circadian rhythms. Drs. Dorothy and Howard Krieger of Mt. Sinai Hospital in New York City, have reported that sodium pentobarbital, given at any time of day, suppressed the rhythm of adrenal hormones in animals. The short-acting barbiturate, sodium thiamylal, blocked the morning rise of hormones only in animals who received the drug in the evening prior to the expected rise. Blood hormone levels, in their rhythmic rise and fall, are altered by seemingly unrelated drugs. Perhaps, these effects of barbiturates upon adrenal steroid rhythms can help to explain the hangover they leave, and some of the mental blunting and confusion that attend it.

Conversely, drugs may be used deliberately to produce desirable phase shifts. It has been hoped that the discomfort of east-west travel can be ameliorated by drugs. The Syntex Corporation has been attempting to discover whether certain adrenal hormones might be given to travelers to speed up their phase shifts and thus make adjustment to new time zones easier. Because different physiological systems adjust to phase shifts at different rates, suggesting that there is no one "clock" in the nervous system, the prospect appears to be complicated.

Nonetheless, Drs. Michel Jouvet and Jacques Mouret, of the School of Medicine, Lyon, France, feel that physiological rhythms may be altered by monoamine oxidase inhibitors in a manner that may shift the phase, or even extend the period of activity and sleep to a cycle as long as 48 or 72 hours. Monoamine oxidase is an enzyme that continually breaks down certain fundamental brain substances known as monoamines, thus preventing them from accumulating. Drugs that counter this enzyme have been effectively used to fight depression. Dr. Mouret has used a monamine oxidase inhibitor to shift the circadian activity-sleep-cycle in rats. After the drug, rats sleep less during the daylight hours when they would normally be sleepy.

Dr. William W. K. Zung of Duke University, has shown that the drugs known as tricyclic antidepressants actually extend the period of the sleep cycle.

Knowingly used, drugs may be beneficial in altering rhythms, either by shifting the phase, by changing the amplitude, or even by changing the period. Unfortunately, many of the effects of drugs upon the body's time-structure are incurred unwittingly. For instance, most laymen do not realize that they may feel the so-called withdrawal symptoms of a hypnotic or tranquilizer many days after they have stopped taking it, or that some drugs will leave subtle traces upon the timing of body functions.

Unexpected Temporal Effects

Somewhat at random, throughout the waking hours, people take tranquilizers, hypnotics, stimulants, antihistamines, antibiotics, sulfa, and other common drugs. It seems possible that some of these medications could produce long-term effects that defy our current methods of appraisal. Dr. Curt P. Richter has suggested such a possibility after a 40-year study of relationships between stress, drugs, and activity rhythms.

During the mid-1950's he began to use drugs to try and alter behavior cycles. One was a drug commonly used against infection, sulfamerazine. It was added to the diet of adult female rats who had shown a very regular crescendo in their running cycles. Every four or five days, as they went into estrus, they would become more active than usual as they approached estrus, always becoming active at the same time of day. After receiving sulfamerazine, the daily rhythm remained, but these animals began to show very long cycles in running and food consumption. Instead of reaching an apex every 4 or 5 days with estrus, the peak activity occurred every 20 to 35 days. Moreover, the abnormal long-term cycle did not appear until after sulfamerazine was removed from the diet. Then the animals began to look like manic-depressives. In one phase of these long cycles, they became very active and lost weight and showed less appetite. In their long inactive period, they became almost totally inert. They ate huge amounts of food, gained weight, and showed vaginal smears that were no longer coordinated with the four-to-five day estrous cycle.

A relationship between long-term and short-term cycles is a matter of great consequence, but at present remains unstudied. Richter speculated that the changes in his rats were due to the action of the sulfa drug on the thyroid. In experiments on the thyroid gland, he found that radioactivity also produced long and regular activity cycles. Was it possible that a patient could take sulfa or antibiotics and eventually suffer changes due to effects on thyroid? If so, the effects might be interpreted as psychiatric, emotional problems.

Society tends to be moralistic about people who lose and gain weight, who periodically seem to lose control of their appetites, and cannot maintain a steady state. But clinicians have always recognized that thyroid malfunction can lead to problems in maintaining a normal weight, appetite, and activity level. Are some of these problems of uneven behavior and weight regulation actually iatrogenic—created by drugs? Similarly, emotional symptoms may be periodic. In attempting to link abnormal behavior with past training or external circumstances, one overlooked possibility may be that the symptoms evolve from chronic after-effects of medication.

Dr. Richter has tested numerous drugs, among them a thyroid-inhibiting drug, thiourea; a pain-killer, aminopyrine; a sedative, sodium barbital; an adrenal hormone, cortisone; and a female hormone, progesterone. In each instance, he saw no abnormal effects while the animals were taking the drugs. Only later, when the drug had been stopped for quite a while, did an impressively large number of animals exhibit abnormally long activity and eating cycles. These cycles would never have been detected except in a laboratory where the activity rhythms of the animals were being recorded every day by an automatic recorder. Eighty percent of the animals that received thiouracil, a thyroid-affecting drug, showed subsequent cycles. The familiar female hormone, estradiol, left an aftermath of abnormal activity cycles in 66.6 percent of the group of test animals. This study was done in the late 1950's before the widespread use of estradiol and other female hormones in contraceptive pills. Could there be subtle after-effects from these contraceptive pills, even in women who show no symptoms during the long periods when they are taking them?

In relative ignorance of drug mechanisms, it is not possible to predict exactly which compounds will produce abnormal running cycles. However, in the huge number of animals that Richter tested in a decade, one-third did show abnormal cycling after drug intervention. Richter's comment was:

> "On the basis of these observations on rats, the possibility must be considered that the cessation of a drug or hormone after prolonged treatment may also produce lasting effects in man; further, that the existence of such changes may not be detected without the aid of special measurements made over long periods of time."

Long cycles of mood and behavior are beginning to be studied in mental hospitals using computerized nursing notes, which also allow an appraisal of behavioral changes in response to drugs.

The first psychiatric hospital to develop such an intensive method of examining drug responses was The Institute of Living in Hartford, Connecticut. There, in the hospital's research facility, monkeys have been the first patients to receive psychiatric drug treatment that was calculated to normalize aberrant circadian rhythms while improving behavior.

Dr. Charles Stroebel had produced, with behavioral stress in monkeys, both behavior and somatic disorders that resembled those of neurotic and psychotic patients. Concurrent with their abnormal behavior, the animals had exhibited gross aberrations of their daily temperature rhythms. Drug treatments that corrected the behavioral abnormalities also restored normal temperature rhythms. Indeed, the rapidity of the animals' recovery seemed to depend less upon the dosage of psychotropic drugs than upon timing. Pulsed doses of tranquilizers and antidepressants had a far faster effect than constant infusion or a single injection.

Recent research on lithium carbonate has offered another most striking display of the manner in which a psychiatric medication can normalize abnormal timing. Lithium removes the oscillations of manic-depressive psychoses in a manner that shows how the alteration of time structure may sometimes be a goal of medication. Acting through mechanisms not yet understood, lithium subdues the violent alternations between manic and depressive states, leaving the patient the relative stability that his normal peers generally take for granted.

"By-the-bye, what became of the baby?" said the Cat.
"I'd nearly forgotten to ask."
"It turned into a pig," Alice answered very quietly, just as if the Cat had come back in a natural way.
"I thought it would," said the Cat, and vanished again. Alice waited a little, half expecting to see it again, but it did not appear, and after a minute or two she walked on in the direction of the March Hare. . .she looked up, and there was the Cat again, sitting on a branch of the tree.
"Did you say pig, or fig?" said the Cat.
"I said pig," replied Alice, "and I wish you wouldn't keep appearing and vanishing so suddenly: you make one quite giddy."
"All right," said the Cat, and this time it vanished quite slowly, beginning with the end of the tail, and ending with the grin, which remained some time after the rest of it had gone.

Lewis Carroll,
ALICE'S ADVENTURES IN WONDERLAND

Chapter VII. Mathematical Instruments

When Androsthenes marched through India with Alexander the Great, he noticed that the tamarind trees opened their leaves during the day and closed them at night. Plant and animal rhythms caught the attention of many astute observers throughout history. An astronomer, Jacques De Mairan, noticed that the mimosa and other plants opened and closed their leaves in a 24-hour rhythm. Moreover, the leaf rhythm persisted when he left the plants in total darkness, an 18th century experiment that was to be the forerunner of many botanical researches into the metabolic and growth cycles of plants. Many of these have employed straightforward techniques of measurement. In Dr. Karl Hamner's laboratory at UCLA, for instance, some experiments use a strain gauge and camera, attached to the plant leaf, so that measurements of movement and time-lapse photographs may be obtained around the clock. Time-lapse photography shows the 24-hour leaf rhythm with no ambiguity, as do straightforward time-plots of leaf motion. Unfortunately, the simplicity of time-lapse photography and continuous recordings cannot be applied to most studies of physiological and emotional rhythms in man. Simple time charts usually are not sufficient, but spectral analyses and statistics have been a hindrance to communication among many scientists, even within the field of biological rhythms.

Some people continue to doubt the existence of biological rhythms unless they can be made visible to the naked eye. Prior to the invention of the microscope, they would have had to doubt the existence of the cell and its internal structure. Within the last five years, however, powerful electron microscopes have displayed molecules such as DNA. Thus, they have made visible the double helix shape that was originally discovered through years of ingenious experimentation, using inferences drawn from X-ray diffraction studies. Until analogous instruments are available for continuous study of the body's time structure, certain mathematical manipulations are the only microscope capable of making rhythms visible.

Light and dark, sleep and waking, social customs and other external factors may act upon the nervous system like the resetting of a wristwatch, synchronizing internal rhythms with the outside world. Rhythms are influenced by drugs, trauma, and seemingly minor events in animals as well as humans. An example of such sensitivity was seen during some experiments to determine blood-cell and adrenal rhythms in mice. Dr. Halberg's group found that a single drop of blood taken from a mouse's tail created a shock that disturbed the animal's eosinophil rhythm for several days. The measuring process may itself change the very state being measured. Thus, the controlled conditions of the study, the proper duration of the study, and appropriate intervals of measurement are the prerequisites for ascertaining cycles of change. In animal research the subject population is purchasable, but studies of healthy or of sick people usually involve very few available subjects.

In a compromise with the realities of equipment, funding, and human frailty, several researchers have devised what are known as transverse studies. They involve unequal time samples, perhaps observations done by different researchers and in different locations. Ideally, rhythms should be charted longitudinally in an individual, who might be tested every hour for weeks or months. Such studies require spartan researchers, capable of working around the clock for a long period. Even were researchers willing, it would not always be possible to take blood or urine from a person repeatedly at equally spaced time intervals, especially if he were vulnerable and suffering from emotional stress or illness. Even healthy people balk at having their sleep interrupted. Many studies, therefore, omit the sleep period. Transverse studies may combine data, using unequally spaced samples and mixtures of subjects who are observed over brief periods.

Using biotelemetry instruments changes in temperature, pulse, respiration, skin conductivity, and other functions can be recorded without impinging on the individual's awareness. If the miniaturization evolved for space exploration were applied in developing socially acceptable, miniature medical measuring instruments, they could be inconspicuous, perhaps sewn into the waistbands of underpants or worn like wristwatches. Inexpensive devices of this sort have been developed for etiological studies of city populations by Mr. M. Wolff at the Medical Research Laboratories in London. A number of American researchers have been working on such instruments—among them engineers at the Franklin Institute in Philadelphia; Drs. Charles F. Stroebel at the Institute of Living, Hartford, Connecticut; W. Ross Adey and his associates at UCLA; William Zung at Duke University. Dr. Franz Halberg and his coworkers are beginning to use a miniature instrument that can record body temperatures for seven days.

Such instruments will make possible continuous data sampling that can be analyzed by computer, and that will not distress or embarrass the person under study but will allow him to lead his normal life. In a technology such as ours, the lack of instruments in medical research is purely a matter of economic priorities. When such equipment becomes available the arduousness of clinical sampling and some of the complex strategies now needed to decipher and fill in missing data will become unnecessary. For the present, however, time structure often must be inferred from other inadequate data that leave huge gaps. Moreover, the changes under surveillance are likely to be subtle, a few more cells dividing at this hour than at that, a person making a few less errors or omissions at a pushbutton task, or a change in a patient's hormone concentration.

Scientists have solved some of these problems of sampling by treating huge groups of subjects as if they were a single individual. In animal researches, a subgroup of animals is used for blood or tissue samples at each time interval, and animals are not sampled twice. Insofar as they are nearly alike genetically, and in age, weight, sex, and background, they may approximate a single individual displaying a rhythm over a long span of time. Such a study might measure functions in 400 mice over 48 hours. In some sense these studies would tell what periodicities to expect in a single individual over a few days.

Sometimes, however, it is neither possible to conduct long-term observations of several individuals nor to muster a large group for a brief study. As a compromise, a few individuals may be studied for two or more cycle lengths. While this is not sufficient to permit evaluations of an individual, such a hybrid study may be better than no study at all, especially when the subjects are for some reason rare—as are astronauts in space, patients with encephalitis, or boys with periodic fever.

Common sense would dictate that any study of rhythms should encompass enough cycles—whether these be in minutes, hours, days, or weeks—to leave no doubt that the observed variation recurs at the attested frequency. Common sense dictates, but the sheer burdensomeness of time studies has induced many researchers to cut corners just a little. Moreover, the charting of a rhythm necessitates measures at intervals in a proper ratio to the total cycle. A roughly 24-hour rhythm should not be studied by taking samples at

12-hour intervals in the manner of many early clinical studies; nor does it make sense to sample every 10 seconds. Dr. Halberg and his associates have worked out what they consider a reasonable ratio of six sampling intervals to one cycle length: 6:1.

The data may be derived from a number of individuals at unequal time intervals (often researchers forego night samples to allow their subjects undisturbed sleep); understanding the data is a little like trying to watch a baseball game through a picket fence. If there are enough missing slats, equally spaced, it is easy to follow the action. But if part of the fence is blocked, one might see only the action at home plate and at first and second base. It is the kind of view that is obtained when tracking a 24-hour rhythm by taking only daytime measurements.

Statistical methods of analysis do permit a scientist to fill in the trajectory of the ball, so to speak, by estimating periodicities even when the values were obtained from different individuals at unequal intervals and during waking hours. The computational method fills in. But like the baseball spectator, the observer of rhythms shows far more accuracy when the phenomena are extremely clear and predictable, and he improves his accuracy when he has a longer time to watch.

In biological or clinical studies measurements taken repeatedly over long periods of time assure a kind of stability that cannot be ascertained in brief studies. Ten-year studies of nitrogen function in periodic catatonia have divulged a three-week periodicity of illness that might not be so clear in a short study, especially if that study occurred during a time of local irregularity. A basic periodicity must be a picture of the overall temporal pattern. In human and animal life, various shocks and perturbations will cause local deviations in the overall rhythms—deviations that may seem small in the long run, but which would loom large on a short-term study. The advantages of longitudinal study are obvious, along with the costs and difficulties. Drs. Arne Sollberger, Franz Halberg and other authors have emphasized the importance of sampling frequency in doing time-series studies. In order to discriminate between a circadian rhythm of 23.5 hours and a 24-hour rhythm, Dr. Sollberger suggests that one would need something like a sample every five minutes for a month. However, there is considerable disagreement about the exact frequency needed for an accurate description of the cycle period. Since all methods of analysis involve the assumptions and judgments of the researcher, the interpretation of time-series data is riven by the kind of controversy found in other branches of science.

Biological Clocks in the Unicorn

In 1957 a professor of zoology at Cornell University began to poke fun at some of the statistical methods used to decipher time cycles in "noisy" data, particularly the attempt to correlate biological changes in vegetables and other forms of life with geophysical events.

Dr. LaMont Cole wrote a famous article in which he assigned an arbitrary metabolic rate to the mythical unicorn. By using a RAND Corporation table of random numbers and a plausible formula he introduced serial correlations into his data to plot the rhythms of "metabolic rate." By assigning random numbers for metabolic change at each hour during a five-day period, and then averaging the hourly data, he found the plot suggested a rhythm but the periodicity was not quite apparent.

> "It occurred to me that in summer, at 40 degrees north latitude, the hour of the rise of the moon may be retarded by approximately one hour each night. To eliminate any such lunar rhythms we 'slipped' the data one hour per day, aligning hour one on the first day with hour two of the second day, hour three on the third day and so on. This seems to be a standard sort of procedure for analyzing such data."

When the data were re-averaged a daily rhythm came into focus.

> "The results revealed that the unicorn is a very strange beast indeed. He rises to a high pitch of activity at about 3 a.m., but by 3 p.m. he has become inert. A mid-morning dip in activity 'remains unexplained' but may possibly be a subject for future research."

Cole's jibe was not aimed at all of the scientists studying biological rhythms, but it expressed a nagging skepticism that has afflicted many grandstand observers of the research. Biological rhythms have sometimes appeared to resemble the smile of the cheshire cat—faint, taunting, appearing, and disappearing. Mathematical manipulation seems to turn cloudy data into clear rhythms in a manner that is disconcerting to the bystander who does not know the process. Still, changes are hard to perceive in raw data, and not many people have detected circadian rhythms in the leaves of their house plants. Indeed, not many people perceive their own rhythmic changes. The skeptic might remind himself that he rises and sinks every night in predictable sleep cycles of dreaming and other states of consciousness. Probably, man has always ridden a roughly 90-minute cycle of sleep stages, unaware of it until the rhythm became a focus of study in the 1950's and was unmistakably displayed by nightlong EEG and physiological recording.

Lacking such devices for other kinds of round-the-clock observations, researchers have devised methods of "filling" in gaps in the data and applying mathematical methods to coax out of the mass of information clues to what order lies within.

Mathematical Models

Since the 1950's, biologists, physicists, physiologists, mathematicians, and others have collaborated and speculated on the kinds of internal mechanisms or oscillators that might generate our many rhythms. Is the circadian rhythm of waking and sleep caused by a mechanism like that of a child on a swing—constantly pumping and needing a regular input of energy to keep the swing going? Are our rhythms caused by endogenous mechanisms resembling various kinds of oscillators? Or do they follow from cosmic sources, waves of energy from the sun, the barometric pressure changes from the moon, magnetic field changes, cosmic rays? Does the oscillation resemble the pendulum of a grandfather clock or the atomic tuning-fork which, once struck, vibrates endlessly until the expiration of the atom's energy?

All self-sustained oscillations require some energy from the environment to replace the energy lost in the course of change. A pendulum, for instance, requires only a push now and then. When the energy loss is small a system resembles a pendulum, but when a lot of energy is expended much energy must be injected from the environment. Such a system, like that of the emptying heart, bladder, or bowel, is known as a relaxation oscillator. The higher the frequency of a biological rhythm (the heart has a higher frequency than body temperature), the more the fundamental oscillation tends to resemble a relaxation oscillator.

Many researchers feel that circadian rhythms in insects and animals are not relaxation oscillators and require no external energy to set them off. It is not necessary to describe different kinds of oscillators in detail to indicate that each model involves special principles that have been described by abstractions and that lead to certain consequences in living organisms.

Experimental evidence has instigated models that postulate how biological cycles in certain species respond to external influences. Models, such as those of Drs. Jürgen Aschoff and Rutger Wever of the Max-Planck Institute in Erling-Andechs, Germany, do not offer strictly mathematical theorems but specify hierarchies of influence upon internal oscillators.

Dr. Wever has proposed a simple differential equation as a model for circadian rhythms. The model assumes only that the cycles (within the circadian range of 21-28 hours) persist under constant conditions, and that the circadian rule applies. This rule specifies how light intensity influences the activity rhythm—the frequency of activity, the ratio of activity to rest, and the total level of activity—in any particular species. Light is a primary synchronizer of the activity rhythms of most species with the exception of man, insofar as is now known. Man may be influenced by light intensity but can be more swayed by his social context.

Wever assumes the persistence of the rhythm and circadian rule and therefrom derives many general properties of circadian rhythms. The relationships between rhythms and their synchronizers can be clarified by expression in mathematical form. These experimental models suggest what will happen under the influence of light at various intensities, temperature change, food scheduling, social factors, and so forth, and postulate what should happen when activity rhythms are delayed or accelerated artificially. Models attempting to make order out of the currently available information suggest a route of research that ultimately will contradict, confirm, or refine the conjectures.

A number of references in the bibliography offer a variety of mathematical models. These may indicate why, aside from the sheer enjoyment of the game of conjecture, scientists want to test the fit of ready-made theories within the unformed field of biological rhythms, especially in areas relevant to man, himself—those regions where time structure is least understood and hardest of all to study. Dr. Jürgen Aschoff's book, *Circadian Clocks*, Dr. Arne Sollberger's, *Biological Rhythm Research*, and a number of review articles indicate how mathematical models attempt to discover the value of known rules in illuminating the unknown. If one can see the semblance of familiar systems for which mathematics has already been devised, then these may help to understand the biological events under study and to describe them accurately.

Some of these models are cybernetic models of feedback, arising from engineering systems first devised in the 1950's. Human beings are exceedingly vulnerable creatures and, like the machines we invent, we keep a semblance of constancy—homeostasis—perhaps by feedback. The thermostat in our brain is constantly reading temperature in parts of our body and, like the thermostat on our home furnaces, it does not let us get too cold or too hot.

Feedback mechanisms, such as these thermostats, are familiar to engineers, and elaborate mathematics describe their function and enable us to use them. Sometimes, however, a system like this breaks down and sets up a huge oscillation known as negative feedback. All of us have experienced it in our homes. A guest requests that the temperature be raised, and somebody turns the thermostatic system up to 90. Soon the room is too hot, for the furnace has been going full blast. Now the thermostat shuts down, a window is opened, and the room soon becomes unbearably cold, whereupon the furnace is turned up to full blast again. Instead of an even temperature and an efficient use of the furnace, the room alternates between too cold and too hot and the furnace is taxed to the limit.

Some scientists have proposed that negative feedback may be a relevant mechanism in understanding certain periodic illnesses. Drs. Franz Halberg, Alain Reinberg, and Jean Ghata are among those who have plainly stated their opinion that feedback models cannot elucidate biological periodicity. Nonetheless, endocrinologists implicitly accept feedback concepts in talking about the way in which the adrenal cortex releases certain steroids, prompted by a hormonal signal from the pituitary that presumably originates in the hypothalamus. When the levels of adrenal hormones increase, the stimulating actions of the hypothalamus and pituitary cease. This is an example of a feedback system, but its various steps do not explain the timing of the process. Feedback mechanisms are used to explain gonadal function, yet the feedback chain of female hormones does not explain the time period of the ovulation cycle, nor why estrous cycles vary in length from one species to another. Only in special instances in which the time period of each step can be specified, could feedback mechanisms elucidate the period of a cycle. If some have conjectured that certain periodic illnesses might be explained by negative feedback, others have conjectured that periodic symptoms might be the result of two oscillator mechanisms that became related in such a way as to produce beat frequencies. Either interpretation suggests an expression of the data using mathematics borrowed from other branches of science.

Some of the understanding of frequency distributions or harmonics has been based on prior work in physics in the analysis of sound or light.

Diagram of a beat frequency which is the function resulting from the difference in period of two component frequencies. By permission of Franz Halberg.

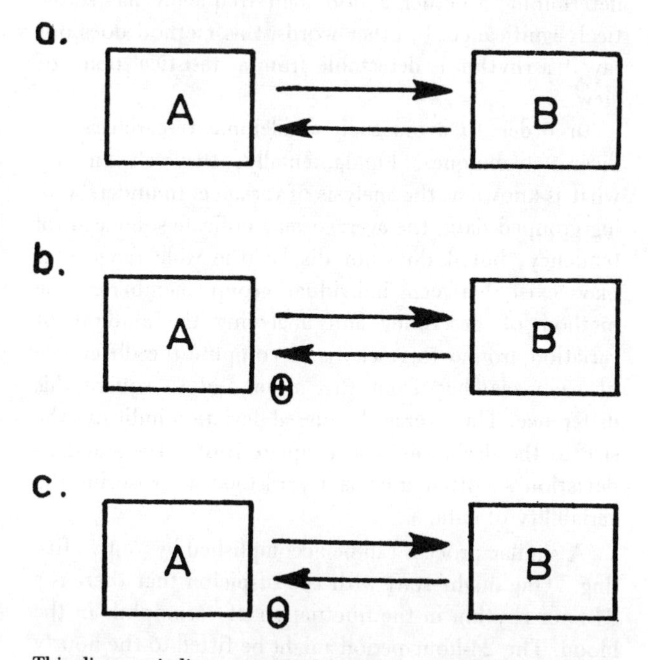

This diagram indicates types of feedback: a. mutual interaction, b. positive feedback and, c. negative feedback. From A. Sollberger.

Wave Forms and Harmonics

When we refer to the "menstrual cycle," we imply an internal process that moves in a kind of circle from a point of origin through various hormonal stages back again to that origin point in predictable repetition. Most fluctuations measured in the body are cycles in an analogous sense. Body temperature moves up and down about two degrees centrigrade each 24 hours, presumably representing a metabolic cycle. Adrenal hormones become more concentrated and less concentrated in the blood. There is more or less potassium in the urine.

Temperature fluctuations appear to comprise a number of frequencies. A very sensitive thermometer would reveal that there are many points on the circadian round at which our body temperatures rise and fall some fraction of a degree in what may be ultradian rhythms of twelve hours, and perhaps three hours, or even shorter intervals.

Any rhythm will be composed of at least one dominant frequency, usually accompanied by random variation that is referred to as noise. Mathematical frequency analyses are used to ferret out the hidden periodic fluctuations while excluding the noise. Frequency analyses are based upon harmonic or Fourier analysis, which represent any given time series (such as hourly temperatures for a week) as the algebraic sum of sine waves having different frequencies, phases, and amplitudes. Related methods, known as periodic regression and autocorrelation, are sometimes used when harmonic analysis is inadequate. A problem not resolved by Fourier analysis, for instance, is the problem of determining whether a dominant frequency has statistical significance. In other words, this method does not say if a rhythm is detectable from a statistical point of view.

In order to avoid such a dilemma researchers use diverse techniques. Fundamentally, they all employ what is known as the analysis of variance. In understanding grouped data, the average may indicate some central tendency, but it does not display the wide range that may exist between individual group members. One method of describing and analyzing the amount of variation around the mean is to compute the difference of each member from the mean and to square this difference. This averaged squared deviation indicates the size of the deviation, whose square root is the standard deviation so often used as a yardstick in measuring the variability of data.

A similar process can be accomplished by "curve fitting." One might start with the suspicion that there is a 24-hour rhythm in the fluctuation of eosinophils in the blood. The 24-hour period might be fitted to the hourly cell count. Then, one might add derivatives of the curve (harmonics), such as a two hourly or irregular peaking of the cell count. This composite curve would be added to the fitted curve until it became possible to significantly decrease the variance of values around the composite curve. By repeating the procedure with a number of basic periods, a series of approximations might determine which base period and harmonics best fit the data.

Another method that has been used by Dr. Gordon Globus at the University of California, in Irvine, in the analysis of sleep records is autocorrelation. Here, one might take a night-long sleep record, composed of alternations between Rapid-Eye-Movement (REM) sleep and non-REM sleep. The distribution of these sleep states would constitute one curve. The curve is now moved by one sleep stage and compared with itself. A correlation coefficient is calculated, and the curve is again moved another step. At each step a correlation coefficient is computed. An autocorrelation function is obtained by plotting the correlation coefficient against the sleep stages over which the curve has been moved. This technique, although it sounds simple, has a remarkable power to reproduce the hidden periodicities in the record and eliminate the noise.

It is useful to talk as if body fluctuations were waves, although it is not literally true. For instance, the daily fluctuation of corticosterone in human blood was long interpreted as a smooth decline to a trough around midnight and smooth steep rise to a peak before awakening. Yet, the close-interval sampling of Dr. Elliot Weitzman and others has indicated that the hormone enters the blood in spurts and puffs that occur in late sleep. It is no process resembling a smooth curve. Over the 24 hours, however, there is a regular maximum and minimum that can be described as a curve. Any fluctuation of levels may be seen as a cycle. The phase of any cycle can be expressed as a point on a circle—360 degrees.

The shape of the curve over 24 hours, its level, amplitude, phase, and slope can be expressed by different mathematical functions. After fitting a curve to the data, so that the rhythm is characterized by its frequency and phase, this may be further abstracted as a sine wave or cosine.

By the least squares method of curve fitting, it is possible to express results either as a sine wave or as a cosine. A cosine has its crest at the point of origin. A sine has its crest at minus-90 degrees. Since at present most scientists are interested in the peak and trough of a rhythm and a clockwise representation is easier to read, cosinor plots are easier to read than sinor plots. (The term "cosinor" was derived from *cosine* and *vector.*) If a cosine is fitted to the data to approximate a rhythm's amplitude as well as phase, the direction of the "hand" on the "clock" will indicate the phase of the rhythm at its peak, or in Dr. Halberg's terminology, acrophase.

Several varieties of analysis have been adapted to interpret biological rhythms, and there has been inevitable controversy over alternative methods. At present NIMH researchers, Dr. Per Eric Bergner and Dr. Nathan S. Kline, at Rockland State Hospital in Orangeburg, New York, are beginning to reassess the available techniques for collecting and evaluating time-series data from human beings.

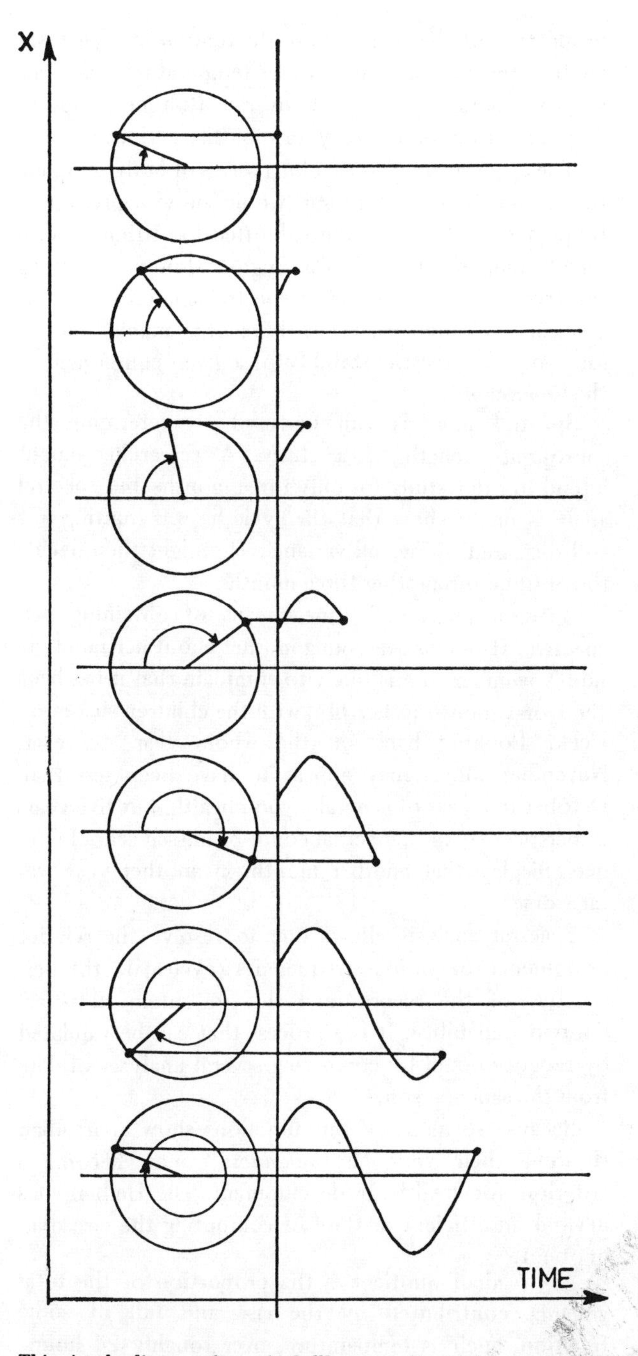

This simple diagram from A. Sollberger indicates how a rotating vector generates a sine wave.

At this writing, however, most biological rhythm studies involve animals or other forms of life, and many of the medically oriented researchers in the field are using techniques developed by Drs. Max Engli and Franz Halberg. By 1970, the carefully developed methods of the Minnesota laboratory were being used by scientists throughout the world, many of whom were sending their data from animal and human subjects to Minnesota for computer analysis. The methods are sensitive enough to ascertain drug effects, and the differential effects of drugs taken at different times of day. They have provided a means of comparing vital functions in sick and healthy persons and are useful in summarizing sparse data and data obtained at unequal time intervals.

Point of Origin: Determining Phase Relations

Trying to determine phase relations without an agreed point of origin is like trying to run a global airline without agreed-upon time zones. In 1883, an international conference met in Washington, D.C., to devise a system that would place all of the earth's clocks in time zones that referred to a single point of origin—Greenwich, England, time. Until then, every locality set its clocks according to a sundial. Catching a train or making a railroad delivery was no mean task. Passengers and railroaders alike had to make elaborate calculations to decide when, for example, in St. Louis' time, the transcontinental express train would actually arrive. In those days, a passing acquaintance with solar astronomy was a prerequisite for the traveler. It is much simpler now that the 360 degrees of the earth are divided into 24 segments, each of them fifteen degrees or one hour wide—with a point of origin at Greenwich.

In the realm of physiology and biology, a similar agreement is needed in evaluating internal rhythms. At present, each scientist chooses his own point of origin—if he specifies any point of origin at all.

When rhythms are expressed in terms of local time of day, anyone out of that time context has to make a translation. Without this translation, for example, it might appear that adrenal hormone cycles of people in New Guinea are out of phase with those of people in Scotland, and researchers have seriously proposed that this is so. By juxtaposing two sets of data and using a single point of origin (the midsleep time), Dr. Halberg and his associates demonstrated that there was really no phase difference. The two groups slept at different local times; but in relation to midsleep, their hormonal rhythms were in phase.

Such studies have now been generalized to other functions and other locations throughout the world, which may reveal whether geophysical factors are important in circadian rhythms. So far, the results of many comparative analyses by Dr. Halberg suggest that circadian rhythms are about the same the world over, with respect to the schedule of sleep and waking.

In studies that are run on a controlled schedule of light and dark or sleep and rising, the reference point might be the middle of the sleep span. For rodents, one might use the onset of light. Halberg's Minnesota laboratory suggests a point of reference: midnight (00), December 31, 1899, the beginning of the 20th century. This point was not idly chosen. It is useful in evaluating a vast amount of already available data, including statistical surveys of birth time, deaths, and suicides in large populations. It was the beginning of a day, a week, a month, and year, although the century actually began at midnight, December 31, 1900. In the future, each individual will probably become his own clock—as, in fact, he really is. By understanding phase relationships among cycles within the body, we should be able to read biological time of day in individuals. It is hard to predict exactly which rhythms will be the most convenient "hands" of the "clock"; probably they will be easy to measure, like body temperature, pulse, and respiration.

Spectral Analysis

A continuous record of body temperature taken with a crude thermometer might show very little change until the person ran a fever. On the other hand, a thermometer capable of registering a hundredth of a degree would show increasing change. Is this change rhythmic or merely noisy? Despite what we have observed in daily and monthly rhythms of temperature change, we know almost nothing about temperature changes over a span of a year or even over a span of three or four hours. A student who recorded hourly temperatures from a volunteer during a year might well gaze in disgust at his time plot over 50 miles of graph paper, wondering whether it contained any perceptible order.

Today a set of programs for a computer can do in minutes work that used to take months. As a first step the computer may printout the spectrum of temperature frequencies; that is, the frequency of occurrence of each temperature sampled. It would also printout the distribution of recorded temperatures along with their mean, total variance, and standard deviation. Looking at a histogram plotted by the computer, one would immediately detect fever or hypothermia by a skewed distribution of temperatures. One would no longer be looking at time, but rather the frequency at which certain temperatures occurred.

If one were to look at the entire picture of temperature change over a year—the thermovariance spectrum—one would see the variance at each frequency unit along the entire spectrum. This would tell what proportion of the temperature fluctuation in a year was contributed by the frequency of temperatures changing a given number of degrees over a 12-hour period, a 24-hour period, or a weekly, or monthly period.

The spectral analysis of a long series of body temperatures from a normal person would show a prominent frequency at about 24 hours. Plotted logarithmically, it might stand out like the Washington Monument among the tree tops. By comparing spectral analyses of short records—say three weeks—with records many times as long, one can see the stability of a given component in the long record.

Spectral analysis can be used to determine the appropriate length of a study. A researcher might intend to run a study for only three months, but spectral analysis might show that the cycle he was hunting was still obscured by overall variance. He might then decide to continue for another three months.

Spectral analysis is one means of obtaining perspective. If one asks a young mother about her family in mid-November, she is likely to complain that it has been the worst month in her life, with the children sick every week. Looking back on the whole year, however, November illness may appear to have been less than October in a year of generally good health. Put five years in perspective and an actual count of missed school days may display that another month, in another year was far worse.

Spectral analysis allows one to resolve the relative prominence of various frequencies (cycles) in the perspective of the record, and also to obtain statistical limits of reliability. It is a process that can be validated by repetition and by comparing several analyses of data from the same person.

Because so many of our functions show a circadian rhythm, their very *"circadianness"* may become a criterion for health or development. Dr. Halberg has devised an efficient method of computing the circadian quotient.

A circadian quotient is the proportion of the total variance contributed by the rise and fall of some function, such as temperature, over roughly 24 hours. Rarely is a frequency absolute. Since temperature does not rise and fall in exactly 24-hour cycles, one would compute a circadian quotient by using a band of frequencies. Spectral estimates for perhaps 23-25 hours would be summed and divided by the total variance.

Using this method, one could compute an infant's progress (if he were raised on self-demand) toward a 24-hour day without straining one's eyes on traditional charts of infant sleep-and-waking spans that show hourly blocks filled in with either black or white according to the baby's observed behavior. In their first six weeks, most babies show no perceivable pattern in their series of naps and wakeful hours. How fast is a baby progressing toward sleeping at night? It is hard to guess by looking at the usual time charts of the baby's activity. The following graph shows the same time chart accompanied by a visualization of the circadian quotient.

The circadian quotient gives an immediate display of the extent to which a child approximates a 24-hour sleeping-waking schedule.

One record of infants on self-demand feeding was computed by the Minnesota laboratory. In the first four weeks the pattern of sleeping and wakins is almost indecipherable by eye. The circadian quotient was less than 15 percent. At seven weeks the record was still confusing: the "circadianness" of the infant's sleep habits was about 25 percent. By twelve weeks, however, his circadian quotient had reached about 65 percent, and he was sleeping long stretches at night.

The Minnesota laboratory has used circadian quotients to depict responses to drugs. In one instance reserpine was given to a young girl, a hospital patient, whose temperature had been recorded at three-hour intervals before, during, and after drug administration. Her temperature, as revealed by the circadian quotient, was somewhat irregular before the drug: a circadian quotient of 25.2 percent. She was even more irregular (11 percent) during the days of reserpine administration and more nearly circadian (33 percent) after the drug treatment was finished.

The variance quotient simply offers a single number to indicate the relative dominance of a certain frequency or band of frequencies in the overall variance. In studies of the heart or of enzyme activity, an experimenter would be examining very high frequencies. In analyzing the rhythms of cell division in human cancers before and after X-ray treatment, the Minnesota team found it necessary to compute what is called a "relatively low frequency quotient"–the dominance of a band of frequencies with cycles around 19-28 hours. They found that the mitotic rhythms in tumors tended to be faster before X-rays and slower after radiation treatment.

In order to track free-running rhythms, or to watch a stable rhythm as it changes in amplitude or phase, it is useful to have an equation describing a harmonic function (such as sine or cosine) that most closely approximates the actual rhythm in amplitude and phase as well as in frequency. The following two graphs from Sollberger (1965) show the total harmonic function.

Such an equation allows one to estimate the period of the cycle, then its amplitude above the adjusted level, and next its phase in relation to some time point outside the body. The least squares method (also known as a multiple regression method) allows the use of data that were not taken at equal intervals, such as data from subjects who could not be tested at night when asleep, data from individuals who were not studied for identical lengths of time, or data from people who lived without time cues in an experimental situation such as a cave. Temperatures measured every ten minutes for three summer months and three winter months might be given to the computer.

The range of assessable rhythms extends from about hourly cycles to weekly cycles, with monthly rhythms on the questionable side. For each trial period the computer would express the temperature values relative to the mean variance for that time interval. For each frequency in the data the computer matches a cosine curve in the form of an equation to the biological values. The difference between the real and abstract curves with various amplitudes, phases, and levels are squared: the best fit curve is the one for which the sum of these difference squares is least.

Using a geographical analogy, the least squares would represent the least wasted area between an actual mountain formation and the curve drawn to represent it. The analogy to the crest of a hill would be the so-called acrophase, or recurring peak phase, of the biological cycle over time.

Phase indicates the relation between the curve and a point of reference. The computer can calculate the extent to which the best fitting cosine or sine curves vary from the original data. It can offer dispersion indices, for example, that are like the standard error estimates of the cosine or sine amplitude.

In the course of refining and testing their least squares computer program, Dr. Halberg and his associates began to reevaluate suitable data that were available. Fortunately, there was an extraordinary longitudinal study of Dr. Christian Hamburger conducted over more than 15 years. Hamburger had been his own subject, interested in fluctuations of 17-ketosteroids, steroids produced by the gonads as well as the adrenal cortex. Dr. Hamburger collected his total urine for each 24 hours and used the same method of chemical analysis for every sample for 15 years. A man of regular habits, he maintained a thorough daily analysis excepting for a few breaks for trips. In 1964, Dr. Halberg and his associates collaborated with him in an analysis of the data.

Records revealed a weekly hormone metabolite (17-KS) excretion rhythm with a peak in the middle of the week. For a decade the phase of the weekly rhythm changed remarkably little. Later, during a span in which

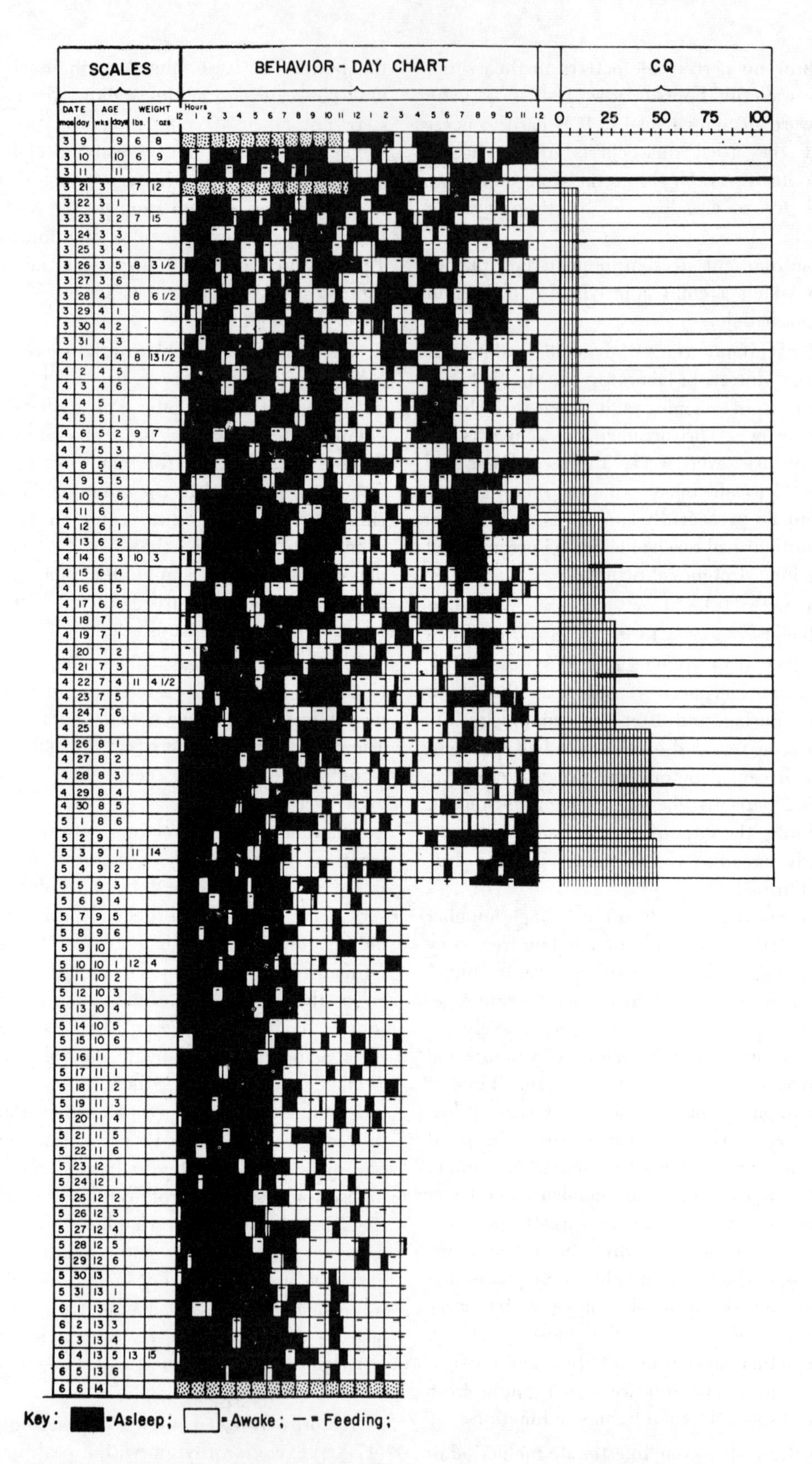

This behavior day chart of an infant on demand feeding illustrates the problems of visualizing rhythms of activity by a non-quantitative representation. The shaded area to the right gives the circadian quotient. By permission of Franz Halberg.

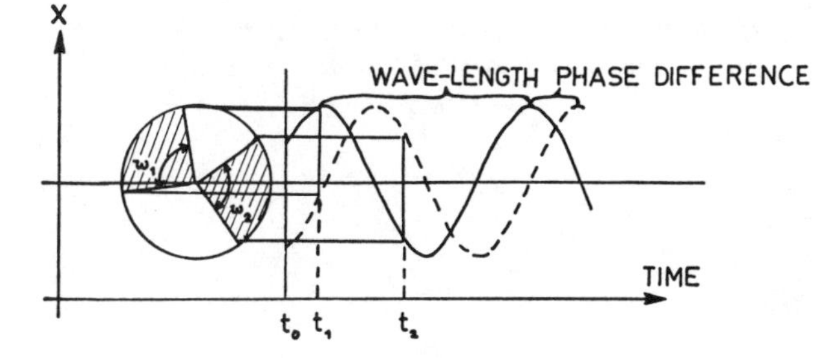

This simple diagram from A. Sollberger visualizes the concept of phase angle.

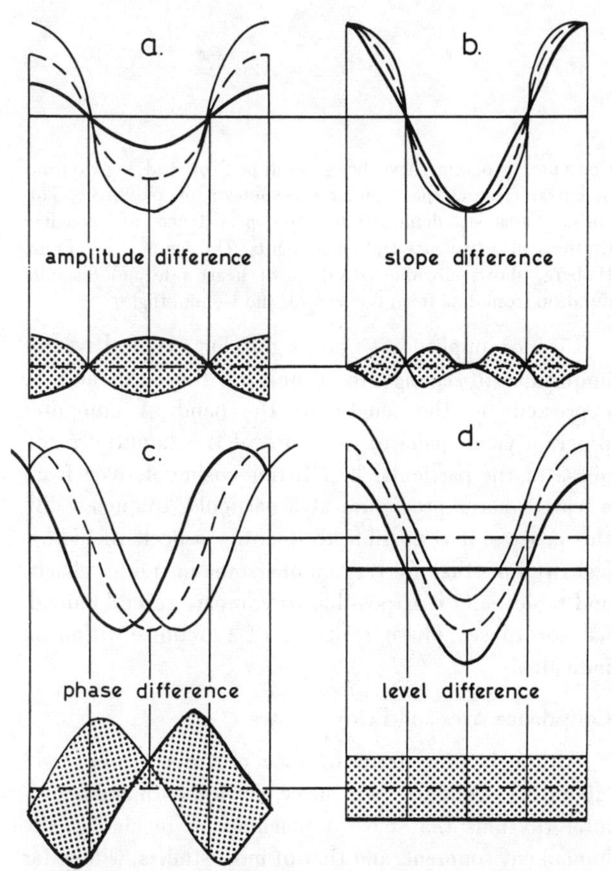

These illustrations from A. Sollberger suggest different ways of looking at data and specifying differences between individual curves.

Dr. Hamburger was taking a male hormone, a computer-prepared display revealed a desynchronization of the weekly 17-ketosterone rhythm. A slow shift seemed to occur over a period of years, vaguely reminiscent of the drift of circadian rhythms in people during experimental isolation. The crest began appearing earlier in the week during a three-year period, first on Thursday, then Wednesday, then Tuesday, suggesting that man has a near-weekly rhythm (circaseptan) with no direct environmental counterpart. Within the same data, Dr. Hamburger found a less obvious 30-day rhythm—suggesting that man's gonadal function may be influenced by cyclic hormonal change in a monthly rhythm.

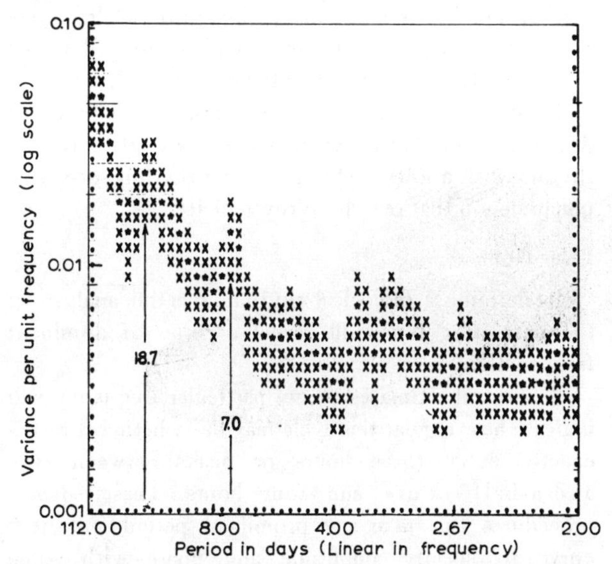

This variance spectrum pools five segments of a time series study of 17-ketosteroid excretion of a man - showing an about-weekly rhythm and a rhythm with a period around 19 days. By permission of Franz Halberg.

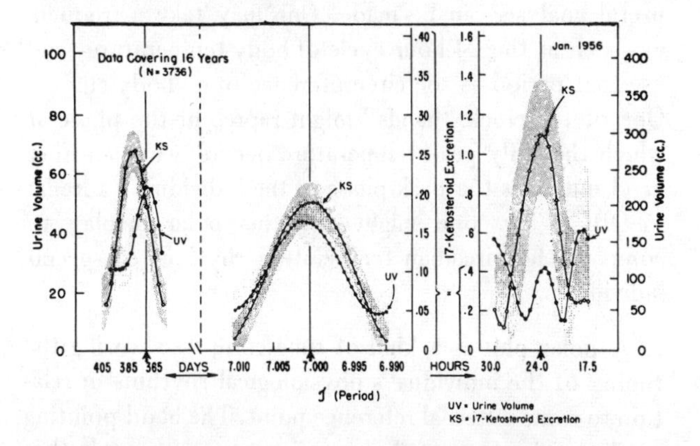

Three spectral windows from a least squares analysis by Professor Max Engli, using urine volumes and 17-ketosteroid determinations taken daily over 16 years. These analyses show sharp ketosteroid cycles with periods of about 385 days, 7 days and 24 hours.

In first applying a least squares spectrum one may not know where the prominent spectral components are located. Usually, there is a dominant circadian period of 22-25 hours. Having ascertained the circadian period, it is possible to define the rhythm more rigorously. With adequate data one can prepare a "window," choosing trial periods linearly, say at six-minute intervals, in the region from 20-28 hours. Using such a magnifying lens on some data collected in the early 1950's, the Minnesota team saw that blinded mice showed circadian temperature periods that ranged from 23.3 to 23.7 hours. Each mouse differed from a precise 24-hour period, and the free-running periods differed from one mouse to another. Differences in phase were detected even when the individuals had the same period.

Least squares analysis also can provide an index of dispersion. One knows whether one deals with a regular rhythm with a low amplitude—a near rhythm—or with a phenomenon that reveals no rhythmicity.

Polar Plots

In defining a biological rhythm, spectral analysis of the data over time will reveal a series of dominant frequencies.

A spectral estimate for any particular frequency will indicate how regular this cycle may be—whether it recurs exactly every three hours or peaks between two-and-a-half hours and four hours. Least squares procedures can take any prominent period and fit a curve to specify amplitude and phase—with some indication of the amount of variation between the actual biological rhythm and the amplitude phase and level of the harmonic function fitted. If the cycle is stable and the frequency does not seem to vary much, then another useful analysis can be made. One may take a frequency—such as the 24-hour cycle of body temperature—and use that period as the circumference of a "body clock." One of the clock "hands" might represent the phase at which the daily peak temperature occurs, while another hand indicates the peak phase of the individual's adrenal 17-OHCS. Or, one might use this polar display to compare the circadian temperature rhythms of several individuals.

A polar plot is a kind of time compass showing the timing of the individual's physiological rhythms in relation to some external reference point. The hand pointing to the peak phase of the circadian temperature rhythm could be read in terms of degrees—on a 360-degree clock. The direction of the hand, relative to the point of origin in the center, would indicate peak phase. The clock of a 24-hour rhythm could be read simultaneously in ordinary time intervals of 15 degrees per hour. The plot would not show exact peaks but rather an averaged estimate or typical phase representing the timing of all peaks that occurred during the collection of data (there being one peak for each function represented). For instance, a volunteer's rectal temperature may have reached its peak nine hours after the midpoint of his sleep span but not precisely at the same minute each day.

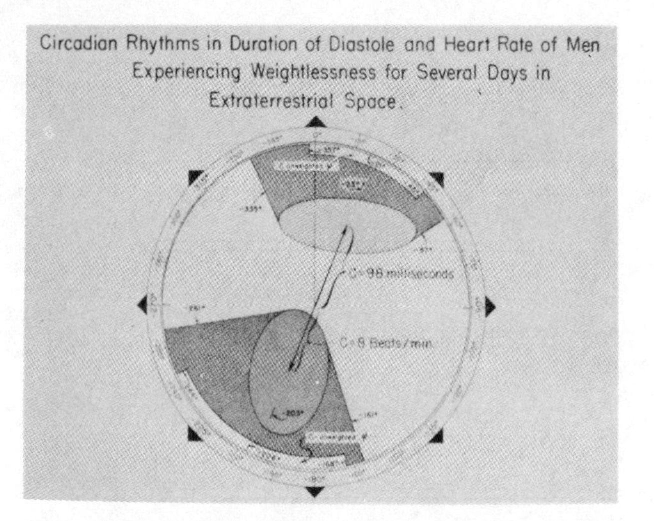

Computer programs have been developed for studying rhythms in extraterrestrial space under the sponsorship of NASA. This cosinor analysis demonstrates the persistence of circadian rhythms in astronauts and cosmonauts. The figure, from Franz Halberg, shows circadian rhythms of heart rate and diastolic duration from data from the Vostok and Gemini flights.

Cosinor amplitudes—again signifying a typical average amplitude of change in a unit of time—can also be expressed by the length of the hand. A computer program yields pairs of estimates of the amplitudes and phase of the particular best fitting cosine, derived from a least squares procedure at a particular frequency. By this process, it is possible to quantify aspects of biological rhythms that are very cumbersome in graphs, charts, and tables, and it is possible to compare several individuals or to see phase relations of functions within an individual.

Confidence Arcs and Error Ellipses

Although some strains of mice, nurtured on very rigid laboratory schedules, will show physiological and behavioral rhythms that have a machinelike regularity, the human environment, and that of most studies, will be far less controlled. People will inevitably exhibit individual differences. This means that a cosinor clock for a group will contain a certain variability. Just as a standard deviation depicts the amount of variation around the mean, a confidence limit estimates the extent to which the data fit the scientists' abstract summary. The cosinor plot reveals the magnitude of error in several dimensions.

The plot may show that a hospital patient, usually but not invariably, exhibited his peak oral temperature around 1 p.m. in a six-month study. Although his schedule was monotonously the same, he was not absolutely consistent from day to day. The peak phase and the amplitude of his temperature values and the slight shifts in peak phase can be summarized by an ellipse. The area within the ellipse denotes the extent of possible error.

This elliptical confidence region is based on the length of the "clock hand," which is analogous to a ruler measuring the amplitude of the rhythm peak. The confidence region describes the range of peak times. The clock hand is making a circular motion. Like a pendulum swinging, its arc has been preserved, representing the "average" peak phase. This point extends to an arc if it includes all recurrent peak times.

Polar plots are still unfamiliar, but they give a quick visual representation of temporal fluctuations in a manner that is more vivid and immediate than linear graphs or periodgrams. When the hour hand of an ordinary "earth" clock points to 12, we know by agreement with the rest of the world that it is noon or midnight. When the phase-amplitude hand points to 360 degrees, we know that the peak of the rhythm falls at the point of origin. This might be the middle of the sleep span or the peak of another readily measured rhythm, such as body temperature. If the phase were to shift to the right to 90 degrees, we would see that peak phase occurred six hours later. If the person were to shift another 90 degrees, he would hit his peak phase at the former time of his daily nadir.

The size of the error ellipse will indicate whether or not the original values were scattered. If they are not dispersed, the area of the ellipse will be quite small. The fewer cycles encompassed by the study, the less reason there is to be confident about the location of peaks and the estimate of the amplitudes—and this is indicated by a wider ellipse.

The error ellipse also indicates if the phase or frequency of the data are dubious. If the plot depicts a 24-hour rhythm and the data contain irregular frequencies, the error ellipse will overlap the pole (the center of the circle). In essence, this indicates that the clock hand representing the purported peak phase (acrophase) could be rotated in every direction and is, therefore, not pointing to any particular location on the clock for peak phase. A confidence arc thus expresses the amount of leeway within which a rhythm has been specified. The cosinor method allows a quantification of statistical confidence for the detected rhythm.

Polar plots have many advantages. For a quick look at the acrophase of any biological rhythm this display is immediate and vivid. The polar plot conveys a great deal of information at a glance—in the manner of a wristwatch. Unlike a wristwatch, which never says whether it is running fast or slow, the polar plot tells the magnitude of its own possible error. A rhythm must be detected with a 95 percent statistical security before the period (τ), the amplitude (C), and the (ϕ) phase are quantified.

By cosinor plots, Dr. Halberg has compared phase relations of rhythms within the body and also among different individuals. This technique was useful in learning that physiological rhythms maintained their usual phase relationships in blinded animals but exhibited a new frequency. The technique was used to decipher the behavior of a rhythm in respiratory function among asthmatic children, as they responded to hormonal treatment; thus, it allowed allergists to discover the acrophase of expiratory peak flow rate in response to drug action in the body and to see how this function changed when the drug was delivered at different times of day or night. The same technique indicated differences between skin and mammary cancer in terms of their mitotic rhythms.

Once a rhythm has been detected and its parameters described by spectral analysis, it is possible to display the stability of the rhythmic parameters over time. Thus, the amplitude, phase, and period can be visualized over a long duration, and it is possible, for example, to observe the effects of a trans-meridianal flight or of drugs.

The cosinor method and other methods now being developed and refined ultimately may allow medical researchers to prepare "physiological clock maps" of patients. As one travels along the 24-hour scale is there a normal phase relationship among the body's many rhythms? Could one construct a clock—showing acrophase for temperature, adrenal steroids, heart rate, eye-hand coordination—and learn the body's time of day by seeing their phase relations?

Time past and time future
Allow but little consciousness.
To be conscious is not to be in time
But only in time can the moment in the rose-garden,
The moment in the arbour where the rain beat,
The moment in the draughty church at
smoke-fall
Be remembered . . . .

T. S. Eliot

Chapter VIII. Physiological Clock Hands for Emotion and Memory

An awareness of circadian rhythms seems likely to generate insight into the way in which we develop emotional and psychosomatic illnesses. Two lines of research that have looked promising are explorations of the mechanisms that may underlie circadian rhythms of adrenal hormones and experiments indicating that biological time of day influences conditioned behavior and memory.

Now that body and mind can be studied as a unity, emotional illnesses and their somatic counterparts may be viewed as the outcomes of particular lives upon particular constitutions. Any individual is a unique composit of genetic subtleties embossed with experiences that distinguish him from even his nearest kin. Recollections of the same events are differently recaptured in different brains. As an experience becomes part of the program of the nervous system and is rendered into tissue, its impact may be biased by the timing of the event. At certain critical stages of development, for instance, the receptivity of the individual appears to be moulded by the events that do or do not occur. In the hierarchy of influences that might predispose an individual to illness, circadian rhythms may not be very important. Nonetheless, the day remains our major social unit, and its cycle is reflected throughout the endocrine system. The endocrine system, and particularly the behavior of the adrenal glands, has been thought to be an important link in the chain between "life stresses" and illness.

Adrenal hormones, with their unexplained bearing on emotion and vitality, show a rhythm that cannot be overlooked. The adrenal system and its possessor give differing responses to events depending upon the phase of the circadian cycle. Vigilance, sensory acuity, drug responses, and even memory and emotional conditioning appear to be tempered by biological time of day. Recent clinical studies suggest that a disruption of the adrenal rhythm may be a telling symptom in a number of illnesses. Indeed, stress has been used experimentally to produce a disruption of basic circadian rhythms, and with this temporal disorder came abnormal emotional and physiological symptoms. Although abnormal adrenal rhythms do not presently explain the symptoms they offer a new way of studying the "stress" precipitated illnesses.

Until the late 1940's, most endocrinologists believed that the body was basically stable (or homeostatic) and that its normal state was one of constancy. This assumption influenced the norms of diagnosis, treatment, and laboratory research and continues to influence endocrinology despite evidence to the contrary. In 1948, Dr. Gregory Pincus observed a rhythmic fluctuation of adrenal steroids in human urine.

During the 1950's, Dr. Halberg and his associates began monumental pioneering research, detecting and displaying unmistakable circadian rhythms in many endocrine functions. Their ambitious studies made it clear that a healthy man on a regular sleep schedule, or a laboratory animal on a controlled lighting and feeding regimen, showed a regular rise and fall of adrenal hormone levels in blood and urine, a cycle that occurred every 24 hours.

For rodents, light and darkness are the most important synchronizers of the adrenal rhythm. In human beings the precise hours of the adrenal peak and trough depend upon the individual's regimen of activity and work, and social factors are extremely important. The rhythm of hormones shifts gradually when a person changes his schedule, eventually adapting to the new routine, but the cycle period in a normal person stubbornly remains about 24 hours, even if the person stays up for three days and nights and undergoes considerable stress.

The stubbornness of the adrenal cycle was first seen in three patients who received electroshock therapy every two hours for two weeks. They were very regressed psychotic patients; that is, they had become so disoriented that they lost all concept of time, had no memory, and were mute and even incontinent. Drs. Franz Halberg and Bernard Glueck had expected that these severe changes might make an impact on the adrenal rhythms, but instead the adrenal hormones continued to approximate a 24-hour cycle. When female mice were placed on a semi-starvation diet so severe that it wiped out the usual estrous cycle, the daily adrenal rhythm persisted.

Human beings are harder to track than mice, for most of us do not live on the exact same sleeping, waking, and eating schedules. This has been a problem in diagnosing certain illnesses. If a sick person shifted his adrenal peak 15 minutes each day, slight but significant abnormalities might never be detected, and he could seem to fall within the so-called "normal range." Such a person's body could be out of synchrony with his life routine. He might, indeed, feel very sick, yet hormone levels taken at the same hour each day over a short period would reveal little. Only if his doctor had infinite patience and made hormone readings at the same time each day, month after month, might he notice that the levels were slowly changing in one direction and then another. An unsophisticated observer might interpret this swing as a sign of improvement or drug response. He would never imagine that it was the track of a repeated cycle with a period just 15 minutes longer than 24 hours.

After examining many normal volunteers, Dr. Halberg and others have shown that there are certain phase relationships among internal rhythms in the healthy person on a controlled schedule. Normally the peak of daily temperature and peak of 17-OHCS in the blood follow one another, about 90 minutes apart. The crests of blood pressure and pulse rate rhythms fall close to the peak of temperature and adrenal hormones, but do not occur simultaneously. If the pulse rate and blood pressure were to reach their peak phase when temperature was at its nadir, one might say the person suffered from internal desynchronization.

The disadvantages of being desynchronized from one's environment are obvious. An animal out of synchrony will be maladaptive. For instance, a blind mouse, whose activity rhythms and body temperature progressively drift out of synchrony with a 24-hour rhythm, might find himself looking for food at the wrong time, when predators were prowling about. A traveler who crosses time zones and feels sleepy by day, wakeful by night, and hungry at hours when local custom provides no food—feels the disadvantage of being desynchronized from the environment. Part of his distress may stem as well from internal desynchronization since physiological systems adapt at different rates to a phase shift and his adrenal corticosteroid rhythm may be out of its usual phase with activity and rest for a week or more. Although the consequences of internal desynchronization are not understood, they might be analogous to a surge of the wrong substance to the wrong organ.

Depression

The phase relationship between the adrenal cycle and activity cycle in man already may have some usefulness in the diagnosis and understanding of various kinds of depression. However, the suspicion that abnormal adrenal hormone levels might be a correlate of depression long antedated any concern for timing. Clinicians have long suspected that depressed patients might be reacting to a kind of inner "stress" which in turn might be visible in levels of adrenaline or of adrenal corticosteroids.

Dr. William Bunney, Jr. and his coworkers at the National Institute of Mental Health in Bethesda, have used adrenal hormone levels in blood and urine to gain some measure of the patient's "stress" and even to predict potential suicides. Although there have been some indications that mood, test performance, a sense of "stress" are correlated with levels of adrenal hormones or with the transmitters that may be related to hormone secretion, the correlations have not been consistent and attempts to understand depression have begun to take account of rhythmic change in the patient's biochemistry.

Researchers have begun to look at the arrhythmicity of sleep among depressed patients, suspecting that adrenal cortical hormone secretion might also show disturbance. Many depressed patients suffer from an insomnia of broken sleep, and many awaken in the predawn hours of the morning. Along with insomnia there are diurnal mood swings; some patients typically feel worse in the morning while others complain more at night. Biochemical rhythms might help to explain these differences among patients. Drs. William Bunney and William Carpenter have surveyed the rhythms of adrenal hormones found in several hundred depressed patients studied by a number of investigators. Unfortunately, no positive conclusion can be drawn since the patients were not on identical schedules, nor were samples taken frequently enough over long enough periods.

A number of clinical researchers have compared healthy and depressed persons for their levels of 17-OHCS at different hours, among them, Drs. George Curtis, Max Fogel, Donald McElvoy, and Carlos Zarate, the Eastern Pennsylvania Psychiatric Institute, Philadelphia. They compared 20 healthy women and 20 healthy men with 10 men and 10 women who were in maximum distress, either suffering depression, anxiety, agitation, or anger. These people were indeed ill, some of them psychotic, all miserable. Another group of 20 patients studied were people who had already improved and were feeling calm and comfortable. Each person was studied for 1 to 4 days. Urine was collected at 8-hour intervals and blood 4 times a day for a 24-hour period. A mood test was given the normal subjects 4 times a day at each blood drawing. This simple test did divulge a diurnal variation in mood parallel to the rhythm of blood 17-OHCS.

The most noticeable result was a sex difference. The distressed men showed a higher amplitude of change in their daily adrenal rhythm than normal men or women. Normal women showed a "flatter" curve than men; the distressed women showed even less circadian change than normal women. The sex differences were somewhat analogous to the differences between hypo-and hyperthyroid patients. The stability of the circadian rhythm was evident in the normal subjects, and variability was considerable among the distressed psychiatric patients with the calm patients somewhere in between.

Until recently, the most enlightening observation of the differences between adrenal rhythms in normal and depressed persons was that the depressed groups showed greater variability. However, a few recent analyses indicate that there may be some evidence of phase differences. In a collaborative study by Drs. Franz Halberg, Per Vestergaard, and M. Sakai, there is the tentative suggestion of a difference in urinary 17-ketosteroids differentiating healthy and depressed patients. The amplitudes of circadian change were the same for both groups in this study, but they differed in phase. The healthy people showed their peak excretion of this hormone around midday, while the depressed patients showed their peak excretion around late afternoon.

In most studies of depression, the adrenal rhythms have been loosely tracked for a day or a few days without reference to other internal rhythms or to quantitative indices of behavior. Patients do not appreciate being disturbed for blood samples, which are often restricted to twice a day. In only a few recent studies, therefore, has internal time discord been detected among depressives.

Dr. Norberto Montalbetti measured temperature, 17-OHCS, and emotional state five to seven times a day in several depressed patients. Three showed a shift in their circadian peak during the worst phase of their depression, but the main difference between the patients and normal control subjects was greater variability in the depressives' adrenal steroid rhythm.

Another study, conducted at the University of Iowa by Drs. Paul E. Huston, Edgar Folk, Jr., and Harold A. Cahn suggested that depressed patients may show some desynchronization among internal rhythms with respect to the normal phase relationships. The Iowa team studied volunteers during 33 hours in an isolation chamber where data could be collected every hour. Heart rate was telemetered by a small radio capsule worn on the pocket of the volunteer's undershirt. Water and food were received at equal intervals and careful records were made of pulse, urine flow, potassium excretion, and body temperature.

Typically, the peaks of temperature, heart rate, and potassium excretion occurred between 5 and 9 p.m. Urinary flow reached its peak slightly later. With some subjects the peak of each rhythm came in the afternoon, while others showed peaks between 7 p.m. and midnight. Since not all these people had been on the same schedule of waking and sleep before their isolation study, they showed their peaks at different times of day. However, the same person was studied after intervals of a day, a week, a month, and in one case, a year later. Individuals were found to have very consistent phase relations, even when tested at widely separated times.

Although the phase relations of physiological function were similar and consistent in normal individuals, a comparable study of depressed patients showed a urine electrolyte excretion that seemed to be out of phase with their peak urine volume. This evidence of what may be a dissociation in timing rather than simply a change in the level of a substance, may become a new aspect in studying the biochemistry of depression.

The secretion of adrenal steroids may be related to levels of epinephrine and norepinephrine (terms that are used interchangeably with adrenaline and noradrenaline). Epinephrine and norepinephrine are secreted by the medulla, the innermost core of the adrenal glands. Norepinephrine is generously produced and secreted by nerve cells throughout the body and brain, and it acts as a transmitting substance across the synapses between neurons. It has been measured in the blood and urine of depressed patients in very few studies. Dr. William Bunney and his coworkers have noticed increases in norepinephrine in patients who were just about to make the transition from depression to mania.

Drs. Julius J. Chosy, Donald T. Fullerton and their coworkers at the University of Wisconsin Medical School in Madison, have studied blood steroid levels and urinary levels of steroids, epinephrine, and norepinephrine in hospitalized patients with depression. Their preliminary study suggests that patients said to have neurotic depression might be differentiated from those with psychotic depression by their biochemical rhythms. Serial measures of epinephrine and norepinephrine were taken during a 24-hour period from 28 newly hospitalized patients. When they entered the hospital the patients were interviewed. They were given the label of psychotic depression only if their symptoms included delusions. They then took personality inventories, I.Q. tests, and other psychological tests during a period of close observation. If the person were entirely free of drugs for a long interval prior to hospitalization and fell within certain criteria on I.Q. and other tests, he became a subject in the study.

During the study, the patients ate a special controlled diet. Every 3 hours for 24 hours, urine samples were taken. Activity was rated every 15 minutes and does not account for the urine profiles. The levels of free epinephrine and norepinephrine differed considerably. The levels were generally lower among the patients with neurotic depression. Moreover, these patients each showed signs of a circadian rhythm, with a daytime peak. The individuals had not previously lived on a standardized schedule which may explain some spread in the peak time. Nonetheless, with just two exceptions, these patients showed a circadian cycle. The two individuals who deviated from the pattern resembled eight of the people with psychotic depression. These patients not only showed much higher levels of norepinephrine and epinephrine, they showed a greater change between nadir and peak. They did not exhibit a regular circadian rhythm but exhibited two daytime peaks in morning and evening.

In a current study, the Wisconsin group hopes to observe the relationship between norepinephrine, epinephrine, and adrenocortical hormone rhythms. Indwelling catheters permit blood sampling at half-hour intervals without disturbing sleep, while continuous EEG recordings may begin to indicate whether there is a correspondence between sleep disturbance and anomalies in the adrenal hormone rhythm in depressed patients.

A comparison of several internal functions with behavior is likely to become a fruitful way of studying depression in the future. There are already new methods for plotting behavior quantitatively, as in the automated nursing notes used by the Institute of Living in Hartford, Connecticut, where behavior can be correlated with physiological measures and analyzed by computer. This technique is already being used in a manner that might be called behavioral chromatography.

Behavioral Chromatography

The automated nursing notes now used at the Institute comprise a checklist of traits and actions, descriptions that provide a comprehensive evaluation of the patient, to be filled out every 12 hours. This checklist, the elixir of many personality inventories and years of research, requires the aide or nurse to observe closely and fill out a printed form with 215 descriptive statements. These include items on the fastidiousness or sloppiness of a patient's appearance, whether he groomed himself, and whether he seemed sad, irritable, angry, tearful, or preoccupied. Did he eat and sleep well, take medication, attend classes? The checklist is analyzed by computer, making it possible to see a patient's profile as it changes over days or weeks.

Analysis of one depressed patient showed that in the course of improvement his acceptable behavior fluctuated in a roughly three-day rhythm. As this patient improved, he soon became less disorganized although he remained anxious and depressed. Next, anxiety decreased. His sense of depression was the last to depart. Like the revelation of the paper chromatogram which separates an organic compound into its components and makes them visible by colors, automated behavior analysis may begin to separate the behavioral components of an illness like depression. As aberrant behaviors of mental illness become disentangled by automated techniques, peeled away like layers of onion skin, one may find that depression is made up of layers without a real "core."

Early in the study of circadian rhythms, the adrenal cycle had seemed to exert a pervasive influence upon rhythms of other functions, such as cell division. The adrenals had appeared to act as a kind of endocrine "clock," and removal of the adrenal glands appeared to abolish circadian rhythms of cell mitosis in renewal tissue such as skin. Subsequent experimental techniques have shown that the circadian rhythms of mitosis did not vanish when the adrenals were removed, but the daily amplitude of change did become very small. Because the adrenals seemed to have a pervasive influence on the circadian rhythms of an animal, researchers sought the "clock" mechanism that guided the adrenal cycle.

Cortisol Rhythms and Sleep Reversal

What would happen to the adrenal cortisol rhythm if a person lived a 12, a 19, or a 33-hour day? Drs. D. N. Orth, D. P. Island, and G. W. Liddle of Vanderbilt University School of Medicine in Nashville, Tennessee, placed volunteers on these odd schedules for 4 to 42 days before testing their blood for rhythmic changes in 17-OHCS. Most subjects were tested after 6 to 10 days. This long preparation was necessary, since a day or two on a new schedule will not alter the circadian hormone rhythm. Indeed, sometimes it takes almost 3 weeks before the rhythms of the body align themselves with a new regimen.

The subjects in this study lived in a metabolic ward in a hospital and wore a permanent catheter in the forearm allowing blood sampling around the clock without disrupting their sleep. The peak hormone levels seemed to come about the time of awakening while the minimum levels occurred early in sleep.

Sleep, itself, had often been postulated to be the synchronizer of the adrenal rhythm. Dr. Orth and his colleagues deprived one subject of his night's sleep and found that his cortisol rhythm looked as if he had slept normally. A similar finding was reported several years before by Drs. Gilbert Frank, Franz Halberg, and coworkers during a 50-hour study in which the researchers became their own subjects, sampling their own adrenal steroids during sleep deprivation as they worked around the clock.

Dr. Orth and his group repeated sleep deprivation in such a way as to prevent any kind of psychological anticipation. They let the subject wash his teeth, make his ablutions, get into pajamas, and settle into bed for a good night's sleep. After he had turned off the lights, they suddenly turned them on and made him get up for the night. The 17-OHCS rhythm was unchanged by this maneuver.

This confirmed what Drs. Franz Halberg, Alain Reinberg, and Jean Ghata and others have indicated—that sleep, itself, cannot be the synchronizer of the adrenal corticosteroid rhythm. In numerous studies of rodents, Dr. Halberg changed the light-dark schedule, and thus shifted the adrenal rhythm. Although animal experiments have suggested that light and darkness must be paramount in synchronizing the adrenal and other circadian rhythms, it is always risky to extrapolate to man.

Drs. Orth and Island conjectured that the circadian rhythms of 17-OHCS may be synchronized by light and darkness in human beings. They studied three blind subjects in whom they found plasma steroid rhythms that were out of phase with the rhythm of activity and rest. They then studied normal subjects for long periods as they slept in illuminated rooms and were awake in darkness. Sometimes a short period of light was scheduled to interrupt the darkness. By dissociating the schedules of light and darkness from the rhythm of waking and sleep, they were able to report that 17-OHCS tended to show a peak around the time of illumination.

The role of illumination as a possible synchronizer of the adrenal corticosteroid rhythm in man may have implications for people who work at night or who suffer from varying degrees of blindness. Dr. F. Hollwich of Munster, Germany, has corroborated in over 200 cases that blindness and severe cataracts affect the endocrine rhythms of the individual.

The role of light in the circadian rhythm of adrenal steroids must not be simple nor the mechanism abrupt. In many sleep reversal experiments, adrenal cortical rhythms have not fully reversed until after two weeks, a transitional period in which the comments of subjects and their diaries indicate a transitional state of distress. Thus, light does not bring about an immediate phase shift of the adrenal rhythm. During transition, the adrenal rhythm has not yet returned to its normal phase with the activity cycle. It has been conjectured that this dissociation of the adrenal rhythm from the sleep-waking cycle might account for the distress of travelers who say they are uncomfortable despite the ability to sleep at will, and thus to adjust to a new time zone behaviorally long before their physiology has come into phase with local schedules.

Clock Hands Behind the Adrenal Rhythm

The very existence of a circadian adrenal hormone rhythm in normal people has prompted research into the possible loci of the "clock" functions that instigate the rhythm. The concept of a "master clock" in the central nervous system has lost some appeal, since the demonstration that in vitro tissue beats with a circadian rhythm, that isolated adrenals secrete hormones rhythmically, and, indeed, that the rhythm exists in single cells. Historically, however, the search for a master clock has engendered much interesting research on the relations between behavior, the endocrine system, and brain. It seems possible that some of the hormone determinations made in studies of blood levels may be reevaluated, in the light of indications that episodic bursts of adrenocortical hormones swiftly rise to high levels and subside.

A number of clinical studies have measured both plasma and urine, the 17-OHCS, and 17-KS levels of patients with diverse illnesses. A group led by Dr. Leon J. Sholiton at the University of Cincinnati College of Medicine, assayed hormone levels in patients with chronic bronchial disease, cancer, and acute but non-fatal illnesses, such as infections. In patients with chronic illnesses, including cancer, there was a significant rhythm of free 17-OHCS quite similar to normal subjects. However, acutely ill patients with symptoms of mental confusion showed an abnormal rhythm of adrenal steroids in the blood and an increase in nighttime excretion of 17-OHCS.

In 1959, Drs. G. T. Perkoff, K. B. Eik-Nes and their associates at the University of Utah College of Medicine, studied the rate of disappearance of injected cortisol at various times of day, as well as the response of blood corticosteroids to doses of ACTH. ACTH (adrenocorticotrophic hormone) is released by the pituitary gland and stimulates adrenal hormone secretion. Plasma levels of corticosteroids seemed to follow the circadian rhythm of ACTH secretion in normal people, but this rhythm was not found in patients who suffered alterations in consciousness.

Another study showed that patients with severe brain damage had no obvious circadian rhythm of adrenal hormones in the blood. This suggested to the investigators that the brain was controlling ACTH, which, in turn, regulates adrenal steroid hormone production.

The same group (G. T. Perkoff, K. B. Eik-Nes, C. A. Nugent, L. Rush, F. H. Tyler, A. A. Sandberg, L. Samuels, and others) compared blood levels of 17-OHCS around the clock in large numbers of patients and normal people. More than 30 patients with non-endocrine disease (rheumatoid arthritis, jaundice, fractured hip, senility, cerebral thrombosis) differed from normal people in a routine test of adrenal responsiveness. This is the amount of the hormone hydrocortisone released after an injection of the pituitary hormone, ACTH. Despite different adrenal responsiveness to ACTH, these patients did not differ from normal people in the rhythmicity of adrenal steroids in plasma. Only patients with altered states of consciousness, such as confusion, coma, or abnormal sleep patterns failed to exhibit the usual circadian hormone rhythm. A number of clinical observations indicate that brain damage or malfunction may alter the circadian rhythm of the adrenal system.

Patients with brain lesions or hypothalamic illness often have shown endocrine symptoms. This has been noted by Dr. Lawrence Kahana and his associates at Duke University Medical Center, and by a team at Columbia University in New York City (Drs. Jack H. Oppenheimer, Leonard V. Fisher, Joseph W. Jailer), who studied patients with hypothalamic or temporal lobe disorders.

One means for gauging adrenal responses has been to inject a patient with metyrapone (a substance that inhibits the enzyme 11-beta-hydroxylase) or with dexamethasone, which suppresses ACTH secretion. Unusual responses to these two substances have been interpreted as signs of endocrine malfunction, yet unusual responses have been seen in patients who showed perfectly normal circadian rhythms in plasma adrenal steroids.

Rhythmic changes in the adrenal system have implications for the administration of drugs to people with arthritis or many other "nonendocrine" ailments. In a classic study, Drs. Thomas Nichols, Charles A. Nugent, and Frank H. Tyler, of the University of Utah, attempted to see whether the daily rhythm of adrenal hormones might be a part of a feedback system. The authors reasoned that the pituitary would release ACTH, which would stimulate adrenal hormone production, but when blood levels of hormones were high, some sensor would inform the pituitary which would stop emitting ACTH. Presumably, one could intercede in this feedback system by injecting cortisol into the blood. The pituitary should interpret this injection as a signal to stop releasing ACTH, which would lead to a decline in adrenal hormone production, followed by a decline in blood levels.

During several weeks, volunteers were given a small amount of dexamethasone. This synthetic analogue of cortisone is often given to arthritic patients because it is effective against inflamation. On two consecutive days each week the person got the drug either at 8 a.m., 4 p.m., or midnight. Then the researchers simply waited and took blood samples 8, 16, and 24 hours later.

The half milligram of synthetic cortisone caused only a transient drop in blood cortisol levels if injected at 8 a.m. or 4 p.m. Apparently, doses at these hours only slightly suppressed ACTH production. However, the same dosage at midnight produced virtually total cortisol suppression for a full 24 hours in blood and urine.

Clearly, no simple feedback mechanism explained the circadian rhythm of adrenal hormones. The phase of the cycle was important in adrenal hormone suppression. In drug therapy, perhaps, a shrewd timing could make a very small amount of cortisol produce adrenal suppression for a whole day. Unfortunately, in subsequent clinical studies the results have not been so clear even when potent glucocorticoids were administered for over a month. Urinary metabolites did not perceptibly differ depending upon the time of drug administration.

It now appears that the adrenal cortex varies in its response to injections of ACTH, and the size of the response depends upon the phase at which the injection occurs in the gland's secretory cycle. Dr. Malcolm M. Martin of Georgetown University, Washington, D. C., has reported considerable circadian variation in adrenal response to metyrapone which inhibits adrenal hormone levels. The greatest response occurred when adrenocortical activity was at its peak, and there was little response at the time of minimal activity in the adrenal cortex. The metyrapone response might reflect how much ACTH had been released in the body. Moreover, one might expect to find a circadian rhythm in the pituitary secretion and release of ACTH. Presumably, this rhythm would reflect circadian rhythms in brain centers that regulate the pituitary.

Drs. Dorothy and Howard Krieger, at Mt. Sinai Hospital in New York, have studied circadian rhythms of adrenal hormones in people with hypothalamic tumors. At the same time they have been attempting to plot accurate records of biological rhythms in normal people in an effort to obtain the shape, level, and variability of each person's daily curve.

According to these authors, normal people on a rigid schedule are exceedingly consistent in their blood hormone rhythms. Dr. Dorothy Krieger feels that an adrenal corticosteroid rhythm for one 24-hour period might be traced on translucent paper and superimposed upon the next day's curve, fitting almost identically. People with hypothalamic tumors, by contrast, showed abnormal plasma cortisol rhythms. Dr. Krieger feels that abnormal circadian rhythms may be as sensitive a detector of hypothalamic disorder as can be used in present-day medicine. The test simply requires round-the-clock sampling for several days.

Since patients with hypothalamic illness have shown abnormal adrenal rhythms, a number of researchers have suspected that the hypothalamus might indirectly mediate the pituitary rhythm that causes ACTH secretion. This ACTH rhythm, in turn, stimulates the adrenal cortex. Does the hypothalamus rhythmically emit something that would be a pituitary corticotrophic releasing factor?

A Critical Time for Brain Stimulation of the Pituitary Adrenal Rhythm?

A number of experiments on rats, cats, and mice suggest that there is a circadian rhythm in the hypothalamus. The Kriegers and their associates have localized regions within the hypothalamus where tiny chemical implants injected through fine guide tubes produce an abrupt and prompt increase in levels of the adrenal hormone, hydrocortisone, in the blood of the animal. The chemicals employed for brain stimulation are basic substances used by the brain in transmitting impulses across the synapses: norepinephrine, acetylcholine, serotonin, and possibly gamma amino butyric acid. The authors believed that these substances, normally released in the natural processes of the nervous system, would activate hypothalamic neurosecretory cells. These cells secrete a characteristic product, perhaps a releasing factor that enters the pituitary portal blood stream and activates the pituitary gland to produce its trophic hormone. This in turn activates the target endocrine gland. If various neurotransmitters could trigger the hypothalamic pituitary-adrenal system, it might be possible to block the circadian rhythm by blocking the action of these transmitter substances.

In preparing for a series of experiments, they had observed ten cats until they were certain that each animal showed a consistent and regular circadian adrenal hormone rhythm with a peak between 8 p.m. and 4 a.m. and very low levels for the rest of the 24 hours. Once the animal showed this consistent pattern, it would be possible to measure change.

Throughout the nervous system there are some cells that respond to acetylocholine and are known as part of a cholinergic system. Other cells respond to norepinephrine and its chemical family and are known as adrenergic. When the Kriegers administered large doses of atropine, which blocks cholinergic receptors, they noticed a drastic alteration in the circadian rhythm of adrenal hormones. They used the drug, dibenzyline, or alpha methyl *para* tyrosine, to block adrenergic activity, but this did not affect the rhythm. However, drugs that interfered with serotonin levels in the central nervous system blocked the adrenal rhythmicity.

The cholinergic system seemed to be critical in the hypothalamic regulation of the rhythm. Moreover, atropine inhibited it differentially at different hours of day and night. The researchers began to wonder if there were a critical period in the 24-hour cycle when the hypothalamus activated ACTH release. If inhibited then but not at other times of day, the neurons might fail to issue the commands that presumably mediate the pituitary-adrenal hormone rhythm. Thus, by interfering with the nervous transmission from the hypothalamus at the right moment, they could eradicate the rhythm of pituitary-adrenal interaction.

Since the rhythm was influenced by manipulations during certain hours of the cycle and not during the rest of the 24 hours, the Kriegers thought that drugs effective on the central nervous system might also affect circadian rhythms. They gave a common sedative, sodium pentobarbital, to animals. It completely suppressed the adrenal hormone rhythm, regardless of the time the animal was injected. Sodium pentobarbital is a long-acting drug. They predicted that a rapidly acting barbituate would affect the adrenal cycle only if given at certain hours. They gave animals sodium thiamylal at 8 a.m. and 6 p.m. Injected in the evening, prior to the normal time of circadian rise, it blocked the morning rise of hormones. It did not suppress the morning peak when given at other times of day. Although the exploration of drug effects was a side issue in this study, the results implied that there are critical times of day in the nervous system, sensitive periods when drugs may alter circadian rhythms in the endocrine system. This may be related to rhythmic changes in levels of biogenic amines in the central nervous system.

Circadian Rhythms in Brain Amines

In 1968, several striking papers appeared by Dr. Donald J. Reis and his associates at Cornell University Medical College in New York City. They had begun a new and arduous form of brain mapping, analyzing levels of brain amines, norepinephrine, and serotonin in nervous tissue, sampled at hours around the clock from cats who had been living on a strict regimen of light and darkness. Twenty-four regions were assayed. Only seven exhibited clear 24-hour rhythms. Norepinephrine rhythms were found in the pineal gland, hypothalamus, lateral thalamus, mesencephalon, pons, and also the cervical region of the spinal cord. Serotonin levels showed circadian rhythmicity in other regions: the red nucleus, hypothalamus, and six regions of the telencephalon, among them, the visual cortex. In only three of these areas was there any overlap of serotonin and norephinephrine rhythms. Usually, they were 180 degrees out of phase, norepinephrine reaching its peak when serotonin fell to its nadir and vice versa.

Although there may be some disadvantages in mapping circadian rhythms in the domesticated cat, which is neither clearly diurnal nor nocturnal, this project clearly showed that rhythms appear in axonal terminals in the spinal cord. A persistent norepinephrine rhythm was seen in cervical ganglia even when a regimen of continuous light had obliterated the usual rhythm seen in pineal tissue.

The pineal gland, in the center of the brain, is a great biochemical factory and storehouse, rich in serotonin, norepinephrine and an enzyme that converts serotonin into other hormones. Some of the biochemical rhythms of the pineal are triggered by light and darkness. These are obliterated by constant light or darkness. However, not all of the pineal rhythms depend upon light; serotonin continues to show a circadian rhythm even in animals that live in constant light.

Reis and his coworkers observed rhythms at surprisingly low levels of the spinal cord. They found 12-hour cycles of change in the thoracic and sacral regions. Thus, circadian rhythms in neurotransmitters do not inevitably seem to be governed centrally, but could exist at various localities in the nervous system. It no longer seems likely that the rhythmicity of a particular group of nerve cells would regulate fluctuations in behaviors so complex as sleep and waking. However, ratios of change among the overall rhythms of masses of cells may provide the significant cues to underlying behavioral rhythms.

The amount of daily change in brain biochemistry is surprisingly large as indicated by Dr. Lawrence E. Scheving and his associates, who analyzed whole brain tissue from rats at intervals around the clock. They found that serotonin showed a circadian peak during the hours of normal sleep. By contrast, dopamine had a faster rhythm with three peaks every 24 hours. The peak dopamine content in brain tissue was 20 to 35 percent greater than at its nadir. Norepinephrine showed two peaks every 24 hours, one at the beginning and one at the end of the animals' quiescent period. The amplitude of the rhythm was sizable; there was 55 percent more norepinephrine in brain tissue at the peak of the rhythm than at the trough. Circadian rhythms in brain biochemistry may begin to indicate how drugs intervene in our many rhythmic processes and may alter their phase relationships.

Memory and Periodicity

While some researchers were discovering a critical period in a neuroendocrine rhythm, psychologists were discovering critical periods in the learning behavior of animals. The unity of biochemistry and behavior have grown even more palpable as this work has progressed; yet the specialization of scientific research is such that these scientists would not be likely to read each other's journals or to know each other.

Studies of the memory process are particularly diverse. A great many of them involve training animals to become proficient at some routine task and then apply drugs or other influences to see if the animals still remember the task. Measures of responses are most indirect measures of memory, yet these studies may begin to illuminate some of the bizarre amnesias that people suffer after trauma, strokes, or accident. Why, indeed, do memories of a specific time span vanish in a person? By interrupting the memory process, researchers have attempted to find the timetable, the location, and the biochemistry of the storage process.

If learning and memory are the results of physiological processes, one might expect that memory consolidation would exhibit the kinds of circadian fluctuations that have been so dramatically demonstrated in rodents. On a fixed light-dark schedule rats and mice exhibit predictable circadian rhythms of metabolic and nervous change. Drs. James L. McGaugh and Gwen Stephens of the University of California at Irvine have looked for signs of such a rhythm in memory consolidation during the extreme phases of the circadian cycle. Would rodents respond with differing memory loss depending upon the phase in the cycle they received electroconvulsive shock?

Mice were placed on a cantilevered platform outside a box. There was a small hole in the side through which the animal could climb into the dark box and escape the bright lights. However, if the mouse stepped through the hole it was immediately shocked. The animal would learn restraint after a while, and would stay on the platform without trying to climb inside. Some animals were given only one trial, others three, while some were placed on the platform time and again, until they managed to sit there for 30 seconds without stepping into the hole. This sign of restraint might be considered learning. After some predetermined interval each animal was given an electroconvulsive shock, and subsequently returned to the platform for a test of its amnesia gradient. Twenty-four hours later, if he were placed on the platform, would he step into the hole at once? If so, one might assume he did not remember having been shocked for that behavior the day before. If he simply waited on the platform for 30 seconds, it was assumed that he remembered. The number of seconds the mouse would stay on the platform was defined as his retention latency.

In one experiment the retrograde amnesia showed a difference between animals that had been shocked at 1 p.m. (the middle of their quiescent period) and animals given electroconvulsive shock at 9 p.m. Electroconvulsive shock caused amnesia more effectively when it was administered during the animals' active period (dark). Further experiments confirmed that mice had greatest retrograde amnesia if electroshock was delivered when they were at their peak of body temperature and activity.

As Drs. McGaugh and Stephens demonstrated, the circadian fluctuations in retrograde amnesia gradients could be shifted by changing the animals' lighting schedule. If the lighting regimen were inverted, the test hours (1 p.m. and 9 p.m.) no longer coincided with the peak and trough of the animals' activity-rest rhythms, or with peak and trough body temperature and hormone levels. The animals were physiologically in transition. The same training and shock procedure did not offer the same retrograde amnesia gradients as did the earlier studies. This strongly suggested that the neurophysiological processes underlying the formation of memory traces and consolidation of memory might, themselves, fluctuate with the 24-hour temperature and activity rhythm.

These memory studies and other studies of biological rhythms suggest that the process of forgetting is not steady. Just as physiological cycles of a week, or about a month, pass unnoticed in most lives, there may be subtle cycles in memory and forgetting. Is there a reason why a sudden onrush of memories or the recall of a missing name invade the consciousness uninvited and unpredictably?

Critical Periods in Fear Memory

If man resembles his stand-ins, the rodents and monkeys used in testing his drugs, and if human memory resembles the artificial memories created in computers, then the timing of experience must be important. This has been an assumption guiding Dr. Charles F. Stroebel of the Institute of Living in Hartford. During the last several years he has demonstrated that biological time of day has an unmistakable influence upon the inculcation of fear. The rapidity with which an animal learns that a signal is the harbinger of something painful, such as shock, and the stubbornness of his fearfulness after the situation no longer pains him, is biased by the particular hour of conditioning. If the depth of a "fear" response is biased by the time of learning, and the unlearning of an inappropriate fear response is also time-locked, then biological time may be playing an invisible role in psychotherapy.

Animal studies have provided many insights into the way in which human behavior is shaped by reward or punishment. In the early 1960's, sophisticated mathematical models of learning had been devised to account for the seemingly random fluctuations in performance that always cropped up in animal studies of learning. Dr. Stroebel and his staff decided to explore fluctuations in learning on the hunch that many fluctuations were not due to chance but to circadian rhythms that changed an animal's reaction to the manipulations of the experiment.

Animal studies have always required some ingenuity. Creatures who cannot talk are coaxed to reveal their inner feelings by responses developed in conditioning. Bar pressing is a convenient procedure with rats and monkeys. The animal learns to press a bar at a certain rate of speed—lured into continuous activity by rewards. Once he becomes a conscientious bar presser, it is possible to measure how drugs, shocks, stimulation, and other maneuvers will affect his bar pressing.

This is an operant conditioning procedure in the sense that it uses some of the animal's own natural motions. Usually, when a rat enters a new cage he will explore hungrily or thirstily with his paws. Ultimately, if there is a lever in the cage he will press it by accident, and if there happens to be a pellet of food or a drink of water as a reward, he will take it and press again. At first, a reward follows each press, but eventually the rewards are spaced out, arriving periodically, and the animal learns to press the lever at a steady rate. New conditions can be superimposed; a light or a sound or shock can be associated with the reward. As the animal becomes conditioned to the added stimulus, a measure of his reaction is the change in his easily quantified, automatically countable, bar pressing.

Emotional Conditioning and Time of Day

In an initial exploration of the possible influence of circadian rhythms on emotional learning, 48 rats were trained to press a bar for a drink of water. During a week of preparation, they lived on a rigidly controlled schedule (12-hours darkness and 12-hours light), in an environment of unvarying temperature, humidity, and background noise. The experimenters used a schedule of light and darkness to biologically align their rats so that all the animals were biologically at the same time of day. This could be judged by the phase of body temperature rhythm or adrenal steriod levels. When the rats became proficient bar pressers, they were divided into three groups.

Within the first four hours of darkness (morning for a human being), the animal was placed in a training cage and would press at the bar for water. Now he heard little clicking sounds as he worked. When the clicks stopped he received a brief electric shock. Then during a short interval of adjustment the rat would be left unmolested. by clicks or shocks and he would resume bar pressing. Then the clicks would start again; he would halt and be shocked. This happened during nine daily training sessions, at the same hour each day. Typically, the animal would press the bar for water until he heard the clicks. Then he would "freeze," defecating, urinating, breathing more rapidly, showing an accelerated pulse, and signs we might call anxiety or fear in humans.

There was nothing the animal could do to avert the shock, and soon the mere sound of clicks would cause a very thirsty animal to stop bar pressing for water and show signs of extreme "anxiety." Human beings may learn anxiety responses in a similar way. For instance, if a parent scolds a child before spanking him, the child may become anxious when he hears a certain scolding tone of voice. Years later a similar tone of voice may evoke anxiety.

The strength of a learned emotion can be measured by repeatedly confronting the individual with the warning stimulus but removing the consequences. How long will he continue to react as if he is about to be shocked? When an individual is repeatedly allowed to experience a conditioned stimulus with no consequence his emotional response eventually dies away or, in the language of learning theorists, becomes extinguished. Extinction is, of course, a major goal of some schools of psychotherapy. Presumably, if the original learning was weak, extinction would occur quickly, but if the original conditioning was strong, then the anxiety would persist long after it had become inappropriate.

Dr. Stroebel explored three extinction schedules, placing the animal in the conditioning cage where he heard clicks but never was shocked. One group remained on the same time schedule during both conditioning and extinction, and each rat was extinguished at precisely the hour he had been trained. The second group remained on the same schedule of light and dark. However, the animals had been conditioned during the first four hours of the dark period but were given extinction trials in the light period. The third group was on an inverted light-dark cycle. This group of rats with its inverted schedule was biologically at a time of day opposite to the other rats.

The measure of extinction was the bar-pressing response. An animal was put in the conditioning cage. At first he would show anxiety and stop pressing the bar after he heard clicks. How many hours did he require to learn that the clicks were innocuous, and he should resume pressing the bar for water? The experimenters counted the number of trials. Then they gave the animals another rest period of three weeks and tested them again. A strong emotional response may seem to disappear during extinction, yet spontaneously reappear later, indicating that the "fear" memory persists and the response was not really extinguished.

Conditioned "Fear" Responses at 8 a.m.

The results of these later tests were startling. First of all, it was clear that the rats showed the strongest "fear" response and seemed most stubbornly resistant to extinction at the biologic time of day that they had been trained. If a rat had been trained at 8 a.m. (during the first few hours of his activity cycle), his "fear" responses were stronger then than at any other time of day. Tested at other hours he would show less fear. Control rats, who were trained and untrained at the same hour each day, took the longest time to extinguish. At first it seemed as though extinction were more difficult at the time of day that coincided with the hour of original training. However, when these animals were retested three weeks later, they showed no resurgence of fear. The other two groups originally appeared to extinguish faster, but on a retest three weeks later they showed fear-responses and clearly had not been extinguished. These rats had undergone extinction trials at biologic times of day different from the time of their original training. They appeared to extinguish very rapidly but later showed a substantial resurgence of fear responses.

Learning at Random Hours

The time-lock of fear responses was too clear and too striking a finding to believe. Goaded by cautious excitement the researchers replicated the experiment with both rats and monkeys. Now they conditioned the animals at random times of day. Animals acquired fear responses very slowly if conditioning took place randomly over the twenty-four hours. This may have some bearing on emotionally tinged learning in humans.

Moreover, animals who learned and unlearned at the same time of day were most rapidly relieved of their acquired fear, while animals subjected to a randomized schedule exhibited an unusual resistance to extinction. Even when trained at random times of day and extinguished at precisely the same hours biologically speaking—as the hours they were conditioned—their responses showed a distinct and stubborn persistence of fear. The investigators concluded that when anxiety is learned at random hours it may have to be unlearned on a random schedule. The possibility that timing might be an element in human therapy has intrigued one Philadelphia therapist, who is currently considering a study of patients under unique schedules of therapy.

Emotional Learning and the Adrenal Rhythm

Rats appeared to be most susceptible to "fear" training at the beginning of their dark cycle—the time when adrenal hormones reach peak concentrations in the blood—the beginning of the activity period. Was emotional conditioning related to the adrenal hormone cycle? Four groups of rats were conditioned at different times of day. The adrenal glands were removed from one group; another received a drug that blocks the synthesis of adrenal hormones; and a control group was subjected to sham surgery but its adrenals were left intact. At each of the hours of conditioning trials, other control animals were killed and assayed to offer a base line for the normal level of blood constituents at these hours.

Rats acquired fear most rapidly when trained at the onset of darkness, the time of peak adrenal hormone levels in the blood. The animals trained at the end of the dark period, or beginning of the light period took twice as long to react fearfully at the sound of the premonitory clicks. The adrenalectomized animals took somewhat longer to exhibit conditioned fear but they, too, showed the same pattern. On the other hand, some animals had received injections of metyrapone, which blocks the secretion of adrenal hormones by its action on the hypothalamus-pituitary system. These animals showed an almost uniform rate of conditioning, whatever the time of day. Thus, the drug virtually eliminated their rhythm of susceptibility, possibly by interfering with central nervous system activity related to the rhythmic release of adrenal steroids and other hormones in the blood. Control animals, whose blood was analyzed at the time of conditioning trials, showed that the rhythm of fear vulnerability paralleled the levels of corticosterone in the blood. The greatest susceptibility came at the peak, with lesser susceptibilities corresponding to lower levels. This correlation does not imply that the hormone level causes emotionality. However, the adrenal rhythm does offer a possible method for predicting when an animal may be most vulnerable or resistant to emotional stress and conditioning.

Man is, of course, a diurnal animal, and if his emotional responses were to fluctuate with his adrenal rhythm, one would expect to observe maximum vulnerability to fear in the hours between 4 and 8 a.m. when human adrenal hormones reach their peak. It has never been demonstrated that man's emotional susceptibility is related to his concentration of adrenal hormones. In 1963, however, Dr. Stroebel and his team showed that monkeys do show fear responses and physiological vulnerability in the early morning hours, coinciding with their peak levels of adrenal hormones.

The Clock, Stress, and Somatic Symptoms

These monkeys, trained, and tested in the same manner as the rats, were also under continual examination for physiological changes. Implanted catheters yielded continuous blood samples without disturbing the animal while blood pressure, temperature, heart rate, and respiration were continuously measured.

During their conditioning trials the animals learned to fear the clicking sound that terminated in electric shock. Not only did they stop bar pressing at the sound of the click, but they began breathing hard and showing a heightened pulse rate along with other physiological changes. Weeks after the conditioning was over, the experimenters noticed that the acidity of the blood rose at the hour of former trials. Even 28 days after conditioning was finished, when they had had a complete rest, one monkey showed disturbances in the acid-base balance of the body at the original times that he had received clicks and shocks. Such acid-base inbalances are usually a sign of pathology. Sometimes they occur in diabetes or in kidney or liver disease. Yet these animals exhibited acidosis only at a particular hour of the day. It was the hour at which they had been conditioned.

In 1965, the experiment was repeated. The monkeys had a very small internal catheter implanted in the right atrium of the heart. They also were implanted with electrodes to measure heart rate, devices to measure respiration rate, and implants to record brain waves. These animals were put through emotional conditioning trials at noon each day for 11 consecutive days.

Later, in a subsequent rest period, the animals would show an increase in pulse rate, rapid shallow breathing, and rising acidosis in the blood shortly before 11 a.m. As the intermission from conditioning continued, they gradually stopped breathing so fast and also so shallowly, but during the next 22 days the monkeys continued to show blood acidosis even earlier and later than the former time of conditioning.

It appears that the body may remember a time of fear and continue to anticipate and react at that same biologic time despite the fact that the provoking situation has been removed. The demonstration that blood acidity could change at a presumably learned time of day may be an important cue in studying psychosomatic and psychiatric illness. The acid-base balance of the blood is a delicate and highly controlled equilibrium. It affects the entire body, and when the equilibrium is thrown off, as in untreated diabetes, or among people who have low blood sugar, they suffer mental symptoms such as depression. A person with acidosis may also feel depression. Hospitals rarely initiate around-the-clock studies of patients for many days in succession, and there may be patients in whom blood acidosis is a vestigial response to a conditioned event that once occurred at a significant time of day, such as the time of a job or class.

In a subsequent study, one implanted monkey spent ten days in his training cage where he heard an ear-splitting noise every three hours. After the ten days, he was monitored continuously for any sign that brain waves, respiration, pulse, or behavior tests would show traces of the three-hourly noise. The computations suggest that for a long time afterward, his brain and body were responding to that unpleasant noise every three hours. Another series of studies now in progress may reveal whether conditioned responses to unpleasant emotional situations resonate like echoes of a bell, fainter and fainter until they die out, or whether the body's resonance to emotional conditioning in youth actually brings on psychosomatic symptoms and psychiatric complaints.

If some kinds of emotional learning fluctuate with the circadian rhythm, there are also many kinds of learning that do not seem to show the influence of biological time of day.

In 1963, Dr. Stroebel and his coworkers looked into the possibility that biological time of day might influence other learning situations. They subjected rats to a number of standard paradigms, among them escape and avoidance situations. In avoidance learning the subject can prevent a noxious stimulus if he discriminates between innocuous and potentially dangerous signals and acts accordingly. Escape learning involves unavoidable punishment. In short, the laboratory found that discrimination tasks and situations involving presumably more cerebral function showed no discernible effects that could be ascribed to the biological time of training.

This adds to the evidence that regions of the brain influencing the circadian adrenal rhythm are located in primitive areas, such as the hypothalamus, where survival functions and rudimentary emotions are regulated. It seemed plausible that fear situations might be oriented to the adrenal cycle, calling upon responses in the autonomic nervous system—a part of the nervous system that may rhythmically cause change in the body's survival functions.

In studying the circadian fluctuations in animal learning environmental controls were crucial. If the animal had taken a drug, for instance, the phase of his rhythm would be shifted. The Hartford Laboratory found that the tranquilizer, chlorpromazine, or the antihypertensive drug, reserpine, could be used to shift the rhythm of "fear" susceptibility in a rat. If rats are most vulnerable at the peak phase of the circadian adrenal steroids in the blood, the injection of reserpine or chlorpromazine shifted the phase of the animals so that they became susceptible at an earlier point on their cycle.

Electroshock produced a different reaction. It delayed the susceptibility cycle, shifting greatest vulnerability to fear toward a later time. After ten shocks, animals behaved almost as if their previous emotional conditioning had been erased. It appeared that the rat's peak susceptibility could be displaced by drugs or electroshock treatment, moreover, displaced in opposite directions.

The laboratory began rapidly switching the light-dark cycle of rats. In this way they disorganized the rhythms. Thus, in a long experiment with rats, four inversions were made at ten-day intervals, and then the animal was shifted again before he became adapted to his new schedule. The effect of this regimen was roughly like a series of ten electroconvulsive shock treatments in reducing conditioned fear. It suggested that a regimen of cycle inversions might be used in psychiatric treatment.

Free-Running and Emotional Learning

In the course of these experiments, Dr. Stroebel had noticed some anomalies in conditioned emotional responses, perhaps because the animals were subjected to an alternation of light and dark. There was no way of saying whether the training related to the animal's own biological circadian clock or to a 12-hour periodicity imposed by the light. In order to study animals without the interpolation of a 12-hour lighting cycle, the researchers decided the animals should live in constant light during some experiments. Using 3 weeks of a rigid light-dark schedule, they aligned the animals so that their 24-hour rhythms should have been in synchrony. Then they were kept in an unchanging environment and constant light. The researchers expected to see the time-of-day effects on learning they had observed in their first study, but under constant lighting they saw very different picture.

Several groups of rats showed a much weaker time-of-day effect than had been observed on the first study. When the experimenters looked closely, it was clear that 15 to 20 percent of the animals were giving inconsistent responses. Their responses did not reflect the biological time of conditioning and subsequent time of extinction. However 80 percent of the animals acted the same way under constant light as they had under an alternation of 12-hours light and 12-hours darkness. If anything, this 80 percent gave a clear demonstration that the extinction of conditioned fear was linked to the biological time at which the fear response was first inculcated. These animals demonstrated forcefully that emotional learning is related to the phase on the adrenal cycle.

Goaded by the anomalous minority, the researchers searched for an explanation for the inconsistent 20 percent. Following the experiment, they took the body temperatures of every animal in the study every three hours. The 80 percent of the animals who had shown a time effect in their emotional responses also showed a near 24-hour rise and fall of body temperature. They remained in tune with their earlier light schedule. The inconsistent 20 percent did not, which may explain their impaired performance. Their temperatures appeared to be free-running and diverged from the 24-hour rhythm by about 15 to 30 minutes each day. Every animal was a little out of phase with all the others, and this inconsistent group was out of step with 80 percent of the animals who maintained the 24-hour rhythm in constant light.

The researchers now wondered whether learned anxiety had interfered with the usually persistent and strong environmentally synchronized 24-hour rhythm of these rats. Was their abnormal drifting rhythm a sign of some oncoming pathology? The question may have very profound implications for the study of pathology in human beings. At present, it appears that a few elderly people and depressed patients have shown dissociation among the phases of internal rhythms.

A number of workers in this field, Drs. Curt Richter, Franz Halberg, Hobart Reinmann, and others who have studied psychosomatic symptoms and periodic disease in man have offered the hypothesis that certain symptoms of illness occur when some of the person's rhythms begin to free-run, moving out of phase and out of synchrony with the sleep-activity schedule. Perhaps emotional stress produces such an uncoupling of rhythms. By deliberately uncoupling rhythms in the body, could one elicit the kinds of illnesses that bring people to hospitals and mental institutions?

I know a Gentleman of a tender Frame of Body, who having once, by over-reaching, strained the Parts about the Breast; fell thereupon into a spitting of Blood, which for a year and a half constantly returned every New Moon, decreasing gradually, continued always four or five days...

Epileptical diseases, besides the other Difficulties with which they are attended, have this also surprising, that in some the Fits do constantly return every New and Full Moon; the Moon, says Galen governs the Periods of Epileptick Cases...

A Discourse concerning the Action of the Sun and Moon on Animal Bodies, Richard Mead, 1704.

It was the hand of Edward Hyde.

I must have stared upon it for near half a minute, sunk as I was in the mere stupidity of wonder, before terror worked up in my breast as sudden and startling as the crash of cymbals; and bounding from my bed I rushed to the mirror. At the sight that met my eyes, my blood was changed into something exquisitely thin and icy. Yes, I had gone to bed Henry Jekyll: I had awakened Edward Hyde.

Robert Louis Stevenson

Chapter IX. Periodic Illness and Stress Induced Illness

Ancient peoples, lacking explanations for events in nature, related changes in their behavior and physiology to the influence of cosmic cycles. Menstruation and fertility were related to phases of the moon, as were many symptoms such as the Italian "*chiodo lunare*" or moonstroke, a neuralgic pain around the eye socket believed to occur when the moon rose and disappear as it set, along with skin eruptions and swellings. Although this folklore seems remote, we are beginning to discover approximately lunar cycles of physiological change in man. We may never consult the stars, but the calendar may turn out to be a most useful modern medical instrument.

One ubiquitous syndrome with a roughly lunar cycle is the premenstrual syndrome. Strictly speaking it is not an illness, for in moderate cases it more nearly resembles a slight failure of adjustment in a normal endocrine cycle. Indeed, as many monthly, seasonal, or annual cycles are documented in man, we may discover that other periodic illnesses are also exaggerations or defects in a normal rhythm.

Menstrual Syndrome

Every woman undergoes major hormonal changes each month that pervade body tissues and inevitably affect the mind. Premenstrual tension is the catch-all phrase for a variety of symptoms that may occur at several phases of the cycle, but generally in the four to five days just preceding the menses. An estimated 60 percent of all women suffer some palpable change at this time, perhaps just mild irritation, depression, headache, decline in attention or vision. A good many women experience a day or so of intense energy, then lethargy that abruptly vanishes with the onset of the menses. Some become jittery, others weep, suffer insomnia, vertigo, or even nymphomania.

The roster of premenstrual complaints includes respiratory ailments, the activation of such chronic illnesses as arthritis or ulcers, along with gastrointestinal complaints, altered appetite, and notable changes in blood-sugar levels. Monthly changes in water retention may account for headaches, blurred vision; and the high proportion of viral and bacterial infections at this time of month may be related to the role of estrogen and progesterone in fighting infection. However, the pervasive social impact of the premenstrual tension comes from psychological, behavioral change, and this is the time of month that women are likely to be admitted to psychiatric wards. In *The Premenstrual Syndrome*, Dr. Katherina Dalton summarizes many studies of behavior change that show a large portion of crimes (63 percent in an English study, 84 percent in a French) are not distributed evenly over time, but clustered in the premenstrual interval along with suicides, accidents, a decline in the quality of schoolwork, decline in intelligence test scores, visual acuity, and response speed. In the United States, absenteeism related to menstruation costs about five billion dollars a year, but accidents, absenteeism, and domestic quarrels are only part of the social repercussions of symptoms that affect everyone. Nonetheless, the premenstrual syndrome has been studied more abroad than in the United States until very recently.

Recently a Stanford University team of researchers (R. H. Moos, B. S. Kopell, F. T. Melges, I. D. Yalom, D. T. Lunde, R. B. Clayton, and D. A. Hamburg) studied the symptoms and moods of the menstrual cycle, along with time estimation and flicker responses. A group of 15 women proved quite consistent in their symptoms: they varied monthly in their perception of a rapid double flash of light, in their anxiety, aggressiveness, and depression.

Dr. Oscar Janiger at the University of California at Irvine, became interested in obtrusive behavioral symptoms among his own women patients, symptoms that could not be satisfactorily explained in a purely psychological manner. Upon inquiry he found that these symptoms clustered in the premenstrual interval. In the early 1950's, monthly belligerence, depression, even psychotic episodes were likely to be explained in psychiatric terms, and it was thought that culture shaped women's menstrual symptoms. Dr. Janiger searched the Human Relations Area Files for anthropological support of this idea, but found none. He and his assistants compiled a questionnaire and made a pilot study of Lebanese, Apache, Japanese, Nigerian, Greek, and American girls. They found that similarities were greater than differences, and premenstrual tension seemed to be universal with women of diverse cultures reporting abdominal bloating, irritability, nervousness, depression, fatigue, allergies, backache, and moodiness.

A questionnaire to zookeepers led to the realization that rhesus monkeys, chimpanzees, and gorillas seemed to show some of the same premenstrual symptoms as women. They were lethargic, irritable, and belligerent in the few days before menstruating. Dr. Janiger had thereby found possible animals for neuro-endocrine research into monthly behavior cycles and their biochemical concomitants. In current studies, he and his associates are compiling a profile of physiological, EEG, and emotional changes in women, and have been reviewing the enormous world literature on this topic for a monograph now in progress.

The monthly cycle of women may have less conspicuous counterparts in men. A monthly rhythm of adrenal 17-ketosteroids has been detected by Dr. Franz Halberg in a 16-year study of daily urines on one man, a rhythm that has been studied by Japanese endocrinologists in attempting to understand periodic psychoses in men.

Emotional Cycles in Normal Men

Emotional rhythms usually pass unremarked in the normal person. One of the few attempts to tabulate emotional undulations in "average" people took place in the unusual setting of a factory, under the auspices of an industrial psychologist. In 1929-1930, the late Dr. Rex B. Hersey of the University of Pennsylvania, began to think that economists and psychologists were omitting an important attribute in describing the working man, whom they treated as if man were invariant. Hersey thought there might be cyclic or rythmic fluctuations in behavior.

He spent a year observing management and workers in industry, concentrating on a selected group of 25 industrial workers who seemed average in intelligence, liked their jobs, and were adjusted and "normal" in overt respects. For 13 weeks he watched and interviewed each man four times a day. He examined each physically and then took an intermission. Meanwhile, he asked each person to rate himself on an emotional scale. Then another series of observations were recorded. Charts plotted for each worker showed that emotional tone varied within each day, but longer trends were typical of each individual. One happy 60-year old, who claimed that he never changed, exhibited a nine-week cycle with a mood decline so gradual that he didn't realize he was refusing to joke with his fellows, withdrawing, and criticizing his superiors.

A 22-year-old man showed a four-and-a-half-week cycle with a variance no greater than a woman's menstrual cycle. During low periods he was indifferent, apathetic at work and at home, and temporarily abandoned his art work. Another, more temperamental person, with a cycle of four-and-a-half to six-and-a-half weeks, tended to be irritable and to magnify minor crises out of all proportion when low. Another man with a five-to six-week cycle had manic periods of great vigor and energy when he felt confident and outgoing; while in his low periods found work a burden, slept more, and was happy to sit quietly. Typically, he weighed less and slept less in his high periods.

Monthly psychotic episodes have been observed among adolescent boys, and there are many indications that lesser undulations of mood may underlie the state we accept as normal. Although modern society presumes constancy of behavior, this may, in fact, not exist in most human beings. A true base line for mood and behavior might reveal cycles of several frequencies: near-monthly, seasonal, and annual. Whenever these fluctuations become pronounced enough to be abnormal, drugs usually are applied at once, masking the symptoms and possibly changing the rhythms themselves.

Unfortunately, patients are rarely observed without drugs over long periods of time; thus, the vast chronicles of periodic illness come mainly from 19th century and early 20th century physicians.

Dr. Hobart Reimann of Hahnemann Medical College in Philadelphia, has summarized an extensive and fascinating literature of case histories in a monograph called *Periodic Diseases*. He suggests that many periodic diseases may represent the effects of sudden excitations within certain regions of the brain, primarily the hypothalamic area. In periodic illnesses there are sometimes changes in the number of certain blood cells, sometimes edema, local swelling of tissues and skin, and recurrent fever, not to mention the familiar oscillation of manic depression, normalcy and catatonia, intermittent psychosis, epilepsy, and migraine.

Some of these are genetic in origin. Periodic edema may be linked with a specific dominant gene. Periodic peritonitis occurs mainly within certain ethnic groups (Armenians, Jews, Arabs) in the Mediterranean, whereas periodic edema affects Caucasians almost exclusively.

Unfortunately, symptoms rarely are recorded graphically or on a calendar and the medical nomenclature is confusing. Unless there is a diary, physicians and victims usually are unaware of the periodic nature of a disease because many parts of the body and sometimes mental symptoms are involved. Many patients may go the round of specialists, seeing various internists, surgeons, dermatologists, hematologists, and psychiatrists, without diagnosis or relief. Lack of adequate records is particularly sad among people with hereditary illness, who cannot be advised against passing on their unwanted genes to another generation. Dr. Reimann's exhaustive compilation of cases and medical lore includes examples of the curious role some periodic illnesses have played in history. Purpura, which can cause internal bleeding and hemorrhaging into the skin, seems to take a form related to periodic edema and may have an underlying neurovascular biorhythm.

Medieval mystics were reported to show the stigmata regularly on Christian calendar days, bleeding regularly on Friday. Some fell into trances, experiencing pain and also communication with their Savior, and were healed after Friday. It is not clear whether self-inflicted trauma or deep trance elicited the appearance of a seven-day cycle, but it is known that blood platelets have a life-cycle around seven days.

Periodic Hypertension

One complex and baffling periodic illness is a form of periodic hypertension. In 1953, a shy, slight Irish nun came to the attention of Dr. Hobart Reimann. She had suffered distressing episodes of fever initially lasting only a few minutes, but later recurring about every five days and lasting for twelve or more hours. Her fevers and headaches were given several tentative diagnoses, among them malaria, migraine, anxiety neurosis, and trichinosis. Finally, it was observed that she had hypertension solely during episodes of fever. Her blood tests, urine, and electroencephalograms revealed no abnormalities. Nothing explained why she had a sudden malaise, chills, and throbbing headache in the back of her head with fever and an increase in pulse rate. Her blood pressure would rise from 130/80 to as much as 170/110, and the headache would become very oppressive. A spectral analysis of her fevers was later prepared by Dr. Franz Halberg and showed that these episodes recurred about every 11 days.

The patient was diagnosed as a periodic hypertensive by Dr. Reimann, and was sent to the National Institutes of Health in Bethesda to be studied by Drs. Frederick Bartter and Sheldon Wolff. A 40-day study showed that she underwent sweeping changes in the concentrations of urinary ketosteroids, of a hormone, aldosterone, which regulates sodium retention, and of other adrenal steroids as well. These hormones play an important role in the retention of water in the body.

Dr. Bartter and Catherine Delea have been studying other patients with hypertension, looking for rhythmic phenomena that may help to explain the sudden increases and decreases of blood pressure that are so torturesome to the victim of this illness. They have found that it is meaningless to collect diagnostic data without taking circadian rhythms into account. This means that no hypertensive patients can be adequately evaluated except by several 24-hour periods of observation within the hospital.

Only occasionally are a patient's symptoms obviously clocklike. Dr. Curt P. Richter has cited the famous case of a star athlete at Cambridge University whose knees would become swollen and painful every nine days. His disability was so regular that his team could schedule games months in advance, avoiding dates on which he would be incapacitated.

Dr. Werner Menzel of Hamburg, Germany, was among the first great clinicians to keep time-charts of patients' symptoms and to search daily rhythms for clues to periodic illness. Guided by the maxima and minima of body functions, temperatures, urinary constituents, he compared the sick and healthy, and elegantly described many cases of periodic disease. In one child with a lymph disease (Hodgkins Disease), he found that body temperature underwent a 12-hour instead of a 24-hour rhythm. He also noticed 12-hourly peaks in urine volume in patients who suffered from arterial tension and over-sleeping. People with liver disease, he observed, often showed their peak temperature and urine excretions at night instead of by day. Menzel looked at the circadian rhythm of primary functions for indications of the nature of the disorder.

Mental Illnesses

The most dramatic of all the periodic illnesses are, in mild form, the most common. These are the recurrent emotional or mental illnesses, some showing rapid alternations of normalcy and symptoms every 48 hours, while others might span weeks or months.

Mary Lamb, the sister of the great English essayist, Charles Lamb, suffered a cyclic psychosis for 50 years, beginning at age 30. During one of her psychotic attacks she killed her ailing mother, of whom she had been very fond. Fortunately, a lawyer friend saved her from prosecution and placed her in her brother's custody. She lived to the age of 83, a long and fruitful life. Between her 38 attacks she lived a normal life. The attacks were regular, and at the first sign of slight irritability, her brother would rush her to the hospital or put her in a straitjacket. Immediately after recovering, she went on her usual round of entertaining literary friends and writing books or stories until the next attack.

This famous case is of interest because Mary Lamb showed no signs of physical or mental deterioration, excepting that of old age. Perhaps some signs of deterioration in mental patients derive from interference with normal development and stimulation. The routine of hospital life deprives a person of his usual modes of expression and fulfillment, and the impairment often observed among mental patients may be not only a result of the illness but also of the way we treat mentally ill patients. Diary information might be used to restrict hospitalization for mental patients in the way we restrict it to the acute phases of physical illness. Chronic patients might lead more normal lives in the interstices between bouts of illness if the calendar of attacks permitted. Sometimes, however, the intermission is too brief and then a hospital setting becomes crucial. A number of patients who alternate between a few weeks of normalcy and a few weeks of psychosis have been studied throughout their adult lives. In what may be the single most thorough longitudinal study of an illness, Dr. Leiv Gjessing and his father, Rolv Gjessing, before him, have attempted to discover the biochemical mechanisms of periodic catatonia.

The Dikemark Hospital is the community hospital for Oslo, founded in 1905. The old, stuccoed, pastel buildings have high arched windows and towers, and the setting, on steep-wooded hills with lawn and moss-covered rocks high above a small lake, is unusual for a mental institution, at least by American standards. Among farm fields and lakes, the hospital grounds have an atmosphere of 19th century New England. Patients are not isolated from the surrounding life of the community. Many work in the hospital farm or its small industrial shops, treating the place as a home where they live, paint, practice the piano, construct furniture, and assist in the laboratories.

In this atmosphere, Drs. Rolv and Leiv Gjessing have made long-term studies of people with predictable swings from a normal state into either hyper-excitability and violence or into the mute and frozen state that resembles paralysis and is known as catatonia.

Periodic Catatonia

In the 1920's, Dr. Rolv Gjessing observed that fortnightly stupors were interspersed with normal periods of behavior, and when behavior changed radically, so did the physical appearance of the patient.

During stupor, one man had sputum so thick it could be drawn out like chewing gum. His skin became very oily. Using the instruments available at the time, Dr. Gjessing measured nitrogen retention and other physiological functions, maintaining his patients on a controlled diet, and training nurses to get exceedingly accurate measurements. Using his own funds, he furnished a biochemistry laboratory on top of one of the men's wards, for he suspected that metabolic flaws underlay the illness and that the thyroid gland was involved. His work has been continued by his son.

Leiv Gjessing has been studying and caring for old men who knew him as a little boy. Exacting round-the-clock studies have won extraordinary cooperation from these patients who once had treated him as a young nephew.

Catatonic patients can be studied as their own controls during normal periods, which also allow a person with periodic catatonia to develop relatively normally in pace with peers. Periodic psychosis often begins in the early twenties, sometimes abruptly. Dr. Gjessing has speculated that stress, brain damage, or perhaps a metabolic shock due to some autoimmune reaction might damage a metabolic regulator and, thus, produce the clocklike symptoms.

When catatonic, a patient may seem out of contact, mute, and immobile for several days at a time, but according to the diaries and reports of such patients they are internally experiencing an implosive intensity. One man, hospitalized since 1935, always began to talk to himself in transition and for a day or two he would babble in a manic fashion, reclining with head and feet raised for several days, in a position that is practiced a few seconds at a time by gymnasts. Eyes open and frozen, his hands clammy, skin oily, he lost his appetite; he had a high pulse rate and blood pressure. He looked like a wax doll until he began to recover.

This paralyzed state masks an intense hallucinogenic state resembling experiences under mescaline, as can be seen in these recollections of one patient.

"In the stupor many strange events enter the soul. The soul is bewitched. [Ordinary experiences, such as being washed, displeased him, and the sensation was strange.] Everything was polar .... In order that the sun should shine, the soul had to have psychic trouble, the trouble corresponding in strength in proportion to the strength of the sun .... Like the Tree of Knowledge, everyone who eats the fruit must die."

"If you ask a simple question, I hear it, but it's as if from outside the room. People help but the people become transformed into words, and from words people are transformed into a kinemagraphic picture. . . .thought stops but for a few fixed points that act as a lighthouse. . . ."

Asked why he did not move in bed, he answered:

"The soul and thinking prevent moving, prevent muscles from doing what I want them to. Impulses are not carried out, and this seems natural. Not to want anything and to have no interest in anything is important. Former interests do not penetrate."

This state suggested a kind of intoxication to Gjessing.

Correlations of Behavior and Physiology

Daily physiological records have been routinely kept. Blood pressure, records of sleep, temperature, and of urine, and in some instances pulse, are taken six to ten times a day, and blood samples three times a week. A controlled diet has been given for periods of time, allowing urine analysis of nitrogen relative to the input of proteins. Over the years it has become clear that nitrogen metabolism is faulty in these patients.

Nitrogen is a key element of all protein and all body tissue. It is absorbed from plant and animal protein and metabolized in the liver. In general, the body maintains a balance, excreting about as much as it takes in. During intervals between stupor attacks, the urine shows a retention of nitrogen, but the balance is shifted noticeably during attacks of illness. In patients with catatonic excitement, nitrogen was over-excreted during the normal interval and retained during the phase of excitement; thus, urinary ammonia was higher at the beginning of an attack than at the end. The electrolytes, phosphate and sodium chloride, are excreted more during attacks.

Ten-year graphs of daily temperature or of urine pigmentation in individual patients show such regular changes that one could predict, within a day, when an attack would fall in the coming year. The oscillation between normal behavior and illness has been found to correspond to major oscillation in many metabolic functions. The periodic catatonic swings between phases resembling certain aspects of patients with underactive or overactive thyroid glands. Indeed, thyroxin, a thyroid extract, prevented nitrogen retention and has successfully ameliorated the symptoms of a number of patients who have been maintained out of the hospital, and have lived normal lives for years. When they stop thyroid treatment, however, they relapse.

An association between thyroid abnormality and psychosis has been observed by a number of clinical researchers. Dr. Jack Durell and his associates at the National Institute of Mental Health in Bethesda, conducted a number of longitudinal studies of thyroid function in psychotic patients, using two standard tests (Protein Bound Iodine, or PBI, and radioactive iodine uptake) for evaluating the thyroid function. One of their patients showed striking motility, changes so swift that, within a few hours, he was transformed. Within 96 hours after his change into a mute state, there would be a rise in radioactive iodine uptake. This was observed in three successive cycles. After studies of another patient with periodic catatonia, Dr. Durell postulated that catatonic episodes may result in the suppression of thyroid function, which might synchronize the development cycle of follicles of the thyroid, resulting in a slow oscillation of thyroid hormone output.

The concept of an internal oscillation as a pathological state, replacing the normal state of metabolic activity has many implications for treatment. By understanding the temporal aspects of the illness, it may become possible to schedule medication so as to restore normal rhythmicity in the metabolic system. Empirically, Dr. Gjessing has found that thyroxin has a different effect if given during the "normal" interval than if given during psychosis. Thus, the success of thyroid treatment in periodic catatonia has an element of timing.

Dr. Gjessing has analyzed the amines, phenolic acids, and other protein products of patients during normal and sick phases, during treatment with antidepressants, and with tranquilizers. Drugs, such as the antidepressant and tranquilizer, haloperidol, help prevent agitation and excitement, but the periodic psychotic person becomes agitated the moment the drug is stopped. Gjessing has postulated that drugs may only mask the symptoms of periodic catatonia, and may actually prevent true remission.

The biochemical mechanisms underlying periodic catatonia resemble a complex jigsaw puzzle with missing pieces. An oscillation stemming, perhaps, from flaws in glandular activity—the thyroid and pituitary—may prevent an even metabolism of the nitrogen products of food, which are the basis for protein building and energy. Dr. Roger Guillemin at Baylor University in Houston, Texas, has reported isolation of TRF, thyrotropin releasing factor, in the hypothalamus, defining another long-sought route of control from the brain, via the pituitary gland to the secretion of the thyroid.

One metabolic flaw has been detected by Dr. Franz Halberg in applying his cosinor method of analysis to urinary excretion data on a catatonic patient. Upon analysis, it appeared that urinary constituents were excreted in a normal, 24-hour rhythm, all excepting sodium—which showed a 22-hour cycle. Such a desynchronization might set up a beat within the metabolism, explaining the man's 21-day cycle of mood and behavior.

Signs of oscillating behavior may parallel oscillating hormone balance in a variety of periodic psychoses. Dr. Hobart Reimann cites the case of a young man, classified as a paranoid schizophrenic, who underwent cyclic alternations between feeling and acting male or female. He would be male for three or four days and then female for three or four days, in extremely regular alternation. Observation of patients with attacks of homosexual feelings every four weeks has also led to the suggestion there must be oscillations in the output of sex hormones or oscillations in the susceptibility of target organs.

Manic Depression

During the late 19th and early 20th centuries in Munich and Vienna, a number of doctors kept careful diaries of mood change in patients. These showed weight and behavior fluctuations over many years, sometimes revealing a manic-depressive trend unfolding over a 15-month period, or two-year intervals. Today, the manic-depressive with swift mood changes attracts most study.

Dr. Curt P. Richter, in his monograph on periodic illness, cites the famous case of a man with 48-hour manic-depression. A foremost salesman in the Washington, D.C., area, he would be so morose and apathetic during his depressed 24 hours that he would drive to a customer's office and find himself unable to move from the car, sitting miserably for hours. Yet, on good days, he was the epitome of the aggressive, talkative salesman. He adapted to illness by accepting appointments only on alternate days.

A manic-depressive with a short cycle is fortunate, for his swings will be detected and treated. Manic-depressives with long swings are likely to go undetected and may do themselves harm in their manic phases by serious misjudgments and grandiose illusions. Such people seem to go through a transformation of character, from normal diffidence and activity to superconfidence and unbounded energy. The ceaseless activity of a person in an elated phase involves an astonishing output of energy.

Dr. William Bunney and his coworkers at the National Institute of Mental Health in Bethesda, have studied what they call the switching point in manic-depressive patients. This point of change from depression to mania is preceded by increases of norepinephrine in the brain. During the 24 hours before the abrupt change, patients have shown a drop in their usual amount of rapid-eye-movement sleep and peripheral increases, both in the precursors and actual amounts of norepinephrine.

A number of 48-hour manic-depressive patients are at present under study at various hospitals around the world. Dr. F. A. Jenner of Middlewood Hospital in Sheffield, England, has been watching a former boxer from Yorkshire who became manic-depressive after a bad accident. For 24 hours he is overactive, talkative, sometimes testy, with grandiose ideas about science and the world and then, typically, at some time in his sleep he changes. He awakens lethargic and bleak, rises reluctantly and late, and falls asleep earlier that night. On his inactive days he urinates and excretes more, but eats and drinks less than on manic days.

For 11 years this man has been studied under controlled conditions in a clinic with a measured diet. He has been weighed everyday, has had his urine and feces collected, saliva sampled, and blood drawn. He has taken performance tests and kidney and thyroid function have been measured. The acid-base balance of his urine and blood has been tested for a variety of hormones including the estrogens, male hormones, and aldosterone, a hormone that influences salt retention. Blood sugars and adrenal hormones and certain amino acids also have been measured.

When kept in bed on a liquid diet, the 48-hour alternation of weight, urine volume, and volume of red blood cells suggested that the amount of fluid held within and around the cells changed with the moods. Fluid retention within and around cells has been thought to be controlled bioelectrically by a balance of charged elements. Indeed, the sodium and potassium content of saliva or urine did alternate in a way that coincided with his alternating moods. On depressed days, the saliva sodium was very low and the saliva potassium was very high, and on manic days it was exactly the opposite.

Was the manic-depressive state a result of shifting electrolyte balance? In 1963, Dr. Jenner took his patient into an isolation room at Lancaster-Moor Hospital, where a team of observers supplied meals and collected urine samples. Lighting was regulated so that the day was 22-hours long. Neither Jenner nor his patient knew the exact day length. The experiment lasted 11 real days in which they lived 12 cycles. Four-hourly urine samples from both patient and doctor were analyzed for sodium and potassium, while mood changes were recorded by Jenner and unseen observers. Throughout this period, Jenner took physiological records on his patient and himself. His companion alternated between morose silence, and an uncontrollable railing, scheming, and incessant shouting, moving about in an agitated manner—almost unbearable for Jenner. However, the results warranted the effort, for the patient's moods adapted to a 22-hour day, and he alternated from lethargy to mania in a 44-hour cycle. Yet, excretion of water and electrolytes largely remained on a 24-hour cycle. Jenner's own urine electrolytes had more nearly adjusted to a 22-hour cycle than those of his patient, whose behavior cycles were evidently modified by social factors. A simple relationship between the electrolyte imbalance and excretory patterns did not explain the behavioral adaptation to a 22-hour rhythm, dissociated from the persisting 24-hour rhythms of renal excretion.

Another approach in attempting to understand the alternations of this 48-hour psychosis has been via the alkali metal, lithium, which seems to obliterate the symptoms. In 1967, this patient received lithium and as long as he took lithium salts he remained essentially normal—an effect which has been seen with lithium throughout the world. The manic-depressive cycle has been obliterated by lithium but restored by doses of sodium. Another patient, who alternated between three days of mania and three days of depression, showed a six-day cycle of water retention which was obliterated by lithium.

Using laboratory animals, Jenner and his associates have been attempting to fit together the pieces of the puzzle by studying the water-retention cycle and hormones known to change during the estrous cycle in rats. Simultaneously, they are searching the manic-depressives' urine for clues to uneven fluid retention, since the amount and composition of fluid inside and outside cells undoubtedly influences the transmission of nerve messages. Why does fluid retention begin to oscillate?

A Shock-Phase Theory of Periodic Illness

In a lifetime of research physiologist Curt P. Richter observed the activity rhythms of thousands of animals, attempting to understand the internal "clocks" behind the regular cycles of activity, feeding, and sexual response in healthy animals. Looking at vitality levels and the distribution of energy in animals, Richter hypothesized relations between age, drive, activity, and diet. He began to manipulate diet, lighting schedules, heat, cold, and many aspects of the environment to explore the nature of the 24-hour rhythm. The function of the adrenal glands seemed crucial, yet removal of the adrenal failed to influence the daily period of activity, merely decreasing the overall level. The next step was to remove part of the pituitary, which governs thyroid and adrenal activity. This did not eliminate the 24-hour rhythms of the rat.

Dr. Richter postulated that a "24-hour clock" must reside within the central nervous system, and he systematically began to assault regions of the brain. He deprived a group of animals of oxygen, gave electroconvulsive shock—even stopped the heartbeat for 50 minutes—but as soon as the animals recovered they resumed their 24-hour alternation of rest and activity. Animals were recorded during starvation, dehydration, and extremes of heat and cold. Blinded animals began to drift by a few minutes each day, showing rhythms a little shorter or longer than 24 hours. Drugs, anesthesia, and toxins had no effect on the "clock." However, when the hypothalamus was partly destroyed, blinded rats began to eat about every 40 to 60 minutes around the clock, and to drink every few minutes—no longer sleeping, as before, in an interval of 12 hours but wandering around in a continuous stupor. Brain lesions in over 200 animals suggested that the hypothalamus or related regions were important in regulating the activity rhythm.

Richter also experimented with behavioral stresses. He forced an animal to swim for its life under a jet of water for 48 hours; or placed two fierce male Norway rats in a cage and precipitated a vicious fight by giving them an electric shock. After returning, exhausted, to their cages, these animals exhibited long, slow swings in their levels of activity, but the 24-hour cycle was unchanged. The slow oscillations in body weight and activity were reminiscent of periodic illness in man, of which Dr. Richter collected case histories, studying many patients with periodic symptoms that often go undiagnosed.

One of his patients was a woman with a 40-day alternation between 20 days of normal life and 20 days of deep depression, cramps, and diarrhea. As it turned out, she had been exposed to ammonia fumes, which caused parathyroid damage, and the abnormal symptoms were entirely removed by calcium therapy. The idea of giving her calcium therapy had been inspired by observing calcium lactate uptake in monkeys whose parathyroid glands had been removed. When offered calcium lactate in free cafeteria style, the animals ingested it in 40-day cycles. Calcium appears to influence nervous transaction, and high levels of unbound calcium lactate have been found in the blood of anxiety neurotics by Dr. Ferris Pitts and his coworkers at Washington University in St. Louis. Moreover, injections of unbound calcium lactate have caused normal persons to show these symptoms. As Dr. Richter had seen, ,removal of thyroid glands produced abnormally long cycles of activity and inactivity in rats, and parathyroid removal appeared to interfere with calcium metabolism in a manner also resembling periodic illness in human beings.

Dr. Richter postulated that "clocks" throughout the body ordinarily regulate functions so that they were out of phase and gave an appearance of overall smoothness. Shocks, such as infection, allergy, surgery, and physical or emotional stresses might upset the overall coordination of phase relations among these multitudinous cycles, setting one metabolic rhythm into phase with another in a manner that would generate a periodic beat and recurrent symptoms.

Although purely conjectural, this "shock-phase" hypothesis of periodic illness inspired interesting research. Dr. Richter had generated long swings in weight or activity level, abnormal cycles that followed physical stresses, drugs, and surgery, but only removal of a large portion of the hypothalamus significantly altered the circadian rhythm. In 1963, behavioral stress was used to generate psychosomatic and mental illness in animals, and to alter the circadian rhythm as well.

Behavioral Stress and Illness

During the early 1960's, Dr. Charles F. Stroebel at the Institute of Living in Hartford, was examining the relationship between the circadian rhythm of adrenal hormones and conditioned fear behavior in rats and monkeys. Two of his rhesus monkeys developed a predominantly 48-hour periodicity in brain temperature. This abnormal brain temperature rhythm began to show up while the animals were being conditioned in unavoidable fear. Stroebel and his associates wondered if behavioral stress, rather than physical stress, might produce abnormal circadian rhythms.

They had also noticed a free-running rhythm in some of the animals, who showed peak temperatures a little later each day after they had been conditioned in anxiety. Did this mean that anxiety interfered with their very strong and resistant 24-hour rhythm?

Dr. Stroebel began to arrange his laboratory so that 24 rats could be monitored simultaneously, with body temperatures and activity transmitted by biotelemetry, recorded on analog tape and analyzed by computer, thus enabling him to see data in statistically viable quantities. He also began to look closely at the temperature cycles of monkeys. These creatures had been implanted with numerous sensors for a series of learning experiments. Prior to training the rhesus monkeys had shown body temperatures with a clear, pronounced peak every 24 hours. It was a rhythm that repeated itself elegantly and predictably each day and persisted under constant light in the early weeks of the experiment. However, after four weeks, one monkey began to show peak temperatures at intervals of 46-48 hours.

There had been no change in his daily feeding schedule, his light-dark schedule, the cleaning of his cage, or the six hours of daily conditioning. At first the shift came about as a drift of 15 minutes or half an hour a day. Soon, his "circadian" peaks were occurring on a 31-hour schedule, then abruptly he leapt to a 46-hour, then a 48-hour temperature rhythm. The experimenters were amazed, but they wanted to protect their valuable subject, and his abnormal rhythm was eventually suppressed by a series of timed injections of a tranquilizer during the subsequent four-week period. When the drug was discontined the normal 24-hour temperature rhythm reappeared.

The members of the laboratory took particular note of the monkey's behavior during intervals of abnormal temperature cycles. It became noticeably neurotic until the tranquilizer was administered, when he grew less agitated. Moreover, when the drug was discontinued and the 24-hour rhythm reappeared, the animal was not at all agitated. The relationship between abnormal temperature rhythms and abnormal behavior demanded a further series of experiments.

Stress and Abnormal Rhythms

Among the many kinds of learning paradigms in the animal laboratory, the one that generated abnormalities of the temperature cycle was unavoidable fear. It seemed to have a singular potency which shook most animals out of synchrony with their environment. This emotional stress disrupted the stubborn 24-hour rhythm of a healthy animal—something that had never been done in the laboratory before.

In 1965, Dr. Stroebel and his associates began using seemingly mild behavioral stress to produce disturbances in monkeys. Each wore scalp electrodes for recording brain-wave rhythms. A microscopic temperature sensor had been implanted in the brain of each. They all lived on a rigid lighting regimen of 12-hours light and 12-hours dark, were fed at regular times of day, drinking water whenever they wanted it.

Each monkey was given a six-week period of adaptation. Each day he was placed in a restraining chair. Before his eyes was a panel with a food hopper that delivered pellets of food, surrounded by four lights. A projector screen, capable of projecting 12 images, was just below and on the left and right were two retracted levers. At the end of six weeks of training, each animal was surgically prepared with 18 external EEG electrodes, two electrodes placed in the intercostal muscles to measure the EKG, a catheter in the right atrium via the carotid vein, temperature sensors on the skin over the right and left frontal cortex, a tiny silicone rubber tube catheter to deliver drugs under the skin, and two electrodes in the muscles at the back of the neck. For two weeks the animals simply recuperated. They were then returned to their chairs in the chamber.

Each chamber was an isolation booth maintained at a constant humidity and temperature. During the experiment, the monkey was allowed to see a human being only at 6:30 a.m. and 4 p.m., but was not allowed to see other monkeys. Throughout this stage discrimination problems were displayed for 24 hours around the clock. By using the right-hand lever the monkey could solve the problems and be rewarded. The left-hand lever was always extended, but pressing it produced no effect. Then began a ten-day program of what might be called *escape training.* The temperature in this testing booth was made uncomfortably warm. At this point, an accidental press of the left lever produced a nice gust of cool air. Once this initial escape task was mastered, the monkey quickly realized that he could escape loud noises, flashing stroboscopic lights, and mild shocks to the body by merely pressing the left-hand lever. As he worked in his problem-cage, the animal could rid himself of any noxious stimulus by pressing the left-hand lever, and for two to four weeks he was irritated sporadically with unpleasant events. The animals soon developed a habit of keeping their left hand on the left lever and indiscriminately pressing it. The researchers affectionately called this "primate hypochondriasis." The monkeys evidently associated the left lever with a feeling of security—like Linus with his security blanket—and held onto it, refusing to let go.

After two to four weeks in this situation, with several hours of trials each day, the experimenters merely retracted the lever into a recess in the wall. There were no further noxious situations. The animal was never again annoyed with high temperatures, loud noise or lights. But now the monkey could see the left-hand lever and not touch it. He became frantic. Initially, the animals spent hours trying to get at the left-hand lever.

From this point on, 12 of 13 monkeys became increasingly disturbed. They fell into one of two patterns, categorized on the basis of their brain temperature rhythms. It should be repeated that retraction of the security lever in an isolation situation constituted the only behavioral stress. By the 14th day after the stress, these 12 monkeys showed a very interesting trend. They were living under a lighting regimen of 12-hours light and 12-hours dark, with regular feeding times, yet 5 animals showed a tendency to desynchronize or to be "free-running." Their temperatures still showed circadian rhythm, but, like men in isolation, they were desynchronized from the dominant 24-hour cycle of the environment, and their peak temperature came 15 to 30 minutes later each day (or in two cases, earlier). The animals continued to perform on their discrimination tasks, but were inefficient. They also developed symptoms that could be described as neurotic and psychosomatic.

Two of the monkeys developed asthmatic breathing. Two of them developed duodenal ulcers and later died of them. All showed gastrointestinal disturbances, eruptions of the skin, and sores that did not improve under antibiotics. They drank much more than their normal amount of water. In addition to their other symptoms, three also began to show asthmatic breathing, high blood pressure, and blood in their stools. The five psychosomatic animals barely managed to function, using the right-hand lever to solve the problems that were continuously presented.

The other seven animals were very different. They began to show a predominantly 48-hour cycle of brain temperature. Fourteen days after the recessing of the lever, their brain temperatures had become significantly abnormal. They stopped showing their usual 24-hour temperature peak. Then the peak came at 16- and 32-hour intervals. As the disturbance in the brain temperature rhythm evolved, this group of monkeys began to show a distinct lassitude and weakness. They no longer groomed themselves and their fur became mottled. They began to lose interest in food. They performed unpredictably, if at all, on the right-hand lever problems. They hardly did their discrimination tasks. Instead, they often paused for naps and their behavior became bizarre. For example, two monkeys spent hours catching what seemed to be imaginary flying insects, and one masturbated almost continuously. Three compulsively pulled out their own hair. All of them tended to show very little interest in their external environment, and sometimes rocked or pulled at their fingers or pulled at their fur in the manic and stereotyped, repetitive manner of some psychotics. There were long periods when these poor animals seemed despondent and enrapt in their own bleak world.

Despite a very strict feeding and lighting regimen and relative isolation, the changes in circadian temperature cycle were obvious and seemed diagnostic. The differences between the psychotic-like monkeys and the "psychosomatic" group became clear. The psychosomatics showed desynchronization from the environment, a free-running temperature pattern like that produced in rats during emotional conditioning and in individuals in isolation. But the animals with "psychotic" symptoms shifted to a cycle of 30 hours or more, and then jumped to 45 and 48 hours in brain temperature rhythms.

The sleep cycle also became abnormal, as often happens in mental illness. Normally, a rhesus monkey will sleep about nine hours at night, about a fifth of that time in a drowsy state, 18 percent in light-spindling sleep, 35 percent in deep slow-wave sleep and over 20 percent of the time in rapid-eye-movement sleep. During their neurotic period, the animals drowsed about 60 percent of the time, spent considerable time in light sleep, and very little in deep sleep or in rapid-eye-movement sleep. These findings are parallel to human studies that indicate a number of abnormal patterns during the onset of mental illness.

The exact relationship between the non-circadian temperature rhythm, sleep, and symptoms was not explored because the animals were too sick and disturbed to be allowed to continue without intervention. A month's study might have made it possible to see how the shifting, free-running pattern changed the rhythmic patterns of sleep. However, the animals with psychosomatic symptoms had severe sores around the implanted transducers. These skin ulcers resisted antibiotic treatment. Monkeys are delicate creatures and require considerable care in the laboratory, so the "psychosomatic" monkeys were studied for about six to ten days and were then restored to health with drugs.

The Effect of Returning the "Security" Symbol

At the end of the second period of the experiment, the researchers put the "security lever" back within reach for two "neurotic" monkeys and two "psychotic" animals. This produced recovery and spontaneous resynchronization of temperature, along with improved behavior in both the psychosomatic animals. It had no observable effect on the 48-hour rhythm or the disturbed behavior of the more psychotic-like monkeys, not even over a five-week period.

Experimental periods had to be calculated very carefully. Not only were the psychosomatic monkeys in danger of infection, but the psychotic animals got sicker and more despondent each day. They refused to eat, and it is impossible to force-feed a monkey for long. Therefore, the potential extent of the abnormal rhythms could not be measured in this experiment.

The very fact that behavioral stress was enough to shake an animal out of harmony with his environment suggested the steps by which trauma might generate physiological illnesses in human beings. Did the free-running animal maintain the synchrony of his internal component rhythms, merely falling out of tune with his environment? Or was his internal harmony disrupted, creating a phase dissonance among physiological rhythms that, in turn, slowly evolved into sickness? A disruption among biochemical rhythms in the blood might be equivalent to having too little or too much of a vital constituent in the place where it is needed, or it might mean that the right molecules go to the right cells at the wrong time. Differences among the animals may have some important diagnostic implications. For example, the two monkeys who died of duodenal ulcers showed body temperature cycles shorter than 24 hours, while the other psychosomatic animals had longer than 24-hour rhythms. Dr. Stroebel and his associates observed a skin pigmentation difference between the monkeys who became "psychotic" and those with "neurotic" reactions, suggesting a possible hereditary predisposition that might be used to predict responses to behavioral stress.

Drug Treatment

The experiments of 1965 also indicated how timing may be effective in the use of drugs. The tiny plastic catheters in the monkeys' abdomens permitted a continual infusion of medicine that the animal would never perceive. For four weeks a phenothiazine tranquilizer, known as TPS-23, was infused into the psychotic monkeys with remarkable impact. The 48-hour rhythm of brain temperature dropped out and, at the same time, there was an astonishing change in sleep patterns. The monkeys began to spend an enormous amount of time in deep slow-wave sleep. For three weeks they showed continuous improvement, but in the fourth week they became despondent and would refuse to eat. This despondency paralleled a stage in recovery from psychosis when people often go through a period of terrible depression.

The depression was so devastating to the monkeys that an antidepressant was added to the infusion (tranylcypromine, or Parnate, an MAO-Inhibitor). Within 72 hours the behavior improved, and a circadian temperature rhythm began to re-emerge. The monkeys started to behave normally again. Since the antidepressant appeared to enhance the circadian rhythm, it was decided that the effect might be further enhanced by administering the drug in pulses every 12 or 24 hours. When drugs were pulsed at 12-hour intervals, the temperature began to show small 12-hour peaks, and then larger peaks every 24 hours. At this point the monkeys again began to press their lever for food rewards.

The investigators are cautious about applying their results to human beings. In 1968 and 1969, they started a replication of the original study. Although it is too early to jump to conclusions, there seem to be many parallels between the patterns of these monkeys and mentally-ill people.

Had they been human, the desynchronized animals would have been classified as psychosomatic or neurotic. Week after week their brain temperatures were slowly drifting off the 24-hour rhythm without regard for external time cues or social regimen. These desynchronized monkeys passed into an inverted phase in which their bodies were prepared for daytime functioning during the night and prepared for nighttime rest during the day. Perhaps these inversions may help to explain the napping of the monkeys and the insomnia and napping of mentally-ill people. A laboratory study of good and poor sleepers has indicated that they differ physiologically and, for example, body temperatures of poor sleepers did not reach daytime levels by the time they awakened in the morning. Moreover, on questionnaires, Dr. Lawrence Monroe of Ohio State University in Columbus, found that the poor sleepers had many more neurotic and psychosomatic complaints. Although his study was only suggestive, it offers tentative corroboration of the expectation that abnormal physiological rhythms may be one aspect of sleep disturbance and emotional disturbance.

Many studies of animals will be needed to indicate whether psychosomatic symptoms result from a phase disruption among internal cycles, or a slight uncoupling from the environment that leaves an animal out of kilter with the demands of activity and eating. It may be possible that the autistic retreat of the psychotic monkey and his apparent disinterest in his surroundings arise from a need to protect himself from the demands of "daytime," which he cannot handle with a "nighttime" body. Only substantial research can indicate whether behavioral stress may exaggerate the indetectable and "normal" emotional and endocrine rhythms of the healthy person, causing the emotional and physiological concomitants called mental and periodic illness.

At the Institute of Living, parallel studies of animals and human patients have begun. Day and night nursing notes have been automated and are analyzed by computer so that patients can be given intensive, behaviorally detailed examinations before they receive drugs. Further studies may specify the optimum timing of medications. Astute clinicians already use scheduling as part of their drug strategy. Dr. Heinz Lehmann of Douglas Hospital in Montreal, has found that agitated patients responded to divided doses of sedative, in afternoon and evening, better than to a larger total nightly dose. Drs. Liev Gjessing of Dikemark Hospital, Oslo, Norway, and Per Vestergaard of Rockland State Hospital, Orangeburg, New York, have used strategically-timed doses of thyroid hormone to obliterate symptoms of periodic catatonia. Drug scheduling is presently an art that will be greatly enhanced by an improved knowledge of biological time structure. Researches, such as the ongoing studies of implanted animals at the Institute of Living, may also help to illuminate the etiology of psychosomatic and emotional illness.

Spectacular reversals of the photoperiodic control of diapause are obtained if females of the Nasonia vitripennis are chilled for four hours in certain light dark cycles. . .

D. S. Saunders

. . .it has been shown that a 75-minute pulse of light per day, when appropriately postioned with respect to the circadian rhythm of the sparrow. . . is sufficient to produce a response normally produced only by long days. . . a strong confirmation of Bünning's hypothesis concerning the mechanism of photoperiodic time measurement.

Menaker and Eskin

Linnaeus Flower Clock:

6 a.m. - Spotted Cat's Ear opens
7 a.m. - African Marigold opens
8 a.m. - Mouse Ear Hawkweed opens
9 a.m. - Prickly Sowthistle closes
10 a.m. - Common Nipple Wort closes
11 a.m. - Star of Bethlehem opens
12 noon - Passion Flower opens
1 p.m. - Childing Pink closes
2 p.m. - Scarlet Pimpernel closes
3 p.m. - Hawkbit closes
4 p.m. - Small Bindweed closes
5 p.m. - White Water Lily closes
6 p.m. - Evening Primrose opens

Carl Linnaeus (1707-1780)

"He appointed the moon for seasons:
The sun knoweth his going down."

Psalm 104

Chapter X. The Impact of Light

Throughout most of history man timed his actions by natural light, accepting this axiomatic part of the environment and its daily and seasonal changes, without imagining that it might play a more than external role in behavior and physiology. However, people did notice that plants responded differentially to seasonal change, and even to the daily alternation of light and darkness.

In 19th Century Europe, formal gardens were sometimes planted to form a clockface, with flowers in each bed blossoming at a different hour. On a sunny day one could tell the time to within a half hour by glancing at the garden. The famous Swedish naturalist, Carolus Linnaeus, was the first to notice that various flowers opened at different hours and to use them in a flower clock. Anyone who has tended house plants or a garden realizes that not all plants thrive where it is constantly light. Some flowers can be seen only at night, while some plants flower in summer, others in winter, guided by some mysterious internal mechanisms that determine when it will bloom, and whether it will bloom at all. Within each plant there seems to be a kind of time sense related to the alternation of light and dark.

Light is radiant energy. A molecule that absorbs a quantum of light may become excited in a manner that can be transmitted as a resonance, re-radiated as fluorescence, or transformed into heat. The action of light involves color (wave length), intensity, and duration, which may stimulate or inhibit plant cells, thus governing rhythms in a manner known as photoperiodism.

Photoperiodism

The changing light of the seasons affects plant and insect rhythms in a way that is better understood because of the vast work of Dr. Erwin Bünning of Tübingen, Germany. A 24-hour rhythm can be observed in many plants; time-lapse photography reveals the rising and falling of their leaves. It was Dr. Bünning who demonstrated that this solar rhythm did not persist exactly when plants were kept in total darkness. Indeed, he found that he could give plants 10 hours of light and 10 hours of darkness, after which they would show a 20-hour cycle of leaf movement. Light and darkness clearly were important in determining their cycles. In a variety of experiments he found that plants could adapt to many different light-dark cycles. However, when they were left in total darkness they would revert to a circadian rhythm of leaf movement that was almost, but not precisely, a 24-hour cycle. The precise period length—whether 24, 25, 26 hours—varied from plant to plant. Dr. Bünning crossed plants and discovered that the hybrid had a circadian rhythm with an intermediate period. Thus, the rhythm seemed to be inherited.

Nonetheless, when he raised seedlings in darkness the young plants exhibited no rhythmicity. A single exposure to light was sufficient to instigate the circadian rhythm of leaf movement. Dr. Bünning performed countless tests of the stability of this rhythm and found, that within limits, it was not altered by temperature variation, or by many kinds of poisons. After some 30 years of research, Bünning concluded that the plant must undergo periodic structural changes in its protoplasm, related to processes of cell division. Plant rhythms and their responses to light suggested that they inherit a kind of time-map of their environment—a time-map giving them flexibility, yet preparing them for oncoming seasonal changes in temperature and light.

One manner in which plants may anticipate seasonal change is by responding to ratios of light and darkness as the days grow longer or shorter. This "timing mechanism" has been illuminated by the classic studies of Dr. Karl Hamner and his associates at the University of California, in Los Angeles. Using the Biloxi soybean, they first demonstrated that this plant would not flower if given too much light, such as 18 hours a day. However, if returned to a short day it would soon bud. Thereafter, switching to long days would enhance budding, but the number of buds was directly proportional to the number of short days it had enjoyed. There seemed to be a critical period in the plant's circadian cycle when light was inhibitory and could prevent flowering.

On various light-dark schedules it became clear that the plants required very small amounts of light every 24 hours. However, light had to be administered on a 24-hour cycle or multiples thereof. The plants seemed to possess a rhythm of light sensitivity. Indeed, it was possible to stop the plant from budding by interrupting the darkness with light at critical periods. Light in one 12-hour period inhibited flowering, yet during the next 12 hours would stimulate flowering. The plant seemed to be "counting" by an inherent rhythm of light receptivity, perhaps within its daily cycle of enzyme manufacture, metabolism, and photosynthesis.

Time counters in plants and insects enable them to anticipate the seasons, initiating intervals of dormancy when reduced metabolism would offer resistance to cold, heat, or drought, and preparations for a cycle of growth, followed by flowering and breeding on a "schedule" that affects the entire cycle of plant and animal ecology.

In more complex fashion, birds and mammals also seem to exhibit light-sensitive cycles. More than heat, food, magnetic field changes, and other geophysical changes, light appears to be a synchronizer of animal activity and physiology. Without internal timing many species of bird and fish would not show simultaneous reproductive development that permits them to breed, or to gather at spawning grounds in huge numbers. Migratory birds react to the sequence of seasonal light change by increasing fat deposits and becoming nocturnal. Then they migrate and breed and once more go into moulting and inactivity to accumulate fat for the return migration. It is thought that seals, whales, and other migrating mammals may show a similar sequence.

Several aspects of light-responsiveness have been explored by biologists, and some of these are relevant to seasonal mating. Activity and physiological cycles are oriented around the daily alternation of light and darkness. The extent to which light, entrains, or guides these rhythms, has been indicated by the work of Dr. Jürgen Aschoff and his associates.

Finches are diurnal, and if kept in constant dim light with only 15 minutes of bright light a day, their activity cycle would be influenced by the timing of the bright light. When given it early in their activity period the birds would grow active earlier, accelerating their cycle. If bright light came in the afternoon they delayed their most intense activity. In the course of many experiments, Dr. Aschoff found that light accelerated the cycles of diurnal creatures and delayed the cycles of nocturnal creatures. The intensity, wave length, and phase of light in the 24-hour cycle all influenced the rhythm. This formulation came to be known as "Aschoff's Rule." In general, nocturnal animals do delay activity in constant light while diurnal animals accelerate. In exceedingly intense light, however, human beings and some mammals have shown delayed activity, suggesting that there are exceptions to the tendency. Generally, however, as the days lengthen in spring and the light intensity increases, nocturnal animals increasingly delay their activity period, while the diurnal animals awaken progressively earlier. Changing ratios of light and dark, along with changing wave lengths and graded temperature changes, give cues that prepare living beings for the coming season in the latitude in which they live. Dr. Aschoff has called these external influences "*Zeitgebers,*" or time-givers. (Dr. Halberg prefers the term, "synchronizer," and Dr. Pittendrigh uses the phrase "entraining agent.")

Light and Adrenal Hormones in Man

After many studies of the effects of light-dark cycles on rodents, Dr. Franz Halberg speculated that the onset of dark (in rodents) or light (in diurnal animals) might trigger the circadian increase of adrenal hormones that normally precedes activity. In this way, perhaps, levels of blood hormones would rise in anticipation of awakening and action, rather than rising as a response.

Now there is some evidence that man's circadian rhythm of adrenal corticosteroids in blood may be shifted by manipulations of light and darkness. Drs. David N. Orth and Donald P. Island of Vanderbilt University School of Medicine in Nashville, Tennessee, studied five normal subjects in a light-proof suite of a metabolic ward, where each person slept the same 8-hour period in each 24 hours. At first they adapted for 10-14 days to sleeping in well-lighted rooms and being awake in darkness. Hourly blood samplings from an indwelling catheter indicated that the usual peak of plasma 17-OHCS was delayed, and occurred after waking, with a second peak around the onset of light.

In another schedule, the subjects adapted for 14 days to a prolongation of darkness for 4 hours after awakening. Now blood samples indicated a 4-hour delay in the peak phase of the 17-OHCS rhythm, which coincided with the onset of light. When the same individuals lived in 23 hours of darkness with 1 hour of light, the usual 17-OHCS increase did not occur as it normally does in late sleep, but only after awakening. The peak was of lower amplitude than in controls, and there was a second peak coinciding with the hour of illumination. Thus, the rhythm appeared to be phase shifted by a change of light and dark schedule, and even seemed to be dissociated from the cycle of sleep and waking. Other time-series studies of plasma 17-OHCS suggest that the hormones rise and drop in the blood in episodic, fountain-like bursts that may bias samples taken at longer than 20-minute intervals, since researchers would not know what portion of the fountain burst was caught in their measures. If this observation breeds caution in interpreting hormone levels in the blood, the weight of evidence suggests an effect upon the endocrine system. People who are blind or who have severe cataracts have, indeed, shown abnormal endocrine rhythms, as Drs. Orth and Island have found. Dr. F. Hollwich of the Universität-Augklinik, in Münster, Germany, has made extensive studies of blood cell rhythms (eosinophils and thrombocytes) and of adrenal hormone metabolites in the urine of blind patients and experimental animals in constant darkness.

In 50 people blind with cataracts, he found a reduction of the amplitude of the adrenal and blood cell rhythms and in the basic activity in the regulatory system involving the hypothalamus, pituitary, and adrenal cortex. After a successful cataract operation restored visual acuity, the amplitude of the adrenal rhythm and basic activity returned to normal. Quite a few metabolic disorders may attend the loss of sight.

People lacking vision from early youth have shown disorder in water balance (nycturia) and glucose balance. Often they have had lower than normal insulin tolerance and lacked the usual circadian rhythm of eosinophils in the blood. Dr. Hollwich demonstrated the physiological importance of light in a study of 250 nearly blind people, using repeated determinations of water balance, electrolyte levels, carbohydrate metabolites, fat, protein, and adrenal cortical hormones. A derangement of metabolism was noticeable before surgery, but metabolism returned to normal after successful cataract operations.

The impact of light upon hemoglobin formation, thyroid activity, and the liver's ability to detoxify foreign substances has been studied by Dr. Hollwich and his associates in experiments with mice, birds, and frogs. Since the visual perception of light and darkness appears to have a great deal to do with synchronizing the circadian rhythm of circulating adrenal cortical hormones, Dr. Hollwich feels that light may play a role in synchronizing the hypothalamic-pituitary–adrenal mechanism that underlies the hormone rhythm.

The far-reaching effects of light and darkness upon the neuroendocrine system may begin to explain how birds and other animals know when to migrate, and why man may really "feel" different as the seasons change.

Seasons for Mating

In the early 1920's, Dr. Vernon Rowan captured migrating juncos in Saskatchewan, Canada, as they headed south. He held them in outdoor aviaries in the zero cold, but at sunset he turned on a few electric light bulbs for a few minutes longer each day. By mid-December the birds were singing the mating calls of their species, and their testes had fully developed. When released in mid-winter in Canada the birds headed north instead of south. Dr. J. Benoit, another pioneer in exploring the impact of light upon sexual reproduction, removed the eyes of immature drakes. These eyeless ducks showed no sexual maturation until a light was shined directly into the hypothalamus.

A succession of subsequent studies indicated that visual light had to be primarily in orange-red wavelengths to stimulate gonadal growth, but all wavelengths stimulated growth when the light was shined directly into the hypothalamus.

Inner hormonal states, mediated by the environment, appear to help guide the migratory directions of birds. Dr. Stephen T. Emlen of Cornell University in Ithaca, New York, captured indigo buntings in the autumn and gave some a laboratory environment corresponding to the natural light changes of the season. Others were put through an accelerated long daylight, which accelerated their cycle of fat deposits, and pre- and post-nuptial moulting. By May the control birds were in spring physiological condition, while the accelerated creatures were already in autumn condition. Now they were tested in an artificial planetarium spring sky. The birds in spring condition flew northward while those in autumnal condition headed south.

In trying to ascertain the physiological route by which light might influence mating, Dr. Robert Lisk of Princeton University studied finches whose eyes were left intact, while an implanted optical fiber shined light directly into the hypothalamus. The birds lived a short day as if it were winter, with only six hours of room illumination, but then received six additional hours of light by fiber directly to the brain. In 20 days the gonads had matured and the birds were ready for mating.

As Dr. Joseph Meites of Michigan State University, once said; "In spring a young squirrel's fancy turns because the days are getting longer, and exposure to longer light periods sets off a chain-reaction involving the brain and pituitary gland, resulting in releases of hormones that affect sex hormone levels and in turn cause the sex glands to enlarge and produce their sex hormones."

Meites and his colleagues demonstrated in rats that constant light increased the hypothalamic production of FSHRF (follicle stimulating hormone releasing factor). This in turn stimulated an increased secretion of FSH by the pituitary. FSH regulates follicle growth and estrogen secretion by the ovaries and also regulates the male's production of sperm. The chain of events from light to gonadal development was becoming clearer.

Dr. Lisk and his associates planted optic fibers into the pituitaries of some blinded rats, and the hypothalamus of others. When light was applied in cycles estrus remained fairly normal, but constant light resulted in constant estrus. This was another one of many examples indicating that the cyclicity of light may be important to normal functioning.

For rodents and for man, the alternation of light and darkness appears to trigger activity within the nervous system that in turn helps to regulate physiological cycles. Are there light sensitive structures in the brain that would respond to nervous information by biochemical responses influencing the reproductive system? The pineal may be one organ that gives biochemical information to the brain and endocrine system about the alternation of light and darkness. Researchers on this gland suggest it may play a mediating role in a relationship between light and sexual function.

The Pineal Gland

The gland is shaped like a tiny pine-cone situated deep in the middle of the brain between the two hemispheres. Many people think of the pineal body as a vestigial third eye, mentioned by Indian mystics and Yogic practitioners. This curious gland, which protrudes on the skulls of lizards like a skin covered eye, indeed, responds to light. Much of our current knowledge about the pineal body and the possible role it plays in biological rhythmicity was instigated by the work of Drs. Julius Axelrod and Richard Wurtman ten years ago at the National Institute of Mental Health in Bethesda, Maryland.

In 1916, pineal tissue was found to blanch the skin of a tadpole or frog. It had a powerful effect on the pigment granules in the skin and was thought to be responsible for discolorations in the skin and blemishes. More than 40 years later, this skin-blanching substance was isolated and identified (1959) by Dr. Aaron Lerner and his associates at Yale University. Because of its influence upon melanin, which darkens our pigmentation, it was called melatonin.

Melatonin

Melatonin is a complex molecule related to a chemical family known as indoles. Serotonin is another indole concentrated in the intestines and blood and considered a central substance in many of the brain transactions involved in mood and sleep. Dr. Axelrod was intrigued by the close relationship between the molecules and wondered if an enzyme transformed serotonin into melatonin. As biological catalysts, enzymes are middle-men of our metabolism, continuously breaking down one chemical molecule and transforming it into another, so that food becomes tissue and tissue performs its functions. Enzymes are abundant, but each one is limited in what it can do.

Since melatonin existed in pineal tissue, that seemed a reasonable place to search for a melatonin-building enzyme. Dr. Axelrod and his associate, Dr. H. Weisbach, took some pineal gland from a cow and incubated it with a form of serotonin and a radioactive amino acid containing a methyl group. Soon the radioactive amino acid had relinquished its methyl group to the serotonin, making it radioactive and highly identifiable. However, paper chromatography showed it was now transformed into melatonin. This suggested there had to be an enzyme in the cow pineal that could transform serotonin into melatonin.

Dr. Axelrod and his associates discovered the enzyme which they called hydroxy-indole-O-methyl transferase (HIOMT). Did this enzyme exist throughout the brain and nervous system? A thorough assay of body and brain tissue gave a startling answer. In mammals the enzyme was found *only* in the pineal gland. (In lower forms of life, such as fish and amphibians, this enzyme is also present in the eye and the brain.)

In 1960, pineal biochemistry and physiology began to attract attention. Here was a gland that secreted a strange substance capable of blanching skin and perhaps influencing sex hormones. It was the product of an exclusively pineal enzyme that converted serotonin into melatonin.

In 1962, Dr. Axelrod collaborated with a clinician and endocrinologist, Dr. Richard Wurtman, who had been studying the effect of pineal glands on sexual function in rats. About this time Dr. Virginia Fiske of Wellesley College, observed that animals exposed to light had smaller pineal glands than animals kept in darkness. Biologists found that light stimulated the development of gonads in birds. Was there a relation between pineal function and sexual development in mammals?

Clinically, it had been observed that young boys with pineal tumors came to a precocious puberty. Infant rats deprived of their pineals also have been observed to show precocious sexual mounting and copulating. This suggests that the pineal might in some way regulate sexual development. Perhaps the pineal might slow down sexual development through its secretion of melatonin.

Drs. Charles C. Rust and Roland K. Meyer of the University of Wisconsin, implanted melatonin in weasels and saw an arrest in gonadal development. The animals, who were already beginning to wear their brown spring pelage, developed white winter coats and showed inhibition of gonadal development. The amount of melatonin injected into animals in this and other experiments exceeds the amount known to be secreted by the pineal gland. However, if pineal melatonin does not directly put brakes on reproductive capacity and influence seasonal change of fur color in animals, it may exert an indirect influence.

A recent study by Dr. A. J. Kastin and his associates suggests that the pineal may act as a brake upon the pituitary production of MSH (melanocyte stimulating hormone). MSH activity in the pituitary was heightened by removing the pineals from young rats. MSH is one possible arbiter of the seasonal change in gonadal activity and fur color in animals. Perhaps, pineal melatonin indirectly regulates MSH production.

Light and Darkness: Pineal Size and Melatonin Production

In early studies of the pineal gland, one stumbling block had been the fact that melatonin and serotonin do not remain at constant levels in the gland but fluctuate. As Drs. Axelrod and Wurtman soon found, rats in constant light had small pineals compared with those maintained in constant darkness. They had less HIOMT activity, while rats in darkness showed heightened enzyme activity. HIOMT and melatonin production seemed to be inhibited by light.

Since rats are nocturnal animals, the researchers expected light to enhance HIOMT activity and melatonin in diurnal animals. Indeed, hens kept in constant light did have heavier pineals than hens kept in darkness. Dr. Robert Y. Moore at the University of Chicago, recently demonstrated similar effects in rhesus monkeys. Animals kept in constant light produced more pineal melatonin than animals maintained in darkness.

The pineal sensitivity to light and the variegated photoreceptors seen from species to species are discussed by Drs. Richard Wurtman, Douglas F. Kelly, and Julius Axelrod in their book, *The Pineal.* It is interesting to know that light, beginning at birth, also affects the development of the gland itself. Postnatal transformations of the pineal gland have been shown on human autopsies by Drs. N. A. Kerenyi and K. Sarkar, at the Pathology Institute of the Nova Scotia Department of Public Health in Halifax. Transformation begins at birth and is complete by two to three weeks of age, producing a typical mosaic pattern. This mosaic pattern becomes less prominent by about six months and disappears completely by about nine months. The sequence of tissue development occurs in a newborn infant whether or not the child is born prematurely, which suggests that light may play a role in the development of the gland which, in turn, may later dictate a rate of sexual development. Dr. Kerenyi used rabbits to study the role of light in the development of the pineal gland. If rabbits were kept in darkness from birth to two months, their pineal glands showed a retarded rate of development.

Ordinarily, we do not think of light and darkness as potent in the development and well-being of infants. Such evidence as this may at least stimulate some thought about illumination levels and lighting schedules in nurseries of hospitals and in homes. Moreover, the color of light may be physiologically important. Biologists and botanists have pointed out that plants respond to short-wave lengths differently than to long-wave lengths. Recently, a team of Japanese researchers at Kyoto University has employed an ingenious new technique which again indicates that the neuroendocrine system must be color sensitive.

Drs. T. Oishi and M. Kato in the department of zoology at Kyoto University, painted the heads of Japanese quail with a radioluminous paint. The inorganic paint, excited by the radiation from its tritium binder, either emitted orange or green light. The quail had been kept in continuous light, which had the effect of increasing their testicular size. Ordinarily, if male quail were returned to a routine alternation of light and darkness, testicular activity would decrease, as it did in the untreated control birds. However, the creatures with orange light on their heads maintained their testicular size for two weeks on a "regular" light-dark alternation. Creatures with luminous paint that transmitted green light showed a tremendous reduction in testicular activity and size. Thus, it seemed that perhaps the pineal gland was color sensitive, and at least in this species, responsive to long-wave lengths.

How Light Could Reach the Pineal in Mammals

One might wonder how light and darkness could affect a gland so deeply buried in the center of the brain as the pineal in man.

The optic tract resembles a complex cable system, consisting of a bundle of nerve fibers that run from the eyes to the back of the head, carrying light messages to the visual portions of the brain. A small bundle of fibers branches off from the main cable system and diverges downward to the nerve cell clusters (ganglia) in the upper segment of the neck. This second optic tract is known as the inferior accessory optic tract and its fibers terminate in the superior cervical ganglia.

As Drs. Axelrod and Wurtman demonstrated in a complicated series of experiments, interruption of the inferior accessory optic tract or removal of the superior cervical ganglia obliterated the effects of light and darkness upon pineal biochemistry in rats. They measured the effect by the rate of HIOMT activity. An elaborate process of elimination showed that light was entering through the eyes and then traveling a circuitous route through the inferior accessory optic tract, transmitting its light messages to the cervical ganglia, presumably activating the sympathetic nervous system, which somehow relayed the information to the pineal.

Three hundred years ago the philosopher, Descartes, proposed that images from the eye were carried by a "string" to the pineal. The light stimulated the gland to tilt so that it poured its "humors" down through hollow "tubes." Nerve fibers were then envisaged as small, hollow pipes. This ingenuous notion seems to have been closer to physiological reality than anyone might have supposed. However, the pineal gland does not function the same way in each species.

Birds do not rely on their eyes to give light to their brains to quite the same extent as do mammals. For instance, blinded chicks exposed to continuous light showed twice the melatonin production of unblinded chicks kept in darkness. Somehow the light was penetrating into the brain of the blinded chick. Photocells placed within the brains of birds have suggested that light must enter through the skull.

Most glands are influenced by changes in internal chemistry or by chemical messages in the form of hormones. The pituitary, thyroid, and the adrenal glands are triggered into production by messenger chemicals transmitted through the blood. The pineal appears to be somewhat different, emitting neurohormones after receiving bioelectric messages about the environment through nerve networks. Dr. Wurtman has described the gland as a kind of transducer, an organ that receives neural messages and gives off biochemical, hormonal responses.

About 1963, Drs. Julius Axelrod and Solomon Snyder and others at NIMH began to focus upon the serotonin content of the pineal gland, wondering if its rhythm might be the reciprocal of melatonin, since melatonin is constructed out of serotonin. Serotonin is far more ubiquitous than melatonin. Dr. Wilbur Quay of the University of California, had shown that—in the rat—pineal serotonin levels are several times greater at noon than at midnight. Unlike melatonin, serotonin levels seemed to decrease in the pineals of animals exposed to constant light.

Levels might decrease without abolishing the rhythm. In tracing the serotonin rhythm Drs. Snyder and Axelrod compared pineal tissue from animals who lived in constant light with tissue from animals who had spent the same amount of time in constant darkness. They focussed upon the enzyme from which serotonin is constructed—5-hydroxytryptophan decarboxylase (5-HTPD). Biochemists often measure enzyme activity in order to gauge the rate at which an end product is formed, especially when there is no convenient way of making a direct assay of the end product. In this instance, however, pineal serotonin fluctuations did not parallel enzyme activity.

Drs. Snyder and Axelrod began a round-the-clock study with animals that had been a week on a schedule of 14 hours of daylight and 10 hours of darkness. Every 4 hours they took pineal tissue for analysis. Even with the help of direct and sensitive assay techniques, they could not find any significant changes in 5-HTPD or MAO activity that would differentiate daylight tissue from night tissue. Puzzled, they began to ask what possible steps in serotonin production might be light sensitive and rhythmic.

Serotonin is formed from the common amino acid, tryptophan. This substance is found in most proteins, especially milk and cheese. Could you inject animals with tryptophan or 5-hydroxytryptophan at different times of day and night and perceive differences in the rate at which the amino acids were converted to serotonin? Some rats were injected by day, others by night, but when pineal tissue was assayed for serotonin content, the increases were the same. It didn't matter whether the rats had been injected by light or dark. This experiment yielded no evidence for the influence of light upon serotonin synthesis. In another laboratory, a scientist had meanwhile noticed that the nocturnal fall in the pineal serotonin level was truly precipitous. The level dropped very rapidly once the lights were off. Perhaps, serotonin was stored in nerve ends in a bound, unusable form and was released suddenly at night by some signal from the darkness (or absence of light). Perhaps, if this release occurred at nerve ends, some of the "free" serotonin would be destroyed by monoamine oxidase (MAO). In this case, drugs that blocked the action of MAO might also block the serotonin rhythm.

Drs. Snyder and Axelrod treated animals with MAO inhibitors. This did not alter the daytime levels of serotonin. However, these drugs did prevent its nocturnal drop. The scientists inferred that more serotonin was released by sympathetic nerves at night, and this caused the rhythmic variation.

Pineal serotonin fluctuates, rising and falling in a daily rhythm. Its persistence after blinding suggested that this rhythm is endogenous but synchronized by light and dark. Like melatonin, its levels in the pineal seemed to respond to light entering through the eyes, transmitted as nerve messages along the inferior accessory optic tract. But unlike melatonin, its increases and decreases were not explained by enzyme activity. The story was evidently more complex and interesting than anyone had forecast, for pineal serotonin seemed to follow an endogenous, circadian rhythm that was entrained by light and darkness in a manner not thoroughly understood.

The pineal gland, itself, could not be understood without exploration of several other compounds, all of them important in the nervous system. Along with serotonin and melatonin, the pineal contains histamine, dopamine, and norepinephrine. Norepinephrine (also known as noradrenaline) is a close relative of adrenaline. The pineal is rich in norepinephrine, which is stored in the vesicles of nerve endings, little saclike cell containers. The sympathetic nerves may act through norepinephrine, and if so, this chemical has a major influence on the biochemical activity of the pineal gland. If norepinephrine is needed to relay light messages, then indirectly it helps to manufacture melatonin and to alter the gland's level of serotonin.

In 1966, Drs. Richard Wurtman and Julius Axelrod began to study norepinephrine in the pineals of rats. As they sampled tissue around the clock, they found the pineal content varied by threefold each 24 hours. The norepinephrine content of pineal tissue was highest at the end of the night, the rat's active period, and reached a nadir at the end of the light period, when the rat was usually still resting.

This rise and fall of norepinephrine in the pineal was striking. Like serotonin and the melatonin-making enzyme, HIOMT, this neurotransmitter showed a cycle related to cycles of sympathetic nervous activity. The norepinephrine rhythm of the pineal suggested that entire segments of the sympathetic nervous system must be influenced by the alternation of light and darkness, by time of day.

Norepinephrine seemed to exhibit a diurnal rhythm. In their subsequent experiments, Drs. Wurtman and Axelrod and their associates attempted to find out whether this rhythm was inherent or was directly related to environmental changes, such as light and dark.

Perhaps the levels rose during darkness because it was not being destroyed so rapidly by the enzyme MAO. Perhaps, on the other hand, it was being produced more rapidly. Rats are nocturnal animals and darkness may stimulate certain kinds of nervous activity and biochemical production.

A series of careful and delicate surgeries was necessary to prepare experimental animals for the test. One group of rats was blinded. Another group had similar surgery in such a way that it did not interfere with their vision. Sham operations were used to provide a control group that had undergone the same trauma as the other animals but not the functional defect.

Blinded and sham-operated rats were kept for five days in a normal alternation of daylight and darkness. They were then killed at different hours of the day and the pineals were analyzed. The sham-operated controls showed the highest levels of norepinephrine if they had been killed just before the end of the dark period. They showed very little at the end of the light period. The blinded animals showed no change in levels of norepinephrine, regardless of the time of day their pineals were excised and assayed. It was strong evidence that the 24-hour changes in norepinephrine levels were related to the alternation of light and dark. When intact rats were kept under 12-hourly alternation of light and darkness, the pineal norepinephrine was three times higher at 7 a.m. than at 7 p.m. However, when rats were kept in constant light or darkness no rhythmic change was observed. Dr. Halberg has reported that the rhythm does, indeed, persist but at a very reduced amplitude. Throughout this field of research there have been many instances of this sort, for physiological manipulations often reduce the amplitude of a rhythm to such a degree that it requires more sensitive methods of detection than the instruments in use.

Dr. Wurtman has speculated that the 24-hour rhythm of norepinephrine might reflect a response to light and darkness that could "drive" other pineal rhythms of melatonin and serotonin. Changing levels in the nervous system would, indeed, suggest that response to drugs might change rhythmically. Working with Dr. Donald Reis of Cornell University Medical School, he helped to survey tissue from many regions of the brain. They used cats, keeping them on a schedule of light between 7 a.m. and 7 p.m. and darkness the remaining 12 hours. The cats were studied for a period of six months, during which animals were dissected at different hours of day and night.

There were, indeed, differences in norepinephrine levels in different parts of the nervous system, depending upon the time of day the tissue was taken. For instance, the upper part of the spinal cord (the cervical region), which is continuous with the brain, showed a 24-hour rhythm in norepinephrine content. Certain parts of the hypothalamus also showed a diurnal variation in levels of this catecholamine. The rhythm was, however, the reverse of the norepinephrine rhythm in the pineal gland: in these regions the highest levels were found at the end of the daily *light* period. They were exactly out of phase with the pineal norepinephrine.

The regions of the hypothalamus that exhibited diurnal periodicity in this neurotransmitter happen to be the regions that control body temperature and glandular activity. Body temperature, brain temperature, and the levels of certain adrenal hormones are known to follow a circadian rhythm—rising and falling about every 24 hours. Perhaps, indeed, norepinephrine is needed to drive these rhythms. Since light entrains the rise of pineal norepinephrine and it is so ubiquitously crucial to nervous activity, it may be an intermediary that causes an individual to follow a rhythm of changes each day—in vitality, hunger, alertness, sleepiness, and wakefulness. The world around us is constantly changing, and each day brings an alternation of day and night, so that we may need an internal manager to keep us in synchrony with our environment.

If this is a role of norepinephrine, perhaps one might see it better by determining what the chemical does in the pineal gland.

Throughout the sympathetic nervous system, nerve endings usually contain a single transmitter chemical. The pineal contains many sympathetic nerve ends, but these terminals contain two transmitter substances, serotonin and norepinephrine. Why does the pineal need two possible transmitters in its nerve endings?

It became important to determine whether one or both of the amines (i.e., serotonin and norepinephrine) in pineal sympathetic nerve endings functioned as neurotransmitters bringing environmental information into the pineal sensory cells. To this purpose Dr. Wurtman began a study with Drs. Axelrod and Harvey Shein of McLean Hospital, utilizing a series of organ cultures. Pineal glands were incubated with a radioactive compound (tryptophan) which would be converted into the pineal hormone, melatonin, if norepinephrine was added to the incubation solution. The pineal responded by converting considerably more tryptophan to serotonin. The addition of serotonin produced no such effect. This suggested that norepinephrine is a sympathetic neurotransmitter that helps to control pineal glandular function. The presence of serotonin in pineal sympathetic nerve endings probably results from the very high concentrations of serotonin in the environment of these neurons.

Melatonin injections have been shown to influence the function of the thyroid glands, to alter the size of the adrenal glands, to modify electroencephalographic tracings, and to induce sleep. Melatonin appears to have the ability to do something that serotonin cannot do—permeate cells in both body and brain. It can enter the brain in a manner that serotonin cannot, for serotonin does not pass through the blood-brain barrier, that biological sieve which protects the brain from bombardment of crude molecules. Melatonin seems to be concentrated in the midbrain and hypothalamus, where it may influence sexual function. Norepinephrine, by influencing the production of melatonin, has an indirect influence upon glandular systems and notably on gonadal or ovarian development.

Melatonin may influence the gonads, the thyroid, and adrenals by a primary action on the brain. Drs. Fernando Anton-Tay and Wurtman have suggested this. They injected serotonin intraperitoneally. Brain tissues analyzed 20 to 60 minutes later were found to show the action of melatonin upon serotonin-containing neurons in the brain. This action of melatonin upon serotonin-containing neurons may explain the endocrine effects of melatonin and its effects upon sleep, since brain serotonin is implicated in the induction of sleep.

There is a chain of biochemical influence, one of whose consequences may be the alternation of sleep and waking, another the output of various glands, and sexual function. Norepinephrine seems to play a role in all of these. It influences the synthesis of melatonin, which influences glandular function, and influences the amount of serotonin in the brain.

The pineal gland has been referred to as a kind of biological clock, but a more appropriate image may be one suggested by Drs. Suzanne Gaston and Michael Menaker, at the University of Texas. They have conjectured that it acts as a kind of coupling system perhaps maintaining phase relations within a multi-oscillator system. Ordinarily, when sparrows or other animals are kept in constant light or darkness, they show a circadian rhythm of activity and perching. When Drs. Gaston and Menaker removed the pineals from sparrows, their circadian activity rhythm vanished in a constant environment (although sham-operated controls continued to be active in a rhythmic manner). Pinealectomized birds became arrhythmic. However, they were again rhythmic when returned to an alternation of light and dark. This suggests that the pineal gland is not necessary for the transmission of light-dark information to the brain, nor necessary for entrainment to a light-dark alternation, although it may be a coupler for endogenous oscillators.

Photoperiodism in Man?

Pineal research has suggested that a system of neuroendocrine structures may act as a kind of integrator that may absorb and transmit information about changes of environmental light. This may be one of many mechanisms that play into the seasonal changes shown by human beings—the well-known "spring fever" increase in suicide and seasonal symptoms in illnesses such as ulcers and psychoses. All of these may have detectable endocrine components. Current research on the pineal gland at least suggests that light and dark ratios might influence the brain and its consequent secretion of hormones. Do human brains utilize cyclic alternations of day and night as a kind of unit counter that influences longer rhythms, like the menstrual rhythm?

One suggestion that there might be a ratio between light levels, activity cycle lengths, and the menstrual cycle comes from a study by Dr. Alain Reinberg and his associates in Paris. They recorded physiological functions and behavior in a young midwife who volunteered to live in isolation in a deep cave for over three months (88 days). A week before and a month after this period of isolation, Josy Laures was studied in a hospital routine for a month, but in addition she took her daily temperature for about a year before and a year after her stay underground. During her underground sojourn, she communicated with a ground station by telephone, informing researchers when she intended to go to sleep and ringing them when she awakened. Through telephone communication and her records, it was possible to reconstruct the calendar of activity underground.

Before she went underground, Josy had lived the 24-hour day demanded by society. In the cave, however, cut off from the imperatives of the world, she lived a 24.6-hour day (differing from 24-hours with statistical significance). While living on a 24-hour day, her monthly cycle of rectal temperature was 29.4 days, and she had a 29-day menstrual cycle. However, when she lived a 24.6-hour day underground, in an atmosphere lit only dimly by the miners' lamp she wore, Josy had a 25.7-day menstrual cycle. Once she returned above ground to a 24-hour day and bright light, her menstrual cycle returned to about 29 days over the course of a year.

Dr. Reinberg and his associates became interested in the possible role of light in the menstrual cycle. Reinberg searched the literature on first menstruation in young girls and found a study of 600 girls from northern

Germany, showing that menarche most frequently occurred in winter. A study of young girls in Prague also showed that a girl was more likely to enter menarche in winter. Throughout the literature, it appeared that dim light—or perhaps lack of light—may act as a kind of stimulant to the system of reproductive hormones in women. Perhaps, the intensity, duration, and quality of light could modify the menstrual cycle.

Girls born blind reach menarche earlier than girls with sight. Many children with pineal tumors have been known to reach sexual maturity grotesquely fast, arriving at puberty while still in kindergarten while overactive pineals have been known to delay puberty. Dr. Richard Wurtman has suggested that light may be as potent as any medicine or surgery. Perhaps, through changing light and a variety of photoperiodism, we, too, show seasonal influences in our physiology and behavior. We now know that there are prominent circadian rhythms in sex hormones in men as there are in women.

A Circadian Rhythm of Testosterone

In 1967, a daily testosterone rhythm in men was discovered by a medical group in Paris and by doctors in New York City. Drs. Fernand Dray, Alain Reinberg, and Jacques Sebaoun, working in Paris, had made round-the-clock studies of five volunteers. Shortly after the publication of the Paris group, Dr. Louis A. Southren and his associates at New York Medical College, published their results on two men and two women. Using indwelling catheters the doctors sampled blood every 90 minutes for 24 hours without discomfiting the volunteers. Normal women, they found, had too little variation in testosterone (a male hormone) to detect any circadian rhythm. Men, by contrast, showed a clear sharp rhythm. The peak came around 8-9 a.m., followed by an abrupt decline in testosterone between 9 a.m. and 12 p.m., which appeared to be a true decrease in the production of steriod rather than an alteration in the metabolic clearance rate of the hormone.

Sex hormones are frequently used to treat cancers of the uterus, breast, and prostate. A knowledge of secretion and dispersion rhythms might prove extremely useful in testosterone treatment. Certainly, it will change diagnostic procedures. In most hospitals the male hormone concentrations are measured in blood taken when the patient awakens in the morning. This measurement is presumed to exhibit the regular or average blood content of the hormone, a procedure as misleading as measuring the average depth of water when the tide is high. The early morning concentration of testosterone runs 35-40 percent higher than the minimum around midnight.

It is not known whether the testosterone rhythm recently charted may be influenced by the alternation of light and dark. During the last decade and a half, classic experiments have begun to show that the cycle of sexual reproduction in rats can be practically altered by the use of light and dark at critical moments in an animal's rhythm.

Controlling Estrus by Light

During the late 1940's, Drs. John W. Everett and Charles H. Sawyer, while at Duke University, began a classic series of experiments with rats. The female sexual cycle consists of various hormonal stages. The hormone that triggers the estrous cycle is known as follicle stimulating hormone, or FSH. It is secreted by the anterior lobe of the pituitary gland at the base of the brain. This hormone incites development in the small follicles of the ovary in which the eggs develop. The follicles release estrogen to prepare the uterus for the coming ovum, and develop the environment for the fertilized egg. FSH is sensitive to environmental changes—which may be a source of irregularity in menstrual periods.

FSH is followed by another pituitary hormone, known as luteinizing hormone, LH, which makes possible the final maturation of the ovum and its release through the rupture of the follicle. The ruptured follicle then becomes transformed into the yellow body—corpus luteum—that secretes progesterone to complete preparation of the uterus for the coming ovum, and develop the environment for the implantation of the fertilized egg.

In 1949, Drs. Everett and Sawyer discovered that there was a 24-hour rhythm in the neuroendocrine apparatus that releases LH in rats. Using a barbiturate (Nembutal) to depress the nervous system and prevent ovulation, the investigators found that the sedative-hypnotic drug prevented ovulatory action when delivered at 2 p.m. on the day before estrus. It did not have the same effect after 4 p.m.

Using a number of barbiturate drugs, they showed that ovulation and estrus could be blocked by injections that were delivered at appropriate periods during the sex cycle of the rat. There was a critical period of about two hours on the day before estrus in rats who lived on controlled lighting regimens. The onset of this critical period is related to the beginning of illumination. If, just before the critical period, anesthetics, blocking drugs, or other depressants are applied to the brain, it is possible to interfere with the neural mechanism releasing LH: the signal that would ordinarily result in surges of ovulating hormone. This means that ovulation will not occur as it should on that night. Moreover, if the same blocking is applied on the next day at the critical time, ovulation is again blocked, and so on, ad infinitum, until the ovarian follicle degenerates. Perhaps there is also a critical time of day and month in which the drugs influence the cycle of the adult woman.

Light, too, can be critical. When rats were maintained under constant light they went into constant estrus. More recently it has been shown that rats in darkness will not ovulate if given two hours of light during the critical period before ovulation. By turning light on and off, people are now able to turn on and off the hormonal mechanisms controlling reproduction in the rodent just as they can control flowering in the soybean plant.

Light affects the male animal, as well. When deprived of light the extraordinarily fertile hamster shows gonadal atrophy. In fact, in total darkness the gonads shrink to about a quarter of their usual size and the animal's fertility is significantly reduced.

Birth-time statistics suggest that there might be some underlying rhythm related to light and darkness in man. Dr. Jeri Malak and coworkers of Charles University in Prague, examined 92,000 births and found that 60 percent began at night, 40 percent by day, regardless of season. Drs. Irwin H. Kaiser at Albert Einstein College of Medicine in the Bronx, New York, and Franz Halberg also found that 60 percent of labor began at night with a peak around 3 a.m. Most stillbirths or neonates with fatal complications were born in late afternoon. The Czech team noticed a possible relation between morning or afternoon births and the circadian phase of the onset of menstruation. In a study of some 800 girls, they found that menstrual periods usually began between 4-6 a.m., a few between 8 a.m. and noon, and fewer in afternoon or evening. A biphasic menstruation curve might be related to a biphasic curve of labor and birth.

Dr. Irwin Kaiser conjectured that the gestation period also might be an integral multiple of 24 hours. If mice were raised on a 28-hour day instead of a 24-hour day, would the duration of pregnancy be changed? Several different strains of mice were put on a 28-hour day. They showed a startling response. The estrous cycle lasts four-six days at most in a normal mouse but now the estrous cycle stretched to eight days. Pregnancies lasted about the usual time, but infant mortality due to maternal cannibalism reached surprising proportions. Infant mortality rose from the usual rate of 6 percent to 45 percent. Maternal cannibalism usually means that the animals are disturbed, but in this case it was not clear whether the 28-hour cycle or human handling was to blame.

In general, we tend to think of the 24-hour solar day of society as our basic unit, although the lunar day of 24 hours and 50 minutes may be quite a fundamental in man especially in reproductive cycles.

Lunar rhythms have been especially interesting to biologists. Drs. Frank Brown, Franklin H. Barnwell, and Margaret Webb of Northwestern University, have studied the fiddler crab (Uca) found throughout the world in tidal marshes. This strange creature blends in with his surroundings by darkening his skin every 24 hours at sunrise. The same rhythm of skin darkening would persist in a laboratory tank, in constant darkness, and under conditions of severely low temperature. The activity rhythms of these crabs follow the tides, which are synchronous with a lunar day, while the skin darkening rhythm follows a solar cycle. Within the laboratory, under constant conditions, the peak daily activity of the crabs will slowly shift from the times of tides on their home beach to the local times when the moon is in its upper and lower transit, possibly guided by some cues from atmospheric tides.

It may be no coincidence that the menstrual rhythm averages 28-30 days. The synodical month is 29.5 days. Once in each synodical month the sun and moon rise and set at roughly the same time—at the dark of the moon. These regular movements of sun and moon may be responsible for atmospheric cycles that help to synchronize the mating of many sea creatures, and may have become part of the time structure of mammals. From man's earliest history, moon and sex were related.

In ancient civilizations, spring, moon, and fertility were connected. The ancient Zimbabwe of Southern Rhodesia, painted and incised on the stone faces of caves their belief in the cosmic forces that governed fertility. There is the story of their rituals: the king evidently regulated his public appearances according to the phases of the moon, which also governed when he married, when he conducted his sex life, and the time of his death since ritual regicide was practised by this tribe. A variety of cosmic moon rituals permeated Africa and ancient Egypt and still dictate some tribal rites of circumcision and initiation. Recently, scientists have again looked seriously at the possibility that lunar rhythms may be relevant to an understanding of human ovulation cycles.

Controlling Ovulation by Light

Some time ago, Dr. Edmond Dewan, a physicist at the Cambridge Research Laboratories in Bedford, Massachusetts surveyed the literature describing the biological synchronization of animal rhythms by light. Could an appropriate use of light be timed at a critical period in the human menstrual cycle to regularize the cycle? Could a woman use light to become so regular that she would predict exactly when menstruation would begin, and when she would ovulate?

Dewan reasoned that all biological oscillations can, at least theoretically, be synchronized to an appropriate environmental periodicity. The similar lengths of the menstrual and lunar cycles might be more than coincidence. As ancient and modern observers have reported, moonlight may synchronize sexual cycles among some fish and other marine animals.

Dr. C. Hauenschild of Freiburg, Germany, has demonstrated how moonlight, rather than atmospheric changes of barometric pressure, might influence these lower forms of life. For 6 nights out of every 30, he shined a very dim light on worms. This approximated the lunar cycle, and moreover, the worms showed a sexual ripening in a rhythm in parallel with the light.

If moonlight were important in mammalian reproduction cycles, one might expect to see a clear case among primates at the equator, but primate observations have been sparse and casual. The most compelling evidence for a role of light has come from the experimental work of Drs. John Everett and Charles Sawyer, with rodents, in which the strategic use of light and dark has controlled the estrous cycle.

On the average the period of a woman's menstrual cycle is 29.5 days, but women vary and some have cycle periods ranging from 16-75 days. Moreover, emotional stress and other factors may cause the period to be irregular. Most women are not as regular as clockwork but can predict the onset of their period to within a couple of days. They could learn to become even more accurate, if given a marker, and might thereby learn to feel and anticipate the hours of ovulation.

In 1965, Dr. Dewan studied a 26-year-old woman whose cycle had varied between 23 and 48 days for the previous 16 years. All night she slept in the indirect light of a 100-watt bulb from a lamp placed on the floor at the foot of her bed, so that light reflected from walls and the ceiling would shine on her face. The light was kept on during the 14-15th and 16-17th nights of her cycle, counting the first day of menstruation as the first day of the cycle.

At first the volunteer was unaware that she was a subject in an experiment, and during the regimen her first three menstrual cycles shortened to 29 days. After the second cycle she became alarmed at the dramatic change in her cycles and was informed about the pilot study. However, the fourth cycle was not so regular, perhaps because the light had been turned off a few hours before dawn.

The second subject, a woman of 38, found that she could sense a slight pain at the time of ovulation, the 14th night of the cycle, or the first night of light treatment. We do not know that light enters the central nervous system of man when the eyelids are closed. Excepting for the indirect evidence from studies of blind people before and after cataract operations, there is no evidence that light affects man's brain as it does the brains of animals. However, the inferred effects of light upon the adrenal corticosteriod rhythm and hypothalamic-pituitary activity suggest that the impact upon the endocrine system may be far reaching. Dewan conjectured that light might influence the release of LH and FSH in human beings, and if so, it would be an instrument for making ovulation predictable and controllable.

Although considerable work must be done before the use of light could become a practical method of regularizing cycles and of birth control, the technique offered great promise. A study was begun at the Rock Reproductive Clinic in Boston, where 17 women (between 16 and 35 years) volunteered to be subjects. All but two complained of infertility as well as irregularity. The first regimen (light on the 14th-16th nights of the cycle) allowed a comparison of volunteers on experimental lighting. They were compared with themselves, as well as with a control population for the variance in cycle period. Here the group statistics showed no significant difference because of two subjects whose cycles were 49 days during the light regimen. If these two women are excluded, there is a reduction in variance that points to the strong effect of light. Looked at this way, the results encourage a larger investigation.

A second regimen has been run with most of the same subjects. Originally, it was designed as an attempt to decrease the average cycle length. The light was turned on beginning the 11th or 12th night instead of the 14th. This regimen led to no reduction in variance of cycle period. This could be interpreted as corroboration for the importance of light at a critical time in the cycle. If there were a critical time for the impact of light upon human females, as there is on rat females, then the premature onset of light might have no effect at all.

So far, the most effective regimen has been three nights of light (the 14th-15th and 16th days of the cycle). One Washington newspaper reporter has used the system for three years to maintain a regular 29-day period, after an irregularity causing her cycle to vary up to 45 days. The regularity of her period persisted even when she forgot the light for a couple of months, but if she neglected the light for more than two months she became irregular again.

Attempts to find an autoregulation mechanism for human ovulation must encompass women with cycles of varying lengths. For this reason future studies may prescribe a timing of light that is based upon the person's menstrual history. Light will be begun just before the probable day of the particular woman's ovulation. Furthermore, since Drs. Edmond Dewan and John Rock have found temperature charts to be unreliable, ovulation will be determined by serum LH tests in the future. By using a marker, such as light or temperature, it may be possible for many women to learn the feeling of their own cycles so that they can anticipate and sense their fertile periods.

Light appears to act upon the brain in a complex manner with the potency of a pharmaceutical, yet we do not think of light as medically potent. Most people are quite unselfconscious in their use of light of different intensities and color. It appears that light may influence the rate of maturation of the pineal gland in newborn babies, which might be a consideration in the light-dark cycle of hospital nurseries. Now that we have learned that light, rather than waking, may trigger the circadian rhythm of adrenal hormones found in man's blood, we may also think of light as it influences the neuroendocrine system and thereby influences responses to drugs. Rather than using it haphazardly, we may find out how light acts upon the nervous system and whether the timing of light can be used to regulate biological processes in man as in plants and animals.

Although the biological ramifications of light and dark cycles have not become factors in medical thinking, light would appear to be an important synchronizer in man's activity cycle and time-structure. Thus, we may begin to learn how jet travel, habits, and schedules of work may—through the phase of light and darkness—have a biological impression to make upon the nervous system and endocrine system of man. Well-being in technological societies might ultimately require that we determine whether there are critical phases in our neuroendocrine cycles when light would play a critical role as it does in the flowering of plants.

He has called together legislative bodies
at places unusual, uncomfortable and distant...
for the sole purpose of fatiguing them....

Declaration of Independence

Chapter XI. Work-Rest Schedules and Isolation

In 1775, fundamentals of human timing were already a practical and political issue, although they were not described in such words. As American colonists were called to assemblies at odd hours and distant places, they construed the orders of the British king as an attempt to disrupt their sleep schedules and render them psychologically vulnerable. With this among their grievances, they issued a declaration of independence. Later, in time of war, the military need for 24-hour vigilance and round-the-clock production of defense industries made the scheduling of work and rest central to the outcome of crises. Night work and rotating shifts in factories and transportation inspired safety precautions limiting the extent of the work week. More recently, schedules of work and rest have become an issue in police departments, among air-traffic controllers, and among a population of diplomats and businessmen who travel by jet plane, as well as among the planners and pioneers of space exploration.

At the beginning of the 1970's, there had been many attempts to define the impact of various schedules and phase shifts on man, and to determine what man's "natural" cycle of activity and bodily function would be were he isolated from time cues as in laboratory capsules or deep caves. Like all human researches, these have been hampered by the difficulty in studying people humanely for sufficiently long periods while controlling a multiplicity of factors—sleep, light, exercise and motivation—not to mention the many emotional and physiological factors involved in isolation or laboratory situations. The difficulties were underscored by discrepancies between the results of the laboratory, such as adaptation to inverted sleep schedules, and the adaptations seen after real inversions in east-west jet travel. There is no way of resolving such discrepancies without knowing the hierarchy of geophysical influences to which man may be subject, and without achieving some basic comparability between the hospital situation and experimental flight.

The demands of shift work, air travel, and space exploration, however, have required practical answers and prompted empirical studies long before there is basic insight into the nature of man's time structure. Real control and predictiveness await more fundamental research. Then it may be possible to determine biological time of day in an individual, evaluate the health of his time-order, and his strength or vulnerability to psychological stress. Before we know whether man's activity-rest cycle is profoundly inherent, and what timing system governs our physiology, it is impossible to know whether human beings must adhere to a circadian cycle, or to predict the effects of light and other geophysical periodicities. For these answers and the implicit control they will give us, we must look to basic research with lower organisms.

Shift Work

Nonetheless, applied studies of work-rest schedules, phase shifts, and isolation are yielding a bank of information that should be useful in screening people for rotating shift work, predicting whether or not a person might withstand irregular schedules, and for learning what penalties may accompany such schedules. Useful data on such practical questions were accumulated before there was any agreement about nomenclature, and before some of the basic issues had been clarified by animal studies. In early surveys, for instance, researchers repeatedly uncovered signs of a circadian rhythm in performance, but did not yet know that all manner of behavioral and mental events, vulnerability to toxin, shock, infection, and drugs might be biased by biological-time-of-day. They saw signs that phase shifts and non-circadian schedules might be unhealthy, but they did not yet know that sleep reversal might be followed by a transition period as long as three weeks, in which physiological rhythms were out of phase with the environment and each other. Even now, many people confuse phase shifts with fatigue and blame lack of rest for the problems of hospital or transportation workers, of travelers, and people on night shifts.

Many of the early studies of accidents, ill health, and performance errors among night workers were attributed to night work, itself, while the effects really came from the workers' phase shifts on weekends and time off. Thus, it was not just the effect of working at night, as Soviet scientists saw in nightworkers on the Moscow subway system, but their abnormal physiological rhythms could be blamed on a poorly organized routine of daytime rest. Despite confusion, the early studies have had an enormously beneficial effect upon industry, engendering safety regulations on railroads, and generating an awareness that both human accidents and human ability are not distributed evenly over the 24 hours.

In Sweden, for instance, Dr. B. Bjerner indicated that meter readers in a gas works made most errors on night shift, less during afternoon shifts, and least in morning. In 1949, Dr. R.T. Browne timed telephone operators as they answered incoming calls, and found they were at their slowest around 3-4 a.m. In 1950, Dr. W. Menzel of Hamburg found that industrial accidents occurred most frequently between 10 p.m. and 2 a.m. which other studies narrowed to the hours 2-4 a.m. However, after an analysis of over 3,000 industrial accidents, Drs. P. Andlauer of Lyon, France, and B. Metz of Strasbourg, suggested that accidents were less frequent among night workers than in evening or morning, and seemed to be related to the nature of the job as well as to time. There is a vast literature of studies, uncontrolled for effects of monotony and phase shifts.

People have long suspected that rotating shifts might detrimental to health and efficiency. In 1957, Dr. E. Thiis-Evensen found an unusually high incidence of ulcers among shift workers. Air-traffic controllers usually rotate shifts, as often as every few days in some airports, or in intervals of two weeks in others. Some controllers say this rotation is preferable to a steady shift, in which one man consistently juggles the peak traffic, the landings and take-offs that occur every few seconds in major airports between 7-9 a.m. and 4-8 p.m. Men who cannot endure the schedule are screened out, but ulcers and hypertension are not too uncommon among those who remain.

Recently, the public has heard about the symptoms known as "jet fatigue" that afflict flying crews and some travelers. One TWA pilot describes "jet syndrome" as a gradual progression from headaches to burning or unfocussed eyes, from gastrointestinal problems to appetite loss, shortness of breath, sweating, and, occasionally, nightmares. Stewardesses also have reported GI problems, insomnia, elusive mental problems, and menstrual irregularities—for which medical examination may reveal no source, and the cure may be to stop flying. Individuals vary in their adaptation to east-west flight and rotating shifts, but adaptation may be lower than has been assumed.

Dr. Andlauer, in a recent survey of over 1,000 industrial workers in the Rhone valley, indicated that 45 percent of the workers could not adjust to a seven-day rotation, while 34 percent could not tolerate a two-day rotation. Body temperature rhythms did not adapt to either regimen; and Dr. Andlauer suggested a re-arrangement that offered stable work hours. Even if a pilot or industrial worker adapts with reasonable comfort to phase shifts in the short term, the long-term effects on health may be worth considering.

Phase Shifts and Non-Circadian Schedules

In the mid 1960's, a doctor, speaking to members of the Flying Physicians' Association, remarked that there might be signs of premature aging among pilots on east-west runs. Preliminary studies by Drs. Walter Nelson and Franz Halberg suggest that it is possible to shorten the life span of mature rodents by inverting their light-dark cycle just once a week. Two groups of over 90 mice were observed on a schedule of 12 hours darkness and 12 hours light. At age 58 weeks, however, one group experienced a weekly inversion of the lighting schedule analogous to a jet flight halfway around the world. In many ways, however, it was an easier schedule than that of a transatlantic pilot, who might have to make two phase shifts in a week with less regularity. The control mice lived an average of 94.5 weeks, while the group subjected to weekly phase shifts had an average life span of 88.6 weeks—a decrease of 6 percent in the mean life span.

In attempting to discover how phase shifts differentially affect efficiency and health, both English and American scientists have studied men in the classic rotating watches of the Navy. This remnant from medieval times, when men did hard physical labor and could not work too long at a stretch, split the day into shifts of four hours. On the average a Navy man would put in eight hours of work each "day," and in 72 hours would have worked each of six different watches. In 1948, Dr. Nathaniel Kleitman took temperatures of crewmen on the submarine, USS Dogfish. These men were on a three-watch system, showed 12-hourly temperature peaks, and essentially lived a 12-hour day in which they never slept for eight hours at a stretch. But their unequal temperature peaks indicated that their bodies still preserved a 24-hour day, and they were most alert at midday. By test criteria, the watch system had dubious value in emergency conditions: 44 of 72 men would have required 20 minutes to awaken, while 14 would have needed an hour or more to become alert to an alarm.

Drs. Robert T. Wilkinson and Robert S. Edwards, of the Applied Psychology Research Unit of MRC in Cambridge, England, compared decision, coding, and vigilance performance in men living a stable 12-hour work interval and 12-hours rest, with men working an 8-hour rotating shift. The stable 12-hour work shift was clearly superior, and decision making improved with stability of schedule.

One problem with rotating shifts is that they often cause some sleep loss. Drs. Peter Colquhoun and Robert T. Wilkinson studied a pool of Navy men over a six-week period, while they underwent varying degrees of sleep loss along with daylong vigilance and other testing. The first night's sleep loss produced little impairment, but two nights even of moderate deprivation (five-hours sleep allowed) showed up in impaired performance.

In a number of studies sponsored by the U.S. Air Force, Drs. O.S. Adams, J.T. Ray, W.D. Chiles, and E.A. Alluisi tested Air Force volunteers in a mock-up crew compartment on a 16-hour work day. They alternated four hours duty and two hours rest. Tests of performance have involved measures of attention, response speed, perception, code-lock solving, communication, memory, and decision-making, often delivered and recorded by sophisticated electronic apparatus. Over the course of 15 days, men on four hours work and two hours rest showed a decline in performance, displaying circadian fluctuations of performance that were slightly out of phase with heart rate and body temperature. Out of 11 men only 2 maintained high quality performance for 15 days. A group of Air Force cadets were now motivated to sustain performance on the same schedule, but even with high motivation certain tasks showed a decline. Cadets performed creditably enough on this schedule unless stressed with further sleep loss—when the weakness of the schedule became apparent.

This research team tested other schedules, searching for those that would elicit optimum work, efficiency, and reliability from a crew, while resisting stress—a search prompted, in part, by the needs of manned space exploration. In a subsequent study, the volunteers were confined 30 days on a schedule of four-hours work and four-hours rest, while controls followed the same regimen without confinement. Although circadian fluctuations appeared in the performance of certain tasks, the schedule looked viable, and was the regimen of Gemini IV Astronauts, Edward H. White and James A. McDivitt. However, the astronauts complained of sleep loss and exhaustion.

Drs. Adams and Chiles had found that 20 percent of their volunteers could not adapt to an altered work-rest schedule, and that the initial four days involved some sleep loss. They had therefore advised NASA to use a seven to eight-day screening test in selecting astronauts, and a five-day preadaptation schedule. For various practical reasons neither recommendation had been followed.

Since then, a variety of studies have been conducted in hope of ascertaining how various factors, such as physical exercise, stress, or simply confinement, may affect performance on different schedules, and many of these are described in a monograph by Dr. T.M. Fraser. Some of the problems of work-rest scheduling have been displayed elegantly in animal studies. The persistence of a circadian rhythm in non-circadian schedules has been one recurrent finding. Drs. D.L. Holmquest, Klaus Retienne, and Harry S. Lipscomb of Baylor College of Medicine, in Houston, Texas, compared rats on a 28-hour cycle (14 hours light and 14 dark) with animals given identical proportions of light and darkness in a randomized schedule. On gross physiological comparison of pituitary weight, or tests of adrenal and thyroid, the animals differed little. But the activity patterns of the animals in randomized light showed a pronounced circadian rhythm. Although phase shifts and non-circadian cycles may not cause gross damage, they create an inner physiological transition that may render a creature more vulnerable to all manner of stresses. The many pioneer studies of Dr. Franz Halberg, mentioned in earlier chapters, indicate how such schedules affect rodents.

Drs. Halberg, Erhardt Haus and their associates kept mice on non-circadian schedules and tested their subsequent susceptibility to alcohol. One control group had 12 hours light and 12 hours of dark, a second group had an inversion of this schedule, while a third group lived an eight-hour day with four hours of darkness and four hours of light. Five days after the regimen began, subgroups from each group of these mouse populations were injected with alcohol at short intervals around the clock. Mice on a 24-hour light-dark schedule were most susceptible at their time of alerting—the beginning of darkness. Mice on an inverted cycle died at the same phase and at the same rate. However, animals on an eight-hour day were far more susceptible to alcohol, and their hours of maximum vulnerability did not seem correlated with the hours of light and dark. It almost appeared that they now showed a free-running, constantly shifting peak of vulnerability that made them most vulnerable as a group at all hours.

The nature of this transitional vulnerability, perhaps arising from a desynchronization of internal rhythms, remains one of the interesting questions of work-rest scheduling. No explanations are likely to be found in human studies, but it is worth singling out a few results from Soviet studies that focus upon possible health impairment from work-rest schedules in confinement.

Soviet Studies

Long before their manned space program Soviet researchers had been studying the physiology of work, correlating pulse, galvanic skin response, and other conveniently measured functions with the performance of workers undergoing the normal stresses of actual labor. With this background they judged the dynamics of efficiency by particular criteria. For instance, they found that the effects of physical exercise could be detected in a person's diurnal test performance, in EEG responses, and general psychological tone. Lack of exercise or hypodynamia thus enters their evaluation of the drowsing and fatigue of subjects on abnormal cycles of work and rest, or simply confinement. Many Soviet researchers have expressed openly an experimenter bias that is a little different from the American outlook. They assume that variations of the normal circadian activity-sleep cycle will have a detrimental effect. They consider the cycle of sleep and waking a profoundly conditioned set of reflexes, so deeply habitual that stability requires perpetuation of the rhythm. Moreover, the fact that the rhythm may be conditioned does not imply that it can be altered. This view has been reflected in the fact that Soviet cosmonauts were maintained on their earthly sleep cycles while in orbit. Our astronauts have been exposed to a variety of schedules, including alternations of a four-hour sleep with four hours of duty. This indicates a difference of "bias" on the part of some American scientists who have postulated great flexibility in at least some human beings, and who do not rule out the possibility of living a three-hour, or perhaps a 48-hour "day."

In studying possible work-rest schedules for space, the Soviet researchers have used a wide range of tests of memory, of psychic and mental and muscular efficiency, of balance, pulse pressure, and other cardiovascular functions. In addition, they have looked for signs of pathology, analyzing peripheral blood and urine constituents, and examining clothing for microbes, along with the microbial content of mucus from the nose and mouth, and lysozyme activity of saliva. Oddly enough, men sealed in "hermit chambers" emerged after 15 days with more skin microbes than they showed before entering, suggesting that isolation may cause changes in immunity.

Lysozymes are the enzymes in tears and other body fluids that help to destroy certain bacteria. Subjects on unusual schedules were evaluated by Dr. O.G. Aleksyeva's method of determining saliva lysozyme activity. For instance, Drs. N.I. Andreshyuk, A.A. Vesolova and others, tested men on a normal routine for 15 days, then after a month's rest, on an 18-hour day, and following another month intermission, on a schedule of 6 hours work and 6 hours rest. Around the clock they were filmed, measured for temperature, weight, mental function, fatigue, physical fitness, EKG, and other cardiac functions, along with urine steroids and measures of ascorbic acid in urine and blood constituents. Judging from neural and muscular activity tests, the 18-hour day was most detrimental. This was clear on tests of saliva lysozyme activity as well. Lysozyme activity was high before the experiment. It declined after the first day on all three regimens, but most pronouncedly on the 18-hour day. Enzyme activity resumed its normal level a day after the subjects left the chamber.

Drs. N.N. Gurovskiy, B.A. Dushkov, and F.P. Kosmolinskaya found in their analysis that the 18-hour day left people somnolent, restless, and emotionally tense. Subjects performed physical exercises abnormally fast yet with impaired coordination. One man declined in muscle power and all became inaccurate on tests. Their difficulty was apparent in their immunoreactive and endocrine systems. Rising ascorbic acid excretion, for instance, was interpreted as an index of adrenal reaction that might be characterized as stress.

Individuality in response to phase shifts and isolation suggests that people are differentially vulnerable to temporal disruption, and there may be means of predicting individual responses. Soviet researchers have suggested that when a person shifts from a 24-hour period to a free-running period in isolation, the length of this cycle and his speed of resynchronization may be an index to the stability of his physiological functioning. Dr. F.P. Kosmolinskaya suggests that exposure to phase shifts and isolation would screen out vulnerable people. Under stress, for instance, a person may drift away from the 24-hour cycle of light and social activity, showing a longer or shorter cycle period. This was demonstrated with monkeys by Dr. Charles Stroebel at the Institute of Living in Hartford, Connecticut, and has been observed in many capsule studies.

Individual Differences

Drs. Harry Lipscomb, John Rummel, Lawrence Dietlein, and Carlos Vallbona studied many physiological functions in a group of students who lived for

30 days in an isolation chamber. The urines, EKGs, EEGs, and other functions were automatically collected and processed by computer, while the subjects took vigilance and time-perception tests. During the first week the lights were on for 16 hours and off for 8 hours. By the end of this week there were sizeable individual variations in physiological rhythms. Then they spent a week on a roughly 23-hour cycle, followed by a week in constant light. One student showed all possible combinations of circadian rhythms, a 24-hour period, a longer, and a shorter period, and total disruption. Another showed a potassium excretion cycle adapted to a 23.5-hour day, which lengthened well beyond 24 hours under constant light.

In a study of three men in a simulated lunar mission capsule, Drs. Thomas Frazier, Rummel, and Lipscomb found surprising variability of physiological function during two weeks on a normal schedule of daytime work and nighttime sleep. The men maintained a 24-hour rhythm of work and sleep, but some rhythms showed a longer cycle period. One man began to complain of great somatic discomfort, and the authors concluded that socio-emotional factors might alter circadian rhythmicity even when the environment remained stable. They conjectured that confinement stress might alter circadian physiological and behavioral rhythms and lead to emotional instability.

In a study of shift workers, Dr. W. Menzel suggested that epileptics, anxiety neurotics, ulcer, and diabetic patients, as well as anyone over 50, would make a poor risk. In 1958, Dr. H.J. van Loon took temperatures of men adjusting to night shift after daytime work. He found that one of his subjects never adapted in two weeks. Variation in adjustment has been witnessed in sleep reversal studies and studies of people on night duty.

Dr. K.E. Klein and his associates at the Aerospace Medical Institute, Bad Godesberg, Germany, found great individual differences among pilots tested at a simulator around the clock for several days in Germany, then during and after transport to the United States, and on return. Each man sat at the panel of a simulator and "flew" a supersonic jet on a course with programmed sudden winds, near accidents, and other "real" contingencies. Circadian fluctuations were pronounced: the simulated readjustment of flight that was done in 53.4 seconds in early afternoon took 103.3 seconds at 3 a.m. Individuals varied in the amplitude of their day-night difference: some of them changing as much as 50 percent. After east-west flight the group showed transient decrement of up to 9 percent in performance, but again individuals differed a great deal.

Drs. George Curtis and Max Fogel at the Eastern Pennsylvania Psychiatric Institute in Philadelphia, have sought behavior correlates that might predict how a person would adapt to phase shifts. In a study using six control subjects and six volunteers, they placed two subjects on a sleep-waking schedule resembling the pattern of a newborn baby. Four lived on a less fractured randomized schedule that often permitted only two hours of sleep at a time. Four-hourly urine samples (excepting during sleep) and hourly blood samples (for stretches of 50-60 hours) were analyzed for 17-OHCS, dopamine, creatinine, adrenaline, norepinephrine.

At the end of a week on a "scrambled" sleep schedule, one subject showed moderate disorganization of his blood cortisol levels, yet recovered circadian rhythmicity by the end of the second week. Despite the "random" schedule, adrenal steroid levels seemed to reach a circadian peak in early morning, with secondary peaks after sleep periods. Each of the volunteers felt worst around 4-6 a.m. whether or not he had slept. One showed a phase shift in the iron content of his blood, and there were other traces of a cycle change, yet no signs of disruption on the mental and personality tests. All the volunteers had been screened for the experiment on a long personality questionnaire known as the California Personality Index, which may be a helpful test for displaying strength rather than weaknesses. Curtis found a correlation between dominance, intellectual efficiency, and good parental relations, and the ability to adapt to a random sleep-waking schedule.

It may become imperative to screen people for temporal stability in advanced societies, where astronauts, pilots, hospital personnel, and many others undergo phase shifts and unusual schedules. A meaningful definition of temporal stability will have to wait for more basic researches, when a person's speed of adaptation, his degree of phase stability among internal rhythms, and his tendency to respond to stress by free-running may become useful measures.

Flight Studies

Among the many studies of responses to phase shifts, a few have been conducted under life-like circumstances. A number of foreign airlines and the Federal Aviation Agency have sponsored studies of crews and of volunteers in flights across time zones. The USSR Ministry of Civil Aviation has studied crews on nine-hour flights from Moscow to Khabarovsk, near China, finding changes in EEG and other physiological functions. The Soviet researchers have encouraged scheduling that maintains a crew on a stable work-rest cycle. Indeed, rumor has it that the Soviet pilots who fly to Cuba are kept in a special hotel run on Moscow time. Studies by the Deutsche Versuchsanstatt für Luft und Raumfahrt investigated fatigue among Lufthansa crews, using many physiological measures. Drs. K.E. Klein, H. Bruner, and S. Ruff found steady deterioration in crews during their round-trip between Frankfurt and New York, along with physiological signs of internal desynchronization.

Dr. E. Gerritzen and his associates have studied crews flying from Amsterdam to Anchorage and Tokyo, noting that it took six days before the urinary excretion of electrolytes adapted to local time. A French medical group led by Dr. E. La Fontaine studied urine constituents after the flight from Paris to Anchorage, Alaska, finding that adaptation of urinary electrolytes took five days. Dr. Jean Ghata, who measured crew members on one such study, reports that experienced pilots maintain stability by remaining on their own "home time." If it is noon locally but midnight in Paris, they draw curtains and sleep. Abroad, they try to eat the same foods they would eat at home, minimizing the need for adaptation.

In 1957, Drs. Edmund B. Flink and Richard T. Doe at the University of Minnesota, looked at adrenal hormones and electrolytes in urine after east-west flights. One investigator flew from Minneapolis to Tokyo and on to Seoul, Korea, taking urine samples every three hours. It took nine to eleven days before the urinary steroid and electrolyte rhythms adjusted to Seoul time—an adaptation rate of about an hour a day.

In the early 1960's, the Civil Aeromedical Research Institute of the FAA sponsored a series of studies of human adaptation to jet flights in easterly and westerly directions.

After a pilot study, Drs. George T. Hauty and Thomas Adams examined men on three trips that were similar in flying time; one westerly to Manila; one easterly to Rome; and a control flight south to Chile. Rectal temperatures, palmar water loss, heart rate, respiration rate, reaction time, critical flicker frequency, decision time, and subjective feelings of fatigue were recorded. The volunteers, indeed, rated their subjective fatigue in a way that reflected their physiological adjustment. They occasionally remarked that when they "got on schedule" they would perform better at the psychological test-tasks. Actually, their performance seemed to improve faster than they realized and faster than their physiological functions adapted to local time. Moreover, their adjustment was not the same going east as it was going west. Body temperature took six days to come into phase with activity and sleep in Rome, whereas, adaptation occured by four days in Manila. Since not all physiological functions shift at the same speed, heart rate and temperature rhythms became temporarily dissociated. Only on the eighth day in Rome did heart rate come back into phase; yet it was in phase with local time by the fourth day in Manila. A discrepancy in adjustment to eastward and westward flight was observed in psychological tests, but oddly, these were more pronounced in Manila than Rome. The same asymmetry of adjustment to east-west flight was observed in FAA subjects, flying back and forth to Japan, and in a European scientist who recorded his adjustment on frequent trips to Minneapolis. Why would adjustment to easterly travel differ from westerly? Is there a difference between advancing and delaying the circadian cycle? Experimental results suggest differences, but the evidences of research are themselves in conflict.

In one experiment Dr. Halberg and his coworkers discovered that the direction of phase shift might influence the rate of adjustment. It happened accidentally. They had been planning to invert the lighting schedule of a group of rats by resetting the automatic lighting system in the morning which would have given the animals an 18-hour span of light, as if they were traveling west and lengthening their rest period by six hours. The experimenters were so eager to start the study that they decided to perform the shift that night—giving the animals only six hours of darkness, tantamount to giving man only a six-hour "day." This was an advance in the lighting cycle instead of the usual delay. To the surprise of the research team, the animals' body temperatures remained desynchronized from the laboratory routine for a much longer time than usual.

Advance and delay affect the behavior of birds differently, as Dr. Jürgen Aschoff has shown. Moreover, inverting the day-night cycle of a finch not only shifted his activity rhythm, which took several days of adjustment, but also resulted in a dramatic decrease in the amount of overall activity, a depression of drive that lasted about six days. Using an underground bunker, equipped so that the inhabitants' movements can be recorded from a control room, and otherwise resembling a very comfortable modern efficiency apartment, Dr. Aschoff and his associates have begun to study people under phase shifts.

Drs. Aschoff and Rugter Wever have predicted a more rapid adjustment in eastward travel, shortening the day. Their experiments with birds and people seemed to corroborate their hypothesis. The rodent experiments of Dr. Halberg point to a diametrically opposite conclusion, which appeared to be borne out by an FAA study. A cosinor analysis of data from these transmeridianal flights, summarized by Drs. Alain Reinberg, and Halberg, suggest that the adaptation period for any given function, such as adrenal steroid levels or heart rate, differs from individual to individual. Moreover, adaptation takes longer after easterly than after westerly flights. The human flight studies of Dr. K.E. Klein seem to confirm these observations. At this time the data on the rate of human adaptation related to cycle advance or delay splits about 50-50.

Since rapid travel has become a way of life for a sizeable population, it would seem urgent to learn how to accelerate resynchronization. Drs. F. Gerritzen, of the Netherlands, and T. Strengers and S. Esser attempted to abbreviate the adaptation time on a test flight from Amsterdam to Alaska and Tokyo, by exposing the seven volunteers to an inverse lighting schedule. However, the passengers did not adapt according to expectation, and it was conjectured that their excretion of androsterone, dehydro-isoandrosterone, and to some extent, etiocholanolone, had been increased by the stress of the flight itself.

The British airlines have approved extended rest periods for crews, and a number of companies—among them Timken Roller Bearing Company, Phillips Petroleum, Continental Oil—are beginning to worry that traveling executives will make business blunders far exceeding the cost of an extra day or two for rest. These companies insist upon a 24-hour rest period after coast-to-coast or international flights. An international Civil Aviation Organization, based in Montreal, now has 116 member nations to evolve policies for sane international travel schedules.

Meanwhile, the Syntex Corporation has undertaken research expected to result in a drug that would phase-shift endocrine rhythms. Hormone therapy for travelers may not be a distant prospect although the mechanisms involved in circadian rhythms are not yet understood enough to allow a predictable manipulation of rhythmic functions.

Quite a few human studies were undertaken in the 1960's in the hope of ascertaining whether circadian rhythms are endogenous, what period they show in the absence of synchronizers, and what man's synchronizers are. In point of fact, while most researchers would agree that circadian rhythms persist in isolation, and appear to be inherent, even that cannot be absolute, since experiments conducted on earth are subject to geophysical rhythms that we may not ordinarily perceive.

Cave Studies

In attempting to determine man's spontaneous rhythm of activity and sleep and of physiological function in isolation, scientists have studied men and women in subterranean caverns, deep beneath the surface of the earth, or in isolation compartments, sound-proofed and insulated against the revealing signs of social rhythms. Researchers above ground have studied reactions of volunteers through bioelectric measuring devices, urine samples, and telephoned reports made when the person intended to go to sleep, when he awakened, when he ate, carefully preventing any time-of-day cues. While American and German researchers have preferred to construct highly instrumented isolation chambers, French and British scientists have also taken advantage of natural caves—not exactly comparable environments. The subterranean caverns tend to be chilly, damp, and treacherous, with dripping walls and loose rock. No sound from above nor light penetrate the dark silence. Within caverns, tents have been made habitable with gas heat, light, books, phonograph equipment, and the accoutrements needed for living for several months. Each volunteer could communicate by one-way telephone to a crew on duty above, but was never contacted by the crew.

If caves seem dank and hazardous places for research, volunteers seem to have been challenged by the difficulty of survival, and were happier underground than in the sterile settings of a hospital during base-line studies or in compartments. As one speleologist put it, people in caves feel they are free, while in a capsule or hospital room they become experimental prisoners. High motivation was essential so the volunteers would be reliable about phoning base camp before going to sleep, upon waking and meals, taking pulse, temperature, and tests, and collecting all their urine in numbered bottles, which were set on a tray outside the cavern where they were collected unobtrusively by research crews.

The precedent for studying man's rhythm in caves was set by Dr. Nathaniel Kleitman in 1939 in Mammoth Cave, Kentucky—but modern isolation studies got their impetus from a courageous geologist and spelunker, Michel Siffre. In 1962, he spent two months on a subterranean glacier in the French Alps. On rising and before sleep, he phoned a surface team who tested him and recorded his time-estimation, temperature, and radial pulse. When the data were analyzed by Drs. Franz Halberg, Max Engli, Dewayne Hillman and Alain Reinberg, it was clear that Michel Siffre had maintained a sleep-activity rhythm of 24.5 hours, while his pulse rhythm was predominantly 24.62 hours.

In 1964, when a young woman, Josy Laures, and man, Tony Senni, lived in separate underground caves for three months, their records of body temperatures, activity, and urinary 17-OHCS rhythms were put to spectral analysis by Drs. Halberg, Reinberg, and Ghata. They too, lived a "day" that was longer than 24 hours: Tony had lived a 24.8-hour day, and Josy lived a

24.6-hour day—periods statistically different from 24 hours.

Free-Running

In most cave and compartment studies, volunteers have shown a "free-running" rhythm longer than 24 hours. At the Max-Planck Institute, 85 volunteers lived three weeks or longer in luxurious underground efficiency apartments, isolated from obvious time cues. Only one individual spontaneously showed a short period of 23.5 hours. Most individuals have cycled at around 25.05 hours. As judged by psychological tests, these volunteers were more stable than the people whose cycles attenuated to periods of 25-27 hours.

Drs. Reinberg and Halberg have noted that some volunteers tend toward a lunar day of 24.8 hours in isolation. Since gravitational fields and many parameters of earth's natural electromagnetic fields can penetrate any cavern or capsule, the longer rhythm may also reflect other synchronizers. Or it may reflect an endogenous time structure that is compressed throughout life and conditioned socially to 24 hours. There is some evidence suggesting that infants may wake and sleep in a longer than 24-hour rhythm, and perhaps the napping of older people reflects a tendency to break away from the 24-hour day. Certainly, it is not necessary for a person to be isolated from time cues before he will "free-run."

In 1963, Dr. John Mills of the University of Manchester in Manchester, England, studied a young man, Geoffrey Workman, who wore a wrist watch throughout his three month stay in a cave. Workman resolved to keep a 24-hour routine, but found himself going to bed when tired and rising later each day, free-running despite his wrist watch.

This was the pattern of a physicist on a year's research mission in the high polar desert of the Antarctic, remote from the temperate zone alternation of light and dark. Unaware of his own pattern, the scientist recorded his bedtime and rising time, and gave his calendar to Drs. Jay Shurley of the University of Oklahoma Medical School, and Chester Pierce at Harvard Medical School. He had been going to bed 15-30 minutes later each night and rising correspondingly later until he retired by day and awakened at "night." Most curious was the fact that he free-ran for only 28 days. Every 28 days he abruptly shifted back to his original bedtime and recommenced his drift.

Some people have been known to "free-run" despite all the time cues of a temperate zone city. Mr. D.R. Erskine of Philadelphia has been following "body time" for 15 years, retiring 65-85 minutes later each night, and rising that much later in the morning. Erksine feels that the drawbacks are trivial next to his sense of well-being. Since he is out of phase with society, he has to consult his calendar in advance in order to see a play or meet friends or relatives. Only a few days each month does his sleep schedule preclude social activity.

It should be added that free-running activity rhythms in isolation do not always show the kind of stability that Mr. Erskine reports of himself for 15 years. During six months of solitary life in a cave in the French Alps, Jean-Pierre Mairtet was studied by a team of interested doctors, led by Dr. Jean Colin, of the French Aerospace Medical Laboratory in Bretigny-Sur-Orge. Mairtet showed a stable rectal temperature cycle (24 hours and 44 minutes), while his activity rhythm oscillated sometimes to 30 or 48 hours. Under spectral analysis, however, this oscillation might have evened out into a circadian cycle.

At first a person may seem to show a regular circadian activity rhythm, but the drift against sidereal clock time becomes more and more evident as isolation continues. Many volunteers, in caves and comfortable capsules, have shown an initial tendency to oscillate. This tendency to initial instability, which may last for several weeks, has encouraged longer and longer studies.

In 1966, Dr. John Mills studied David Lafferty in a cave in Cheddar for 127 days. Lafferty was totally erratic at first. His day was sometimes 19 hours, in which he would be active for 10 hours and asleep for 9, and sometimes it was 55 hours, in which he would be awake only 18 hours and would sleep for 35. Lafferty slept for 60 percent of his time beneath the ground, living a totally irregular cycle until his last 2 months, when, like other men in solitude, he also settled down and lived a day of about 25.2 hours. Each individual seems to have his own free-running period, but it cannot be determined from just a few cycles. Indeed, few cycles in nature are absolutely regular. A person may lengthen one day, sleeping less that night and shortening the next activity period. However, in analyzing the overall time pattern, these transient irregularities are smoothed out.

During isolation, as the circadian rhythm extends to a longer period than 24 hours, people seem to slow down and to lose their sense of time, usually underestimating long intervals.

In 1962, Michel Siffre emerged after 63 days in Scarasson Cavern, having kept a calendar based on his awakenings. He thought he had been there 25 days less, for he had imagined that his activity periods were very short when in fact he lived a circadian cycle. In 1966, Jean-Pierre Mairtet telephoned base camp to give estimates of his day. For a while he thought himself living a 48-hour day when his activity rhythm was circadian.

Yet, he would call in to announce a nap and then sleep for ten hours. Other cave dwellers announced naps over six hours. And most people have vastly underestimated the total time they spent underground. Hundreds of emerging volunteers at the Max-Planck Institute have been asked: "What day is it and what time is it?" Only one man ever answered correctly. He happened to be an artist who normally lived a free-running schedule, keeping a calendar in his mind in order to meet social commitments.

Light, Exercise, and Social Activity

Light is an important *Zeitgeber*, or time-giver for human beings, and Dr. Jürgen Aschoff, of the Max-Planck Institute, has conjectured that the intensity of illumination might influence the activity and physiological rhythms of man, as it certainly does influence birds. In a series of studies where light levels were varied with human volunteers living in isolation, the results were suggestive but did not give clearcut answers. One volunteer lived in the luxurious underground apartment for eight days in dim illumination (40 lux) and showed an activity-sleep cycle with a 24.5-hour period, yet his urine excretion followed a cycle with a period of 25.1 hours. When the illumination was made five times brighter, his activity and urine cycles shortened. Intensity of illumination is one factor affecting the cycle period, but there are many unanswered questions about the effect of light on human rhythms. Indeed, there is the possibility that light has a different effect depending upon the phase of internal cycles and these may include cycles with long periods—such as a month or season.

When Josy Laures lived in a cave in dim light for 88 days, her circadian period of activity-and-sleep lengthened to 25 hours, but her menstrual cycle shortened from 29 days to 25 days. A year later, Dr. Reinberg found that her monthly temperature rhythm was still shorter by a day than during the pre-experiment period (28.3 days instead of 29.4).

Some of the troubles of night workers may be due to lack of natural light at the time they are awake. Dr. Mary Lobban of the National Institute for Medical Research in Hampstead, London, has studied miners in Spitsbergen, Norway, taking frequent urine samples from both day-shift and night-shift workers. This town was oriented around mining, and mines operated 24 hours a day, six days a week, providing workers with an environment that could support a stable night schedule. Dr. Lobban found a well-defined rhythm of potassium excretion, in phase with activity rhythms. However, sodium and chloride rhythms were dampened during the dark, winter months. In spring these electrolytes went into phase with the external environment rather than the living schedule of the miners. The wavelengths of natural light have not been employed in the illumination of caves or capsules, so that volunteers, even in bright light, may be deprived of some neuroendocrine input related to the color of natural light.

Monotony and relative physical inactivity have been reported to cause noticeable brain-wave alterations in people in isolation, according to Dr. V.I. Myasnikov. A number of American researchers are now incorporating exercycles and other equipment for strenuous physical activity in their studies, since exercise may affect the cycle period and internal physiological synchrony.

The most powerful synchronizer for man may be social. In 1965, Drs. Reinberg, Marian Apfelbaum, Paul Nillus, and Franz Halberg studied seven young women who lived in two tents in a deep cavern for fifteen days. When records of their sleep-waking cycles were later analyzed by computer in Minnesota, it was clear that the girls lived nearly a lunar day in length—24.7 hours. Halberg, who knew nothing of the physical arrangements of the experiment, noticed that the seven women seemed to divide into two groups. One group showed the peak of its activity cycle at one time, while the other peaked somewhat earlier. He was then told that there were two tents: his observation on the timing of peak of the cycle had defined the two tents, whose occupants, on the average, lived slightly different period lengths, slightly out of phase with each other. One group went to bed earlier and rose slightly earlier than the other. Social factors have been observed in animals, too.

Dr. F. Gwinner, of the Max-Planck Institute, influenced the activity phase of the cycle of finches by playing them tape-recorded bird songs. The caged finches began to orient their activity cycles around the hour of the tapes, showing what can only be called a social influence.

There is suggestive evidence, however, that social factors cannot alter a person's basic cycle period. A person with a short free-running period was observed among three others with long periods by Dr. Ernst Pöppel, while he was at Max-Planck Institute in Erling-Andechs. Four young men lived in adjacent suites underground. During the first ten days they were in bright, constant light (1400 lux), and their second week they were in dim light (100 lux). One man initially emerged as a leader of the group. He led because he rose earlier than the others, preparing food, initiating activity.

In the bright light of the first week, three men showed an average 26.2-hour cycle. In the second week of dim illumination, they showed a 27.2-hour cycle. The dominant man showed urine and temperature rhythms that were shorter. He clearly tried to stay in phase with the others, leading them as his faster rhythms demanded. To stay in phase, he had to sacrifice sleep and soon he was fatigued in the morning. Finally he gave up trying, at which point he became completely desynchronized from the group, rising when they slept, having breakfast as they ate dinner. One student filmed the experiment and ended with a shot of two men eating breakfast, one eating lunch, while another ate dinner. It was interesting to note that the man with the very short cycle seemed more neurotic than the others judging by his answers on the Minnesota Multiphasic Personality Inventory.

This experiment suggested that social situations and motivation might entrain a person's circadian activity cycle only up to a point. Evidently, there are some profound and probably inherent individual differences in timing that may be relevent in selecting groups for compatibility in isolation situations such as polar work or space.

Internal Synchronization in Isolation and After

Not only do people show individual tempo in the period of their free-running activity cycle, they seem to vary as well in the stability of physiological phase relations.

Drs. Reinberg, Halberg, and their associates analyzed the internal phase relations between body temperature, urinary 17-OHCS, and activity in Josy Laures and Tony Senni. However, this was the only cave study in which month-long base-line measures had been taken before and were collected afterward. The conflicting results of other researchers were, therefore, based on more fragmentary data and a more approximate method of analyzing their data. Curiously enough, Josy showed almost no change underground in the phase relations of the three chosen parameters, while Tony Senni had shown a considerable modification of his internal phase relations underground. This was odd, in the light of the follow-up study in collaboration with Dr. Edwin Sidi, of the Foundation of Adolphe Rothschild in Paris. If Tony's rhythms showed more phase modification during isolation, he nonetheless showed a more rapid readjustment than Josy to a 24-hour day. Within broad limits, however, Drs. Reinberg and Halberg saw no gross change in their subject's physiological phase relations while he was underground.

Dr. Mills has suggested, on the contrary, that subtle chemical and behavioral alterations occur as a person deviates from the 24-hour rhythm. For two months underground, Geoffrey Workman's potassium excretion rhythm resembled his rhythms of sleep and waking, while sodium and chloride were often out of phase. Perhaps this accounted for high urine flow during his "sleep" period, causing him to get up out of bed and urinate. Each week his potassium and chloride peaks came later in his cycle until chloride lost any regular rhythm.

Dr. Aschoff also noticed that one volunteer in isolation showed two frequencies in the kidney. Water, sodium chloride, and potassium showed periods of 24.7 hours, although the man had an activity rhythm around 33 hours. Calcium excretion followed the activity cycle. About every third or fourth day, activity and calcium rhythms came into normal phase with the other three functions. Later when this man's diary was compared with the researchers' calendar, it was clear that the days when the volunteer mentioned feeling well were days when his rhythms came into phase, and he was literally "together."

Tony Senni and Josy Laures were studied for the month following their emergence into a 24-hour world, and it was found that the rhythm of 17-OHCS excretion did not adapt for 3 weeks. One factor that may influence the speed of a person's resynchronization time may be the point on his cycle when he begins resynchronizing.

Dr. V.M. Myasnikov has suggested that people adapt more rapidly to schedule change if their sleep period is lengthened, and has found that free-running subjects tend to lengthen their sleep periods. This has been true of several men in isolation in caves, but not consistently. Some have exhibited prolonged periods of activity and reduced sleep.

It is not clear whether or how environmental factors have an impact upon the internal synchronization of physiological rhythms: studies with electric fields suggest that circadian rhythmicity may, indeed, respond to periodicities other than those of light and social activity.

Electric Field Changes

The Max-Planck subterranean apartments lie under a rock-garden in Erling Andechs—two modern efficiency units, each with toilet and kitchen. Wall-to-wall carpeting covers wires that transmit the inhabitants' every movement to the control room. One of the experimental rooms is heavily shielded with metal plating against magnetic and electric fields, reducing the constant magnetic field of earth to 40 db, while alternating fields are even more diminished. People living in the bunkers would not have known the difference. When Drs. Jürgen Aschoff, Rutger Wever, and Ernst Pöppel compared data, however, they found that the 33 people who lived in the unshielded bunker had an average cycle period of 24.84 hours, while the 49 people who lived in the shielded room had an average period of 25.26 hours. This suggested that natural electromagnetic fields might shorten the period of circadian rhythms in man. Moreover, the men exposed to natural fields showed less variation in cycle length than the shielded volunteers. Only in the shielded bunker did individuals show a desynchronization of internal rhythms which became pronounced as their activity rhythms lengthened.

One man who lived 24 days under constant light in the bunker had a period of about 25.1 hours, but in the first 8 days his body temperature showed a shorter cycle, thus becoming desynchronized from his activity period. This transitional state seems to be typical of a person moving from a rhythm enhanced by external synchronizers to a state of perhaps autonomous rhythmicity.

Other individuals tended to exhibit an activity period that was twice the period of rectal temperature, yet both rhythms maintained constant phase relationships, suggesting that the interaction between rhythms remained strong. This stable pattern was only seen in the unshielded room, whereas shifting phase relationships were observed in the shielded room. All in all, there seemed to be pronounced differences between the individuals in the two bunkers. Those in the unshielded room, exposed to natural electromagnetic field changes, showed shorter free-running periods and little internal desynchronization. When electromagnetic fields were shielded out, however, some volunteers showed dissociation; urinary excretion moved out of phase with the activity cycle. Earth's electromagnetic fields seemed to influence man.

Soviet and American scientists have postulated that earth's magnetic field changes make it possible for animals to measure time and that they have a pronounced effect upon the human brain. Experiments with pulsed stimulation have changed brain-wave configurations and also subjective time perception. Acute episodes in mental illness, moreover, have been correlated with times of magnetic disturbance by reputable psychiatric researchers.

Dr. Wever explored the effect of artificial fields by introducing into the shielded compartment a constant electric field about 1,000 times stronger than the corresponding natural field. The volunteer showed an immediate effect, although he could not have been conscious of environmental change. His free-running period shortened and stabilized at 23.5 hours. Ten subjects showed a tendency toward shorter cycle periods when the field was in operation. Each volunteer spent a week in the shielded bunker with no electric field and a week in a constant field. During the week in a constant field, their free-running rhythms contracted to a shorter period. Some volunteers who had become desynchronized before the field was turned on showed a synchronization of internal rhythms at the onset of the field. Some volunteers became desynchronized when the field was stopped.

Other experiments used alternating electric fields of ten cycles per second, a frequency basic in the human nervous system. These alternating fields were pulsed into the bunker on schedules with different periods. Most of them were given for 11.75 hours, then turned off for 11.75 hours. For a few days the subjects seemed to be entrained to the periodicity of the alternating electric fields and showed synchronized sleep-waking cycles, but the entrainment did not last. Thus, a periodically changing 10 cps field may have a slight impact as a synchronizer, but the effect is very weak. Since natural fields would be 1,000 times weaker, these studies do not indicate that natural fields are important as a synchronizer for human activity cycles.

Most isolation studies so far suggest that man has an inherent circadian oscillator system that can be entrained to cycles in a range roughly from 23-28 hours. The rhythm seems to be affected by light intensity, by electric fields, by social, and probably other periodicities in the environment, but to what extent and through what mechanisms nobody can presently say. Considerable individual differences may be due to early conditioning or may be genetic. In any event people have shown adaptation to isolation, and have lived a non-24-hour day with a reasonably stable free-running cycle that persists for long periods. Both the period of the cycle and the degree of internal phase stability seem to vary from individual to individual. Do these researches imply that man is limited to a "day" of around 24 hours?

The answer is not clear. In the 1930's, Dr. Nathaniel Kleitman tried, unsuccessfully, to adapt to a 48-hour day. Dr. R. Meddis of Bedford College at the University of London, and his wife tried a 48-hour day for 2 months. He continued to show a 24-hour fluctuation in performance, while during sleep his rapid-eye-movements began to bunch toward the beginning of sleep as if he were REM-deprived. Neither he nor others who tried the 48-hour cycle above ground managed to adapt, but none seemed to suffer detectable harm from their attempts.

In 1966, Jean-Pierre Mairtet displayed a body temperature rhythm that had a period of 24 hours and 44 minutes, while his activity period was sometimes 48 hours during his lengthy sojourn underground. In 1968, Jacques Chabert and Phillippe Englender went into caves to stay for four months, and were initially conditioned by a 48-hour lighting schedule. From then on, Jacques controlled his own lights, and Phillippe lived in constant light. In addition to many performance tests conducted by phone, and urines collected "daily," they put on electrodes for EEG recording whenever they went to sleep. During about 2 of their 4 months underground, Michel Siffre reported that the subjects felt they were living a 48-hour cycle. However, in 1970 the data had not been analyzed or made public. Both men seemed to have stayed awake for 36 hours, studying, exploring their caverns and doing the chores necessary to survive. They slept 12-14 hours, but they later reported that they were drowsy and on the verge of sleep much of the time, and one man described a sense of being in a waking dream that merged with reality.

There have been many attempts to alter the activity rhythms of adult rats. Drs. M.L.R. Goff and F.W. Finger of the University of Virginia, exposed adult rats to 48-hour and 16-hour light-dark cycles; using an illumination level higher than usual in the laboratory. After six months of abnormal cycle lengths, the animals immediately reverted to a circadian rhythm when left in constant darkness. Drs. F.M. Brown and Finger have suggested that after experience with a lengthened day from birth on, an activity cycle significantly longer than 24 hours would persist for several weeks after the adult animal was placed in continuous darkness. They found this to be true of pups whose mothers had been on a 27-hour day since before mating.

The circadian period has never been successfully altered for long in healthy animals, and there is still no indication that man can adapt to a non-circadian cycle. There have been transitional periods in which volunteers in caves, Tony Senni, Jean-Pierre Mairtet, Jacques Chabert, Phillippe Englender, showed spontaneous (and conditioned) activity cycles of 48 hours. Adaptation is another question. To date, there has been no evidence that they also exhibited a 48-hour temperature and excretory cycle. Thus, the extent of man's temporal flexibility is not defined. One can only say that he seems to function best and feel best when living a cycle in which his internal functions, as gauged by temperature and other convenient measures, remain in certain phase relationships with activity and sleep.

Perhaps, in our long evolution on a turning earth, we have acquired a prominent inherent rhythmicity in the range of the solar-lunar day. Perhaps, indeed, earth surrounds us with hidden synchronizers that encourage a circadian rhythm. How electric fields, light, or conditioning might affect cycles in our nervous systems remains unknown. Nor do we know whether our health and efficiency depend upon functioning within the limits of a circadian cycle, with an alternation of light and dark that approximates nature. What is certain, however, in the light of these different studies is that phase shifts and non-circadian schedules, even isolation, usually have detrimental effects on man's functioning and physiological synchrony.

Monotony and inactivity cause a slowing down of nervous responses, and isolation, itself, may cause a diminution of the body's immune reactions to bacteria, while phase shifts can cause man to suffer internal desynchrony and to be exposed to stresses at times when he is physiologically vulnerable. Some people have low tolerance for temporal disruption and deteriorate rapidly in isolation. These may be the people who show high amplitude circadian rhythms while on non-circadian schedules. Their free-running periods in isolation, speed of resynchronization, and responses to psychological tests may help predict how they can withstand unusual schedules. All of these findings indicate how work-shift schedules and personnel selection may be improved. Attempts to engineer man's activity cycles are likely to precede any real understanding of human time-structure, for researchers already are investigating drugs that might enable man to live a 48-hour day, or adapt immediately to phase shifts. It has been reported that Soviet researchers have explored electrical brain stimulation to accelerate phase shifts, and Dutch scientists have tried to phase-shift jet passengers with inversions of the light schedule. Amidst the attempts to alter time structure to suit the convenience of socio-industrial activity, there remains a possibility that the earth silently guides our rhythms by means we have not begun to study. There do not seem to be closed systems in nature. For this reason some scientists express mild skepticism as man attempts to escape from the cycles of day and night, the periodicities of sun and moon and seasons, periodicities visible or invisible, that keep us in harmony with the earth—of which we are part.

This succession of the seasons provides us with the great cycle we call the year. The next most important biological cycle would appear to be the twenty-eight-day lunar month. Though we moderns reckon our calendars by solar months, that is really only a bookkeeping convenience. Many animal behaviors are dominated by the phases of the moon, by the cycles of its waxing and waning; more and more human behaviors and biological processes are now considered to be affected either directly or indirectly by lunar periods. Of course the smallest and most dramatic cycle which dominates our behavior is the diurnal period, the rotation of the earth, the alternation of night and day.

Yet we don't live our lives in accordance with this knowledge of the importance of cycles in our physiological and psychological well-being.

John Bleibtreu, **The Parable of the Beast**

While the hands of a clock move from one position to another, an infinite number of other changes take place in the cosmos. And wherever that phenomenon which we call awareness exists, there is probably a sense of the passage of time, and a sense of sequence. In other words, experience seems to be inseparably interwoven with time sense which as is true of other 'primary' experience, is indefinable. Yet we all know what it is, and we apparently conceive of duration as a magnitude, for we speak of a long or short time, and readily compare time intervals one with another. Our experience of time differs from that of space in a strange way in that it seems to be of us, and inseparable from our very existence.

Lynn F. Cooper and Milton H. Erickson, **Time Distortion in Hypnosis**

Chapter XII. Summary

CIRCADIAN RHYTHMS

(from: Circa dies, "around a day")

The unit of a day is primary in the physiological and psychological health of man. Presumably circadian rhythms originate in our evolution on this rotating earth, with alternations of light and darkness almost every 24 hours—a cycle that must have influenced life for some 100 million years. Rhythmicity may have been one of the first forces of natural selection since organisms that timed their activity and life processes in accord with the changing light, temperature, humidity, and other factors of the environment would have had an edge in survival. Of course there are many other rhythms in our structure, from cycles lasting only microseconds to the 90-minute cycles seen in sleep, cycles of about a week, monthly, seasonal, and even annual rhythms.

These rhythms of the body seem to be influenced by a sea of time-information in which we live—changes of light and dark, atmospheric pressure, and electro magnetic fields. The influence of these time cycles is reflected in the uneven way the primary events of life are distributed, births and deaths clustering at night and early morning, people with ulcers, allergies, and some psychoses suffering more in spring. There are yearly symptoms such as the brief winter mental illness known as arctic hysteria and yearly peaks in the numbers of suicides and accidents.

Only a few of these rhythms have been studied, mainly by biologists, by ornithologists seeking to understand migration and mating, botanists, and farmers eager to control the fruiting of plants, and more recently, by physiologists, cardiologists, and neurochemists. They are beginning to reveal how man might incorporate environmental changes into his body and behavior through such time counters as the changing ratios of sunlight and darkness.

Within each day many of our physiological functions undergo fluctuations as large as 70 degrees. As animal experiments show, this circadian rhythm is a clock of changing vulnerabilities and strengths. The amphetamine dose, the x-rays, shock, noise, or virus that prove fatal to animals at one point on the circadian cycle merely annoy them or make them sick at another phase. So with man—the lack of oxygen that quickly would render a pilot unconscious at 4 p.m. affects him much less at 4 a.m. Many aspects of human variability in job performance, in response to medical treatment, in learning, and in symptoms of illness are being illuminated by studies of circadian rhythms.

Each day, as we know, a person's body temperature rises and falls about 1½-2 degrees, reaching its peak around afternoon or evening, a time of peak performance on tasks requiring close attention or muscle coordination. The hormones of the adrenal cortex (17-hydroxycorticosteroids), which so influence our nervous apparatus, are dropping at this time. They fall to their nadir at night, and reach their peak levels in blood sometime before a person arises in the morning or shortly after.

At this time, levels of the male hormone, testosterone, are at their daily high point in a man, and he tends to show peak excretion of elements such as sodium and potassium. Hormones, such as insulin, follow a daily rhythm as do blood glucose and levels of amino acids. The body utilizes protein differently, depending upon time of day. Protein eaten at 8 a.m. rapidly raises the amino acid levels in the blood, but the same meal at 8 p.m. does not. Some researchers conjecture that foods may be more efficiently utilized early in the day, indicating that a dieter may be well advised to have a hearty breakfast as his main meal of the day.

There are 24-hour rhythms in liver enzymes, in the biochemicals of the brain and the spinal chord, in the division of cells, and throughout the nervous and endocrine system. These multitudes of cyclic functions are in different phases at any given hour, lending an overall appearance of steadiness.

The characteristic circadian rhythms of activity and sleep, of urine constituents, of temperature, and performance do not disappear when men are removed from clocks and time cues, but have persisted in volunteers who lived for as long as six months in deep caves or in simulated space capsules. Isolation experiments have begun to show how an individual veers away from the precise 24-hour day when he is removed from the constraints of society. Ordinarily, people lengthen their cycle a little, going to bed 15-30 minutes later and rising later. In some conditions physiological cycles may begin to uncouple temporarily. For instance, when a person abruptly shifts to sleeping by day and waking at night, hormone and temperature rhythms will be at odds with waking and sleep. It may take three weeks before internal rhythms adapt to the new schedule. At first a person may feel dreadful, walking around with body temperature and hormones at "sleeping" low levels. During sleep he may have to get up and urinate, since the urine flow rhythm takes time to adjust. Each system adapts at its own pace, but a transient internal desynchronization occurs whenever a person undergoes a phase shift of sleep and waking. After an east or west flight, a person might feel "rested" enough to sightsee and might perform creditably on certain tests, but his temperature, adrenal hormones, and cardiac rhythms probably would be out of phase with activity.

Mysterious symptoms known as "jet fatigue" and the exhaustion of flying businessmen, may some day be explained as consequences of excessive internal desynchronization. The recent experiments of Dr. Franz Halberg have indicated that phase shifts have lingering effects upon the vulnerability of rodents to various toxins or stresses. Moreover, by simply inverting the light-dark cycle of rodents once a week, this laboratory has seen a small but significant reduction in the average life span. Schedules, as we have begun to learn, may be physiologically and psychologically potent.

Perhaps it should be no surprise that time is part of our structure, yet it has taken dramatic research to present a new image of ourselves. We have always been aware of our physical structure, the skeleton, fingers and toes, yet remained unaware that we were organized of time as well. Most people are faintly conscious of some rhythmic interaction between respiration and heartbeat when they run, or of sleep and waking, and of fluctuations of hunger or energy each day. If a person were to see a calendar of his life he would be startled to see that many recurrences of his moods and his periods of illness or strength were predictable. Were he to look inside himself he might note that his body, like a most complex factory, operates on something analogous to an elaborate production schedule. How else could millions of biochemicals arrive at the nerve ends or organs that needed them in the right quantities at precisely the right time? Without a surge of activity amongst liver and kidney enzymes, we would be poisoned rather than nourished by the food we eat, but these enzymes have to be manufactured and ready when the food arrives. Some production schedule must govern this manufacturing and inventory in the body, a schedule that is also synchronized by the timing of our sleep and activity each day. Just as some hormones seem to be released during certain phases of sleep, other biochemical functions are triggered by food intake or exercise. Irregularity within a single production line might have ramifications throughout. For instance, faulty timing in the output of hydrochloric acid into the duodenum might result in damage such as ulcers. Thus, social factors, such as jobs or companionship, which dictate hours of eating and sleeping become vital to the synchronization of the individual.

As the reality of time structure begins to inform our behavior, it may become a saving factor in medicine and preventive health programs, touching every corner of practical life.

Diagnosis

Alterations of normal circadian rhythms provide diagnostic cues to some illness. For example, many emotional illnesses are preceded by insomnia, which can be described as a profound disruption of the sleep-activity cycle. Intensive sleep researches have shown that nightly sleep is itself, rhythmic, consisting of 80-120 minute cycles in which a person moves from one EEG state to another, repeatedly throughout the night. Patients with depression often complain of insomnia; many of them suffer from a broken sleep whose cycles are fragmented and shortened. Antidepressant drugs that cause them to feel better also lengthen the sleep cycle.

Changes in a circadian rhythm typify some illnesses, and many other illnesses show exaggerated or minimal symptoms at particular times of day. For instance, a knowledge of circadian rhythms is of especial interest in diagnosing or treating illnesses involving adrenal hormones.

At the time one awakens in the morning, certain adrenal hormones have risen or are still rising from their lowest to highest daily levels. These glucocorticoids, which drop by night, strongly affect the nervous system, so that a person becomes more sensitive to sounds, tastes, and smells when they are low. Perhaps what we know as daily fatigue may have much to do with the rhythm of these hormones, whose sensory effects were discovered by Dr. Robert Henkin and an NIH team. The perceptual keenness, fatigue, and deficits suffered by victims of Addison's Disease, an insufficiency of adrenal hormones, has been related to the action of the hormones on nerve transmission.

This hormone rhythm is relevant in understanding certain allergies. As Dr. Alain Reinberg has shown, dust-sensitive housewives or hayfever victims will suffer their largest histamine reaction at night, just before their usual bedtime. Histamine is one of the substances that causes nasal congestion during hayfever and colds. At night anti-histamine drugs may have their most palpable effect. During these nighttime hours when cortisol and potassium excretion are low, many asthma patients suffer breathing crises.

At the Childrens' Asthma Research Institute, a hospital in Denver, NIMH grantees have collaborated in a study, finding that children have detectable rhythms of respiratory function. Moreover, early morning or midday doses of hormones may have greater efficacy than treatment later in the day. This is a valuable observation since smaller doses could reduce such side effects as retarded growth.

Rhythm studies may also be helpful in treating a variety of ailments whose symptoms occur at a certain phase of the circadian cycle. We may gain new insight into epilepsy from studies by Dr. Franz Halberg and his associates. They found that a high-pitched loud noise would cause audiogenic seizures in mice at the beginning of their activity cycle. During their inactive period of light the same sound merely caused the animals to walk or run as if insulted; whereas at night they reacted with convulsions that were often fatal. Drs. Halberg and Rudolph Engli also studied six epileptic patients and found they almost always had seizures in early morning, or if they had no attack, abnormal EEGs occurred at that phase of the cycle.

Immunity to infection is also rhythmic. Circadian rhythms in metabolism mean that, for each of the many kinds of stresses a person is exposed to, there are periods of maximum strength or weakness, hours of greater endurance for nervous strain, of greater patience, keener perception, greater muscular strength, and even hours of better immunity. Mice, for instance, have been shown to be most vulnerable to pneumonia infection at the end of the activity period of darkness. Circadian rhythms in immunity or susceptibility could mean that the timing of vaccinations might enhance their effectiveness. Healthy volunteers, immunized against a Venezuelan equine encephalomyelitis, seemed better protected if they received their vaccine at 8 a.m. rather than 8 p.m. Animal and human time studies now suggest the hour at which an infection is incurred may have a significant effect upon the individual's intensity of illness.

Clearly, the diagnosis and treatment of illness cannot overlook biological time of day. A person whose blood pressure registers "normal" in the morning may be hypertensive by afternoon, and a person who is "normal" in the morning may be "diabetic" by afternoon. Since blood levels of gamma globulin are different at one hour than another, immunization is likely to be altered by time. Biological time of day is also important for accurate biopsy and laboratory tests.

Because proteins bind differently according to the time the tissue sample was taken from the body, even the histologist needs time-of-day information in making plates. Cell division reaches its peak in various tissues and organs at various times of the cycle. Thus, in biopsy, too, it is necessary to know the phase of the cycle from which the tissue was taken.

In the diagnosis of many illnesses it is necessary to measure levels of substances in the urine or blood, or to gauge a patient's response to a hormone or hormone inhibitor. Here, methods used to ascertain time of day within the body might help to minimize the number of ambiguous readings and repeated medical tests that are worrisome and expensive to patient, doctor, and laboratory. Instruments for reading biological time of day would help make knowledge of time structure useful to medical people everywhere.

Unfortunately, today's clinician may search for biological time cues in an outpatient, but with measures taken daily at 8 a.m., he will never notice if his patient's hormonal cycle is just 5 minutes longer than 24 hours. The slow drift will be too gradual to interpret, even with daily measurements for 6 months. Measurements taken at monthly intervals might show a change of level that would be mistaken for improvement or drug response.

The clinician who wants to know biological time of day in a patient currently has no choice but to place his patient on a rigid schedule of waking and sleep, and obtain physiological samples at intervals around the clock, preferably for several days in a hospital.

Since no test of performance nor measure of any physiological function can be done accurately without accounting for biological time, one might expect that a project of high priority in medicine would be the development of a "clock" that will register biological time of day in the individual who wears it for 48-72 hours. Miniature sensors and biotelemetry instruments developed for manned-space missions could be applied to medicine. Researchers at the Franklin Institute of Living and U.C.L.A. Brain Research Institute are working on the initial steps, but such projects have lacked backing. The development of such techniques ought to be of interest to hospitals for the timing of surgery, X-ray therapy, and administration of drugs.

Drugs

In testing drug toxicity, timing can make the difference between life and death. During the 1950's, Dr. Franz Halberg and his associates showed that laboratory rodents kept on controlled schedules of lighting and feeding exhibited a pronounced 24-hour rhythm of vulnerability or resistance to drugs and toxins as they were delivered at intervals around the clock. When ouabain, an alkali resembling digitalis, was given to mice at 10 p.m. (in their activity period), only 25 percent survived. The same dose had much less effect at 8 a.m., when 85 percent survived. Alcohol, barbiturates, anesthesias, and many other substances were carefully tested by this laboratory, proving lethal or relatively inoffensive, depending upon the hour at which they were administered.

Although drug companies and the Federal Drug Administration judge the toxicity of a drug by ascertaining the dosage that kills half of a group of experimental animals (the famous LD50), this is meaningless without specifying the biological time of the determination. As Dr. Lawrence Scheving has indicated, for instance, amphetamine will kill 78 percent of the test animals at one hour, while the same dosage kills only 8 percent at another.

This extreme rhythm of toxicity is not too surprising in the light of studies of liver enzyme activity. These potent catalysts that detoxify drugs appear to have characteristic rhythms of maximum and minimum activity. Not all of them reach their peaks at the same time. Recently a muscle relaxant had proven lethal to laboratory animals at a certain time of day. Scientists later found that this coincided with the very time of least activity in the liver enzyme that might break it down.

In the future, clinicians are likely to seek time information about drug toxicity—and also about efficacy. Today most medications are given after meals, before bedtime, or at the convenience of medical staffs and patients. Empirical experiments with pulsed doses of sedatives and antidepressants have suggested that timing can be effective. For example, patients who were agitated by nightly sedatives have responded differently if given fractional doses in afternoon and evening.

At present, biological time in the individual is all but unknown, and the administration of drugs must be done without consideration of timing. Occasionally a response observed in psychiatric hospitals or among outpatients will suggest that the patients could be physiologically out of phase with the doctor's clock time. There have been few tests indicating how biological time of day might change a person's response to drugs. Round-the-clock studies by Dr. Alain Reinberg and Jean Ghata have indicated that responses to penicillin, anti-histamine, and aspirin are biased by the time of day or night they are taken. More impressive, however, is their discovery among patients with Addison's Disease, who take hormones as a replacement for their own deficiency. When naive patients were asked to choose a daily dosage schedule that minimized their fatigue and optimized a sense of well-being, the patients intuitively mimicked the circadian rhythm. They took two-thirds of their dose upon rising in the morning, and a third later in the afternoon or evening.

Studies of several patients on self-timed medication suggest what rhythmometrists have conjectured: that a variety of medications, used to replace some insufficient body chemical, should imitate the normal circadian rhythm rather than a steady state. Many illnesses require hormone therapy. For instance, testosterone is administered to patients with certain forms of cancer, and insulin to diabetics.

If a person travels east or west, or goes on night shift, his change of schedule places his physiology in a state of transition. Dr. Halberg's laboratory has shown that phase shifts change the susceptibility of animals to toxins, sometimes increasing resistance at hours of former vulnerability, sometimes making them more vulnerable at all hours.

Drugs, themselves, may shift the phase of an animal's endocrine rhythm. In a recent study of adrenal rhythms in cats, barbiturates given at a certain phase of the day were found to shift the circadian hormonal rhythm. Such effects may account for some of the side effects of drugs, perhaps including the "barbiturate hangover." Barbiturates, other hypnotics, tranquilizers, and narcotics, noticeably alter the usual rhythm of nightly sleep, most often interfering with the phase of rapid-eye-movement (REM) sleep that is typified by dreaming, and which appears to be integral in emotional learning and memory. Here, the drug action on the rhythm of sleep usually is followed by a reaction during withdrawal in which a person goes into REM sleep unusually early and may have nightmares. For many years, there have been hints that it may be important to explore the mechanisms of drug action by looking at the after-effects on observable biological rhythms.

Dr. Curt P. Richter of Johns Hopkins University, long ago observed that the antibiotic, sulfamerazine, and the hormone, estradiol, used in some anti-fertility pills had a pronounced and unexpected after-effect upon female rats. Ordinarily these creatures maintained a steady 24-hour activity rhythm with a crescendo every four to five days in the estrous cycle. Normal cycles were observed while the animals were receiving the drugs, but after the drugs had been discontinued, the rats showed abnormally long cycles of apathy and weight gain, followed by periods of unusual activity and weight loss. Instead of four-to-five-day cycles, these were 25-day cycles. Richter's studies suggest that it might be valuable to do long-term studies of people after antibiotic treatment and of women who have discontinued the contraceptive pills. The use of these hormones raises some interesting questions from the rhythmometrist's point of view. These drugs effectively "reschedule" the menstrual cycles of women with irregular periods, creating a 28-29 day menstrual cycle in women whose periods may vary from 16-75 days. Will the imposition of a 29-day cycle differently affect a woman with a spontaneous cycle period of 28 days than a woman with a spontaneous cycle of 35 days?

Individuality in response to drugs has been studied along a number of modalities, but never before have we had to consider individuality in time structure. Our language acknowledges that we walk to different beats. There are people who rise rapidly in the morning like larks and some who do not get "wound up" until they have been awake many hours. A profile of time-structure would describe fluctuations in a person's activity and intensity each day and over long periods. Some people are steady, while others are overly erratic and may swing from apathy to intensity, gaining and losing weight. A profile of circadian rhythms would probably describe an individual's spontaneous "free-running period" in isolation, the amplitude of his day-night difference in temperature and other functions, with information about the phase relationships among these measures and their resistance to disruption. Volunteers in isolation studies show considerable individual difference in their activity-sleep cycles—with periods from 24.5 hours to 27 hours. A rare individual has shown a rhythm shorter than 24 hours, and this has been conjectured to be maladaptive. Even in some instances when men have lived together in underground compartments, the differences in their circadian periods were so compelling that each lived his own cycle and one would be eating breakfast while another had lunch and a third had dinner.

People differ in their ability to adjust to phase shifts and in the extent of their discomfort. Dr. George Curtis has seen that volunteers who adjust to rapid phase shifts also give certain responses to a psychological test. Other studies suggest that the amplitude of a person's rhythm—the amount he changes from day to night in temperature, adrenal, cardiac and other measures—may predict his adjustment. Some of this temporal profile may be part of a behavior style that is conditioned, but a person's basic time-structure is probably inherited. People with "flat" low amplitude circadian rhythms may be the ones who suffer least from irregular schedules, or from rotating shifts as in medical internship and commercial flying. They may be most resistant to certain physiological and emotional illnesses.

Stress—Psychosomatic and Emotional Disease

Some people may be ill advised to expose themselves to irregular work-rest schedules for phase shifts may be a kind of stress. The ulcers and other symptoms seen among night-shift workers and professionals on rotating shifts may begin to indicate how stress and a desynchronization of circadian rhythms might lead to physical or emotional illnesses. For 30 years scientists attempted to discover how stress might generate emotional and psychosomatic illness. An NIMH grantee, Dr. Charles F. Stroebel, has demonstrated a possible mechanism through some striking new research. By conditioning rodents at different hours around the clock and later attempting to extinguish the responses, Dr. Stroebel found that fear was learned most rapidly at the time of high adrenal hormone levels in the blood, and unlearning occurred best if the procedure took place at the same time of day as conditioning. Man is far more complex, but if biological time of day is related to the impression of a trauma, then fearsome events in the early morning should have more reverberations than the same events by midday or evening.

A replication of fear conditioning was done with monkeys. After the study was terminated and one animal had been left alone for twenty-eight days, he still showed hyperventilation and blood acidosis at a certain hour. It was the hour at which he had been conditioned a month earlier. As Dr. Neal Miller and his associates, at Rockefeller University have shown, the autonomic nervous system can be conditioned. Life events may "teach" blood vessels to constrict or the pulse to race, just as "stress" may induce rapid, shallow breathing. Now it appears that this physiological learning can be related to biological time of day. Thus, one's nervous system might retain a time-memory of a difficult examination, accident, or family disaster. Symptoms of anxiety might occur at a certain biological time that would not necessarily jibe with clock time, since stress may cause "free-running."

Ordinarily the 24-hour brain temperature rhythm of implanted animals in the laboratory is very stable. Dr. Stroebel found that behavioral or emotional stress could alter this rhythm. Monkeys implanted with many sensors to measure brain and body functions, had learned to obtain rewards for discrimination by pushing a lever. They pushed a second lever whenever bombarded with noise, hot air, or flashing lights, and the nuisance instantly stopped. The animals began to keep one hand on this "security" lever. One day it was recessed so they could see it but not touch it. Although there were no more noxious stimuli the monkeys were frantic.

Within two weeks some of the animals showed a brain-temperature cycle that shifted by about 15-30 minutes a day— "free-running." At the same time they began to develop sores, bleeding stools, asthmatic breathing, and skin ulcers that were not amenable to antibiotics. They acted analogously to people with neuroses and severe psychosomatic illness. Other monkeys developed 48-hour brain temperature rhythms. These animals acted "psychotic" showing stereotyped movements, "hunting" invisible insects, and taking no interest in their environment.

When the recessed lever was restored for a pair of animals from each group, it had no effect upon the "psychotic" monkeys, but the neurotic animals slowly improved and lost their symptoms. As they improved, their appetites returned and so did the 24-hour brain temperature cycle. The other monkeys were treated with a tranquilizer that "flattened" the abnormal brain temperature rhythm, but left the animals too depressed to eat. An antidepressant pulsed at 12-hour intervals resulted in rapid improvement, and was followed by the reappearance of the 24-hour brain temperature rhythm. It is not at all clear what relation holds, if any, between abnormal brain temperature rhythms and abnormal behavior. However, mental and physical health must in some way be influenced by our physiological rhythms and their synchrony with our environment.

Some researchers have conjectured that a stress, such as infection or emotional trauma, which catches a person at a critical moment, may jar rhythms that are normally out of phase into synchrony, causing a beat with periodic symptoms, such as fever, or manic-depressive psychosis. Many people do not suspect that they have a periodic and predictable illness because they do not keep a calendar.

The calendar may be one of our most overlooked yet valuable diagnostic instruments, for many people go through undulations of mood and physiological well-being that have undue social ramifications. Their changes may result from an exaggeration of normal endocrine cycles, but behavioral symptoms that can be predicted receive different treatment than behaviors that must be explained causally. A prime example is the premenstrual syndrome of women. The symptoms may be as severe as brief psychoses causing accidents and crimes. Surveys suggest that many women are more vulnerable to viral and bacterial infection, to allergy and certain physical symptoms in the few days before the first day of the menstrum. An estimated 60 percent of women undergo notable mood changes that can result in domestic quarrels and discord. Men also seem to have a less conspicuous monthly rhythm, judging from studies of adrenal hormone excretion and weight. One survey of industrial workers revealed four-to-six-week mood cycles of which the men were unaware. Some fraction of socially harmful errors, quarrels, eruptions of violence, and accidents might well be prevented if each person were to tabulate his changes on a calendar and thus forecast his own propensities.

Studies of mood fluctuations are now being undertaken in parallel with endocrine studies which may permit us to detect emotional disorders that stem from slight exaggerations or inbalances in normal glandular cycles. Surely our picture of normal behavior must include fluctuation. Most people are not absolutely constant from day to day or month to month. Like the proverbially "unpredictable" woman, who needs only a calendar to become quite "predictable," men, too, may find that their bad moods, their aggressions or lethargies should not be blamed on jobs or families. These moods may reflect cyclical change in the person.

One aspect of our environment that we take for granted is the alternation of light and darkness. Botanists and biologists have shown that light may be stimulating or inhibitory at particular phases of a plant's physiological cycle, and that light affects the central nervous systems of birds and other species, prompting migration and mating. NIMH researchers have participated in studies of the pineal gland, which may receive light messages and influence the neuroendocrine system. Indeed at a critical time in the estrous cycle of rats, scientists can delay estrus for 24 hours by exposing the animal to light. This can be repeated at 24-hour intervals, delaying estrus until the animal shows palpable harm.

Little is known about the effects of light on man. Recent studies suggest that it may trigger a rise in adrenal hormones, and that it affects cell maturation in the pineal glands of newborn infants. Recently, researchers have begun to explore the possibility that light may influence the ovulatory cycle in human females. Although the pilot studies are inconclusive, they suggest that light, timed properly, may help women to regularize their menstrual cycles and predict ovulation.

Light and darkness are, at present, experienced haphazardly. Yet some scientists feel that light, and perhaps particular wavelengths, may have a potent impact upon the central nervous system and upon endocrine functions. Light may affect us differently by night than by day and also in different seasons.

Subjective Time

Despite the fact that we live on a globe dominated by the solar rhythm of day and night, by lunar tidal rhythms and cycles of seasonal change, we exclude these rhythms from much of our social planning. Technologically advanced people tend to ignore cycles of day and night with rapid east-west travel and night life. In fact, we may need to know even more about our time-structure because we are no longer in tune with the slow tempo of the natural environment. We may need to learn about the effects of the electric daylight in which we live, as well as the effects of irregular schedules and phase shifts. Individuals may need a sense of the oscillations within, the rising and falling of energy, undulations of attention, mood, weight, activity, sexuality, and productiveness. Because the clocks and calendars of social activity are designed for economic efficiency or convenience, an individual may have to learn to detect his own cycles, and become aware of scheduling to protect his health.

Civilized people need to cultivate an awareness of their time-structure and develop a self-consious "sense of time." The pace an individual keeps in work and recreation, his subjective sense of the duration of time intervals, and what he imagines he can accomplish within any specific interval—all are aspects of time that may be influenced by culture. Some of our cultivated sense of time may not rest upon any physiological "clocks" as presently understood. A number of recent studies suggest that subjective estimates of time intervals, for instance, do not necessarily correlate with metabolism in way that can be measured. For example, people under the influence of psychedelic drugs have felt as if months of experience were compressed into a few hours. Drs. Lynn Cooper, Milton Erickson, Jean Houston, and R.E.L. Masters, all have shown that learning can be intensely focused by using time distortion and trance states. In their experiments they have been known to enable a student to improve graphic art skills in a few hours—skills that might normally take a semester of course work. People in time-distortion experiments make judgments about their experience with reference to some learned sense of time, usually based on the sidereal clock. Indeed, there may be natural units of time and of attention in the nervous system that could be sensed by an individual, and only future research can tell whether or not they approximate the units of our clocks. Studies of time distortion emphasize how limited is our cultural view of "time sense." These studies may offer us ways of enriching the education of the young by compressing more learning into the early school years. A number of scientists have conjectured that any intelligent youngster could have the knowledge of today's college graduate by the age of ten. Children using time distortion techniques might indeed accelerate their own study.

Children acquire what we call a sense of time and learn to pace themselves in the example of their own culture, but usually this learning is haphazard rather than intentional. From an early age, children experience the cycles of change in their environment; for a while they live almost totally in the cycle of a day. But adults rarely help them to anticipate cyclic changes within themselves. They could be taught to anticipate without fear the changes of falling asleep and of dreams, as well as to learn how to anticipate excretion and hunger. At this basic level, youngsters are often taught to ignore rather than attend their inner rhythms, to eat when they are not hungry, to ignore signals of the kidney or bladder that might interfere with convenience, to time themselves by the clock. A child who was able to listen to his inner rhythmicity might acquire a sense of mastery for he could anticipate his hours of efficiency, know when to fall asleep easily, while expecting without chagrin the fluctuations of his moods and abilities. A person trained to observe his own cycles would not expect a machinelike stability of himself, and would not be quick to blame his changes upon the world around him. Such training might allow a person to take his place more comfortably in the ecology of nature.

Not knowing that one has a time structure is like not knowing one has a heart or lungs. In every aspect of our physiology and lives, it becomes clear that we are made of the order we call time. As we look deeper into this dimension of our being, we may find that we, too, are like the plant that flowers if given a little light at the right time every 72 hours. There may be in man a combination lock to his activity and rest, his moods, illnesses, and productiveness. Moreover, by cultivation, he may learn to utilize his subjective sense of time.

REFERENCES

Behavior and Performance

Ader, R. Gastric erosions in the rat: Effects of immobilization at different points in the activity cycle. *Science,* 145 (3630):406-407, 1964.

Ader, R. Early experiences accelerate maturation of the 24-hour adrenocortical rhythm. *Science,* 163:1225-1226, 1969.

Ader, R.; Friedman, B.; and Grota, L.J. "Emotionality" and adrenal cortical function: Effects of strain, and the 24-hour corticosterone rhythm. *Animal Behavior,* 15:37-44, 1967.

Agadzhanyan, N.A., and Rafikov, A.M. Diurnal variations in altitude tolerance and the role of the adrenocortical function. [Sympozium] *Biologicheskiye my i voprosy razrabotki rezhimov truda i otdyka,* 1967. Materialy. Moscow, 1967, 8-9.

Alluisi, E.A., and Chiles, W.D. Sustained performance, work-rest scheduling, and diurnal rhythms in man. *Acta Psychologica,* 27:436-442, 1967.

Alluisi, E.A.; Chiles, W.D.; et al. Human group performance during confinement. AMRL Technical Documentary Report, 63-87 Aerospace Medical Division, Wright-Patterson Air Force Base, November 1963.

Alluisi, E.A.; Chiles, W.D.; and Hall, T.J. Combined effects of sleep loss and demanding work-rest schedules on crew performance. AMRL Technical Documentary Report, 64-63 Aerospace Medical Division, Wright-Patterson Air Force Base, June 1964.

Altukhov, G.V., Belai, V.E.; Egorov, A.D.; and Vasilev, P.V. Federation of American Societies for experimental biology, Washington, D.C. Diurnal rhythm of autonomic functions in cosmic flight. *Izv Akad. Nauk SSSR, Ser. Biol.,* (Moscow), 30(2):182-187, 1965.

Bjerner, B.; Holm, A.; Swenson A. Diurnal variation in mental performance. *British Journal of Industrial Medicine,* 12:103-110, 1955.

Cherepakhin, M.A. Normalization of physiological functions under conditions of hypokinesia caused by a regime of motor activity. [Simpozium] *Biologicheskiye ritmy i voprosy razrabotki rezhimov truda i otdykha,* 1967. Materialy. Moscow, 1967, 69-70.

Davis, W.M. Day-night periodicity in phenobarbital response of mice and the influence of socio-psychological conditions. *Experimentia,* 18:235-236, 1962.

Deutsch, J.A. and Leibowitz, S.F. Amnesia or reversal of forgetting by anti-cholinesterase, depending simply on time of injection. *Science,* 153:1017-1018, August 1966.

Findley, J.D.; Nigler, B.M.; and Brady, J.V. A long-term study of human performance in a continuously programmed experimental environment: *University of Maryland Institute for Behavioral Research, and Walter Reed Army Institute of Research,* November 1963. [A space research laboratory technical report].

Frazier, T.W., Rummel, J.A., and Lipscomb, H.S. Circadian variability in vigilance performance. *Aerospace Medicine* 39(4):383-395, 1968.

Gerd, M.A. Human work capacity at various periods of wakefulness. [Simpozium] *Biologicheskiye ritmy i voprosy razrabotki rezhimov truda i otdykha,* 1967. Materialy. Moscow, 1967, 20-21.

Gramvall, S., and Lundberg, U. Variations in pulse rate at a cycle ergometer test on Swedish Air Force Pilots. *Meddelanden Fran Flygoch Navalmedicinska Naemnden,* 12(1):4-5, 1963.

Hartman, B.O., and Cantrell, G.K. Sustained pilot performance requires more than skill. *Aerospace Medicine,* 8(7):801-803, August 1967.

Henkin, R.I.; Gill, J.R.; and Bartter, F.C. Studies on taste thresholds in normal man and in patients with adrenal cortical insufficiency: The role of adrenal cortical steroids and of serum sodium concentration. *Journal of Clinical Investigations,* 42(5):727-735, 1963.

Henkin, R.I.; McGlone, R.E.; et al. Studies on auditory thresholds in normal man and in patients with adrenal cortical insufficiency: the role of adrenal cortical steroids. *Journal of Clinical Investigations,* 46(3):429-435, 1967.

Henkin, R.I. Presence of corticosterone and cortisol in the central and peripheral nervous system of the cat. *Endocrinology,* 82(5):1058-1061, May 1968.

Henkin, R.I. Auditory detection and perception in normal man and in patients with adrenal cortical insufficiency: effect of adrenal cortical steroids. *Journal of Clinical Investigation,* 47(6):1269-1280, June 1968.

Klein, K.E.; Brüner, H.; and Ruff, S. Investigation regarding stress on flying personnel in long-distance jet flights. *Zeitschrift fur Flugwissenschaften,* 14:109, 1966.

Klein, K.E.; Wegmann, H.M.; and Brüner, H. Circadian rhythm in indices of human performance, physical fitness and stress resistance. *Aerospace Medicine,* 39:512-518, May 1968.

Klein, K.E.; Brüner, H.; Rehme, H.; Stolze, J.; Steinhoff, W.D.; and Wegmann, H.M. Circadian rhythms of pilots' efficiency and effects of multiple time zone travel. *Aerospace Medicine,* 41, 1970.

Miles, G.H. Effects of physiological rhythms on performance. North Star Research and Development Institution. Minneapolis, Minnesota: March 28, 1967. 88 pp. AF 49-638-1604.

Richter, C.P. Psychopathology of periodic behavior in animals and man. *Comparative Psychopathology.* New York: Grune & Stratton, 1967.

Stephens, G.J.; McGaugh, J.L.; and Alpern, H.P. Periodicity and memory in mice. *Psychonomic Science,* 8(5): 201-202, 1967.

Stephens, G.J., and McGaugh, J.L. Retrograde amnesia - effects of periodicity and degree of training. *Communications in Behavioral Biology.* Part A, 1(4):267-275, 1968.

Stephens, G.J., and McGaugh, J.L. Periodicity and memory in mice: A supplementary report. *Communications in Behavioral Biology,* Part A. 2(2):59-63, 1969.

Sterman, M.B. Facilitation of spindle-burst sleep by conditioning of electroencephalographic activity while awake. *Science,* 167:1146-1148, February 1970.

Stroebel, C.F. Behavioral aspects of circadian rhythms. In: Zubin, J., and Hunt, H.F., (eds.), *Comparative Psychopathology.* New York: Grune and Stratton, 1967.

Stroebel, C.F. A biologic rhythm approach to psychiatric treatment. In: *Proceedings of the Seventh Medical Symposium.* New York: Yorktown Heights, IBM Press, 1967, pp. 215-241.

Stroebel, C.F. Biologic rhythm correlates of disturbed behavior in the Rhesus monkey. In: Rohles, F.H. (ed.), *Circadian Rhythms in Nonhuman Primates.* New York, Karger, 1969., pp 91-105.

Stroebel, C.F. (Investigator) and Luce, G. (Writer). The importance of biological clocks in mental health. Mental Health Program Reports-3. National Institute of Mental Health, Chevy Chase, Md. February 1968. PHS Publication 1743. pp 323-351.

Svyadoshch, A.M., and Romen, A.S. The effect of autosuggestion on certain cyclic processes. [Simpozium] *Biologicheskiye ritmy i voprosy razrabotki rezhimov truda i otdykha,* 1967. Materialy. Moscow, 1967, 60-61.

Walsh, J.F., and Misiak, H. Diurnal variation of critical flicker frequency. *Journal of General Psychology,* 75:167-175, 1966.

Wilkinson, R.T. Sleep deprivation: performance tests for partial and selective sleep deprivation. In: *Progress in Clinical Psychology.* New York: Grune and Stratton, 1969.

Wilkinson, R.T. Evoked response and reaction time. *Acta Psychologica,* 27:235-245, 1967.

Wolfe, J.W., and Brown, J.H. Effects of sleep deprivation on the vestibulo-ocular reflex. *Aerospace Medicine.* 39(9):947-949, 1968.

Wyrwicka, W.; and Sterman, M.B. Instrumental conditioning of sensorimotor cortex EEG spindles in the waking cat. *Physiology and Behavior,* 3:703-707, 1968.

Books and Symposia

Ajuriaguerra, J. de, (ed.). *Cycles Biologiques et Psychiatrie,* Paris: Masson and Company, 1968. 422 pp.

Aschoff, J. (ed.) *Circadian Clocks,* Amsterdam, Holland: North Holland Publishing Company, 1965. 479 pp.

Aurelianus, C. *On Acute Diseases and on Chronic Diseases.* (Edited and translated by I.E. Drabkin.) Chicago: University of Chicago Press, 1950.

Bünning, E. *The Physiological Clock.* Revised, Second Edition. New York: Springer-Verlag, 1967.

Cloudsley-Thompson, J.L. *Rhythmic Activity in Animal Physiology and Behavior.* New York: Academic Press, 1961. 236 pp.

Claiborne, R., and Goudsmit, S.A. (eds.). *Time.* New York: Time-Life Books, 1966. 200 pp.

Dalton, K. *The Premenstrual Syndrome,* Springfield, Illinois: Charles C. Thomas, 1964.

Danilevskii, A.S. *Photoperiodism and Seasonal Development of Insects.* Edinburg: Oliver and Boyd, 1965. 283 pp.

Edholm, O.G.; and Bacharach, A.L. (eds.), *The Physiology of Human Survival,* Academic Press, London, England, 1965. 581 pp.

VIII conference internationale de la societe des rythmes biologiques, Hambourg, September 9-11, 1963. *Concours Medical* 40, October 3, 1964.

Folk Jr, G.E. *Introduction to Environmental Physiology,* Lea & Febiger, 1966.

Fomon, S.F., (ed.). Circadian Systems. Report of the 39th Ross Conference on Pediatric Research. Columbus, Ohio: Ross Laboratories, 1961. 93 pp.

Fraser, J.T. (ed.) *The Voices of Time.* New York: George Braziller, 1966, 710 pp.

Fraser, T.M. *The Effects of Confinement as a Factor in Manned Space Flight,* NASA Contractor Report, NASA CR-511, 1966, 176 pp.

Gerathewohl, S.J. *Principles of Bioastronautics.* Englewood Cliffs, New Jersey: Prentice-Hall, 1963.

Halberg, F. Symposium on rhythms. In: *Verhandl. Deut. Ges. Inn. Med. 73 Kongr.,* pp. 886-994, 1116-17. Munchen; Bergmann, 1967.

Halberg, F. Symposium on rhythms. Proceedings 4th Panamerican Symposium on Pharmacology and Therapy, Mexico City. Amsterdam: Excerpta Medica Foundation International Congress Series. No. 185:7-39, 1969.

Hall, C. *The Meaning of Dreamss.* Second edition. New York: McGraw-Hill, 1966.

Harker, J.E. *The Physiology of Diurnal Rhythms.* London: Cambridge University Press, 1964. 114 pp.

Hartmann, E. *The Biology of Dreaming,* Springfield, Illinois: Charles C. Thomas, 1967. 151 pp.

Kales, A. ed. *Sleep: Physiology and Pathology.* Philadelphia, Pennsylvania: Lippencott, 1969.

Kleitman, N. *Sleep and Wakefulness.* Chicago, Illinois: University of Chicago Press, 1963, 552 pp.

Koella, W.P. *Sleep - Its Nature and Physiological Organization.* Springfield, Illinois: Charles C. Thomas, 1967.

Kosmolinskiy, F.P., and Dushkov, B.A. (eds.); or Gurovskii, N.N. (ed.). *Papers on the Psychophysiology of the Labor of Astronauts.* Foreign Translation Division of the Clearinghouse, Department of Commerce, Springfield, Virginia, 22151: AD-684-690. English pages: 182. 1968.

Menzel, W. *Menschliche Tag-Nacht-Rhythmik und Schichtarbeit.* Basel/Stuttgart: Benno Schwabe, 1962. 189 pp.

Ornstein, R.E. *On the experience of Time.* Harmondsworth, England: Penguin Books Ltd., 1969. 115 pp.

Reimann, H.A. *Periodic Diseases.* Oxford, England: Blackwell Scientific Publications, 1963. 189 pp.

Reinberg, A., and Ghata, J. *Biological Rhythms.* New York: Walker and Son, 1965.

Richter, C.P. *Biological Clocks in Medicine and Psychiatry.* Springfield, Illinois: Charles C. Thomas, 1965. 109 pp.

Rocard, Y. *Le signal du Sourcier.* Paris: Dunod, 1962.

Rohles, F. *Circadian rhythms in non-human primates.* Basel/New York: Karger, 1968.

Siffre, M. *Beyond Time,* New York: McGraw-Hill Book Company, 1964. 228 pp.

Sollberger, A. *Biological Rhythm Research.* Amsterdam-London-New York: Elsvier Publishing Company, 1965. 461 pp.

Sweeney, Beatrice M. *Rhythmic Phenomena in Plants.* Experimental Botany Series of Monographs. New York: Academic Press, 1969.

Symposia on Quantitative Biology. Vol. XXV Long Island Biological Association, Biological Laboratory, Cold Spring Harbor, Long Island, New York, 1960. 514 pp.

von Mayersbach. H. (ed.). *The Cellular Aspects of Biorhythms.* New York: Springer-Verlag, 1967. 198 pp.

Wolf, W. (ed.). *Rhythmic Functions in the Living System,* Annals of the New York Academy of Sciences, vol. 98, 1962, 573 pp.

Wurtman, R.J.; Kelly, D.E.; and Axelrod, J. *The Pineal.* New York: Academic Press, 1969.

Brain

Axeldrod, J.; Snyder, S.H.; Heller, A.; and Moore, R.Y. Light induced changes in pineal hydroxyindole-O-methyl-transferase: abolition by lateral hypothalmic lesions. *Science,* 154:898-899, 1966.

Berendes, H.W.; Marte, E.; Ertel, R.J.; McCarthy, J.A.; Anderson, J.A.; and Halberg, F. Circadian physiologic rhythm and lowered blood 5-hydroxytryptamine in human subjects with defective mentality. *Physiologist,* 3:20 1960.

Brown, D.W., and Iverson, D.G. Diurnal variation of intraocular pressure and serum osmolality. *Experimental Eye Research,* 6:179-186, 1967.

Cherniakov Ivanov, D.I.; Malkin, V.B.; Popkov, V.L.; and Popova, E.O. Automatic study of diurnal periodic changes in the human electroencephalogram. [Avtomaticheskii analiz sutochnykh periodicheskikhzmenenii elektroentsefologrammy cheloveka]. *Problemy Kosmicheskoi Biologii,* 4:642-645, 1965.

Engel, R.; Halberg, F.; and Gully, R.J. The diurnal rhythm in EEG discharge and in circulating eosinophils in certain types of epilepsy. *Electroencephalography and Clinical Neurophysiology,* 4:115-116, 1952.

Engel, R.; Halberg, R.; Tichy, F.Y.; and Dow, R. Electrocerebral activity and epileptic attacks at various blood sugar levels (with a case report). *Acta Neurovegetativa,* 9:147-167, 1954.

Frank, G.; Halberg, F.; Harner, R.; Matthew, J.; Johnson, E.; Gravem, H.; and Andrus, V. Circadian periodicity, adrenal corticosteroids, sleep deprivation and the EEG in normal men. *Journal of Psychiatric Research.* 4:73-86, 1966.

Friedman, A.H., and Walker, C.A. Circadian rhythms: rat midbrain and caudate nucleus biogenic amine levels. *Journal of Physiology,* 197:77-86, 1968.

Friedman, A.H. and Walker, C.A. Rat brain amines, blood histamine and glucose levels in relationship to circadian changes in sleep induced by pentobarbital sodium. *Journal of Physiology,* 202:133-147, 1968.

Galicich, J.H.; Halberg, F.; French, L.A.; and Ungar, F. Effect of cerebral ablation on a circadian pituitary adrenocorticotropic rhythm in C. mice. *Endocrinology,* 76:895-901, 1965.

Halberg, F.; Anderson, J.A.; Ertel, R.; and Berendes, A. Circadian rhythm in serum 5-hydroxytryptamine in healthy men and male patients with mental retardation. *International Journal of Neuropsychiatry,* 3:4379-4386, 1967.

Halberg, F.; Bittner, J.J.; and Gully, R.J. Twenty-four-hour periodic susceptibility to audiogenic convulsions in several stocks of mice. *Federation Proceedings,* 14:67-68, 1955.

Halberg, F.; Engel, R.; Halberg, E.; and Gully, R.J. Diurnal variations in amount of electroencephalographic paroxysmal discharge and diurnal eosinophil rhythm of epileptics on days with clinical seizures. *Federation Proceedings,* 11:63, 1952.

Halberg, E.; Halberg, F.; and Bittner, J.J. Daily periodicity of convulsions in man and in mice. Report from the 5th Conference of the Society for the Study of Biologic Rhythms, Stockholm, 1961, p. 97.

Hamberger, A.; Hyden, H.; and Lange, P.W. Enzyme changes in neurons and glia during barbituate sleep. *Science,* 151:1394-1395, March 1966.

Harding, G.F.A., and Jenner, F.A. The electroencephalogram in three cases of periodic psychosis. *Electroencephalography and Clinical Neurophysiology,* 21:59-66, 1966.

Harker, J.E. Internal factors controlling the suboesophageal ganglion neurosecretory cycle in Periplaneta americana. *The Journal of Experimental Biology,* 37(1):164-170, March 1960.

Henkin, R.I. Presence of corticosterone and cortisol in the central and peripheral nervous system of the cat. *Endocrinology,* 82(5):1058-1061, 1968.

Ivanov, D.I.; Malkin, V.B.; Popkof, V.L.; Ye,O.; and Cheryakov, I.N. Automatic Analysis of Diurnal Periodic Changes in Human EEG Rhythms, *Problems in Space Biology,* 4:642-645, Izd. vo, Nauka, Moscow, 1965. Clearinghouse for Federal Scientific and Technical Information, Springfield, Virginia.

Jouvet, M. Insomnia and decrease of cerebral 5-hydroxytryptamine after destruction of the RAPHE system in the cat. *Advances in Pharmacology,* 6B:265-279, 1968.

Jouvet, M. Biogenic Amines and the states of sleep. *Science,* 163:32-41, 1969.

Kahana, L.; Lebovitz, H.; et al. Endocrine manifestations of intracranial extrasellar lesions. *Journal of Clinical Endocrinology and Metabolism,* 22:304-324, March 1962.

Kardzic, V.; and Marsulia, B. Deprivation of paradoxical sleep and brain glycogen. *Journal of Neurochemistry,* 16(1):29-33, 1969.

Krieger, D.T., and Krieger, H.P. Adrenal Function in central nervous system disease. *Endocrines and the Central Nervous System,* 43:400-417, 1966.

Krieger, D.T., and Krieger, H.P. Circadian pattern of plasma 17-hydroxycorticosteroid: alteration by anticholinergic agents. *Science,* 155:1421-1422, March 1967.

Krieger, D.T. and Krieger, H.P. The circadian variation of the plasma 17-OHCS in central nervous system disease. *Journal of Clinical Endocrinology and Metabolism,* 26:939, 1966.

Krieger, D.T., and Krieger, H.P. The effect of short-term administration of CNS acting drugs on the circadian variation of the plasma 17-OHCS in normal subjects. *Neuroendocrine,* 2:232, 1967.

Krieger, D.T., and Rizzo, F. Circadian periodicity of plasma 17-OHCS: mediation by serotonin dependent pathways. *American Journal of Physiology,* 217:1703, 1969.

Krieger, D.T.; Silverberg, A.I.; Rizzo, R.; and Krieger, H.P. Abolition of circadian periodicity of plasma 17-OHCS levels in the cat. *American Journal of Physiology,* 215-915, 1968.

Makarova, L.G. Changes of EEG in a healthy individual with trigger light stimulation. *Bulletin of USSR, Academy of Medical Science, Institute of Neurology, Moscow,* 62:6-11, November 1966.

Meites, J. Releasing factors in hypothalamic control of the anterior pituitary in mammals. (Symposium on Hypothalamic Control of the Anterior Pituitary). American Association for the Advancement of Science, December 1965.

Miasnikov, V.I. Electroencephalographic changes in persons isolated for long periods. *Kosmicheskie Issledovaniia,* 2:154-161, January-February 1964; and in *Cosmic Research,* 2:133-138, January-February 1964.

Mink, W.D.; Best, J.; and Olds, J. Neurons in paradoxical sleep and motivated behavior. *Science,* 158-1335-1337, 1967.

Morgan, M., and Mandell, A.J. Indole ethyl amine N-Methyltransferase in the brain. *Science,* 165:492, August 1969.

Mouret, J.; Bobillier, P.; and Jouvet, M. Insomnia following parachlorphenylalanine in the rat. *European Journal of Pharmacology,* 5:1, 1968-1969.

Nishiitsutsuji-Uwo, J.; Petropolos, S.F.; and Pittendrigh, C.S. Central nervous system control of circadian rhythmicity in the cockroach: I. Role of the pars intercerebralis. *The Biological Bulletin,* 133(3):679-696, 1967.

Nishiitsutsuji-Uwo, J., and Pittendrigh, C.S. Central nervous system control of circadian rhythmicity in the cockroach. II. The pathway of light signals that entrain the rhythm. *Zeitschrift für vergleichende Physiologie,* 58:1-13, 1968.

Nishiitsutsuji-Uwo, J., and Pittendrigh, C.S. Central nervous system control of circadian rhythmicity in the cockroach. III. The optic lobes, locus of the driving oscillation? *Zeitschrift für Vergleichende Physiologie,* 58:14-46, 1968.

Nishiitsutsuji-Uwo, J.; Townsend, R.N.; et al. Day-night variation the enzymatic dephosphorylation of ATP in hamster liver fractions. *Comparative Biochemistry and Physiology,* 22:319-323, 1967.

Pujol, J.F.; Hery, F.; Durand, M.; and Glowinski, J. Increase in serotonin synthesis in the brainstem of the rat after selective deprivation of paradoxical sleep. [Augmentation de la synthese de la serotonine dans le tronc cerebral chez le rat apres privation selective du sommail paradoxal]. *D R Acad Sci,* (d) (Paris), 267(3):371-372, 1968.

Quay, W.B. Regional and circadian differences in cerebral cortical serotonin concentration. *Life Sciences,* 4:379, 1965.

Reis, D.J.; Corvelli, A.; and Conners, J. Circadian and ultradian rhythms of serotonin regionally in cat brain, *The Journal of Pharmacology and Experimental Therapeutics,* 167(2):328-333, 1969.

Reis, D.J.; Rifkin, M.; and Corvelli, A. Effects of morphine on cat brain norepinephrine in regions with daily monoamine rhythms. *European Journal of Pharmacology,* 8(1):149-152, 1969.

Reis, D.J.; Weinbren, M.; and Corvelli, A. A circadian rhythm of norepinephrine regionally in cat brain: its relationship to environmental lighting and to regional diurnal variations in brain serotonin. *Journal of Pharmacology and Experimental Therapeutics,* 164(1):135-145, 1968.

Reis, D.J., and Wurtman, R.J. Diurnal changes in brain noradrenalin. *Life Sciences,* 7:91-98, 1968.

Scheving, L.E.; Harrison, W.H.; Gordon, P.; and Pauly, J.E. Daily fluctuation (circadian and ultradian) in biogenic amines of the rat brain. *American Journal of Physiology,* 214:166-173, 1968.

Schildkraut, J.J., and Kety, S.S. Biogenic amines and emotion. *Science,* 156(3771):21-30, 1967.

Schildkraut, J.J.; Schanberg, S.M.; and Kopin, I.J. The effects of lithium ion on H^3-norepinephrine metabolism in brain. *Life Sciences,* 5:1479-1483, 1966.

Stacy, B.D., and Thorburn, G.D. Neurosecretory cells: Daily rhythmicity in leiobunum longpipes. *Science,* 152:1078-1079, 1966.

Steriade, M., and Iosif, G. Opposite changes in responsiveness of the motor and somaesthetic cortex during natural sleep and arousal. [Abstract] *Electroencephalography and Clinical Neurophysiology,* 25(3):299, 1968.

Strumwasser, F. Neurophysiological aspects of rhythms. In: Quarton, G.C.; Melnechuk, T., and Schmitt, F.O., (eds.), *The Neuro sciences: An intensive study program.* New York: Rockefeller University Press, 1967. pp. 516-528.

Strumwasser, F. Membrane and intracellular mechanisms governing endogenous activity in neurons. In: Carlson, F.D., (ed.), *Physiological and Biochemical Aspects of Nervous Integration.* New Jersey: Prentice Hall, 1968. pp. 329-341.

Tamura, A. Changes of diurnal rhythm of Na, K, and Ca-excretion in urine with disorders in brain. *Psychiatria et Neurologia Japonica,* 5:405-23, 1965.

Tepas, D.I. Evoked brain response as a measure of human sleep and wakefulness. *Aerospace Medicine,* 38:148-153, 1967.

Walker, C.A., and Friedman, A.H. Circadian rhythms in hypothalamic and caudate nucleus ultrastructure of untreated rats and those pretreated with reserpine or 1-DOPA and MAO inhibitor. *Federation Proceedings,* 27:600, 1969.

Wolfe, J.W., and Brown, J.H. Effects of sleep deprivation on the vestibulo-ocular reflex. *Aerospace Medicine,* 39(9):947-949, 1968.

Wurtman, J.R.; Axelrod, J.; and Reis, D.J. Metabolic cycles of monoamines and their modification by drugs. In: de Ajuriaguerra, J., (ed.), *Cycles Biologiques et Psychiatrie.* Paris: Masson et Cie., 1968. pp 373-381, 1968.

Wyrwicka, W., and Sterman, M.B. Instrumental conditioning of sensorimotor cortex EEG spindles in the waking cat. *Physiology and Behavior,* 3:703-707, 1968.

Circadian Rhythms: Animals and Lower Organisms

Adkisson, P.L. Internal clocks and insect diapause. *Science,* 154:234-241, 1966.

Albrecht, P.; Visscher, M.B.; Bittner, J.J.; and Halberg, F. Daily changes in 5-hydroxytryptamine concentration in mouse brain. *Proceedings of the Society for Experimental Biology and Medicine,* 92:703-706, 1956.

Aschoff, J., and Wever, R. Circadian rhythms of finches in light-dark cycles with interposed twilights. *Comparative Biochemistry and Physiology,* 16:507-514, 1965.

Aschoff, J. Circadian activity pattern with two peaks. *Ecology,* 47(4):657-661, 1966.

Aschoff, J. Circadian activity rhythms in chaffinches (fringilla coelebs) under constant conditions. *Japanese Journal of Physiology,* 16(4):363-370, 1966.

Barnett, A. Cell division: a second circadian clock system in paramecium multicronucleatum. *Science,* 164:1417-1418, 1969.

Barnwell, F.H. Daily and tidal patterns of activity in individual fiddler crabs (Genus, Uca) from the Woods Hole region. *Biological Bulletin,* 130(1):1-7, 1966.

Bolles, R.C. and Duncan, P.M. Daily course of activity and subcutaneous body temperature in hungry and thirsty rats. *Journal of Physiology and Behavior.* 4:87-89, 1968.

Bowers, W.S., and Blickenstaff, C.C. Hormonal termination of diapuse in the alfalfa weevil. *Science,* 154:1673-1674, 1966.

Brown, F.A., Jr., and Park, Y.H. Phase shifting a lunar rhythm in planarian by altering the horizontal magnetic vector. *Biological Bulletin,* 129(1):79-86, 1965.

Brown, F.A.; Park, Y.H.; and Zeno, J.R. Diurnal variation in organismic response to very weak gamma radiation. *Nature,* 211(5051):830-833, 1966.

Brown, F.A., Jr. Effects and after-effects on planarians of reversals of the horizontal magnetic vector. *Nature,* 209(5022):533-535, 1966.

Brown, F.A., Jr., and Park, Y.H. Association-formation between photic and subtle geophysical stimulus patterns—a new biological concept. *Biological Bulletin,* 132(3):311-319, 1967.

Brown, F.A., and Park, Y.H. Synodic monthly modulation of the diurnal rhythm of hamsters. *Proceedings of the Society for Experimental Biology and Medicine,* 125:712-725, 1967.

Brown, F.A., Jr. Endogenous biorhythmicity reviewed with new evidence. *Scientia,* 103(V-VI):245-259, 1968.

Bruce, V.G., and Pittendrigh, C.S. Temperature independence in a unicellular "clock." *Proceedings of the National Academy of Sciences,* 42:676-682, 1956.

Bruce, V.G., and Pittendrigh, C.S. Endogenous rhythms in insects and microorganisms. *American Naturalist,* 91:179-195, 1957.

Cardoso, S.S.; Ferreira, A.L.; et al. The effect of partial hepatectomy upon circadian distribution of mitosis in the cornea of rats. *Experentia,* 24:568-571, 1968.

Centre National de la recherche scientifique (C.N.R.S.) Colloque sur le rhythme d'activite section de psychologie experimentale du comfortement animal - Union International de Sciences Biologiques. Institut de Neurophysiologie et de Psychophysiologie, Marseille, October 4-5, 1965.

Chaudry, A.P., Halberg, F.; Keenan, C.E.; Harner, R.N.; and Bittner, J.J. Daily rhythms in rectal temperature and in epithelial mitoses of hamster pinna and pouch. *Journal of Applied Physiology,* 12:221-224, 1958.

Chaudry, A.P., and Halberg, F. Rhythms in blood eosinophils and mitosis hamster pinna and pouch; phase alterations by carcinogen. *Journal of Dental Research,* 39:704, 1960.

Clark, R.H., and Korst, D.R. Circadian periodicity of bone marrow mitotic activity and reticulocyte in rats and mice. *Science,* 166:236-237, May 1969.

Crowley, T.J.; Kripke, D.F., Halberg, F.; and Schildkraut, J.J. Circadian rhythms in monkeys: sleep, EEG, EMG, body and eye movement and temperature. *Psychophysiology,* 6:242-243, 1969.

Duke, M.B. Biosatellite III: preliminary findings. *Science,* 166:492-493, 1969.

Edmunds, L.N., Jr. Replication of DNA and cell division in synchronously dividing cultures of Euglena gracilis. *Science,* 145:266-268, 1964.

Edmunds, L.N., Jr. Studies on synchronously dividing cultures Euglena gracilis Klebs (Strain Z) III. Circadian components of cell division. *Journal of Cellular and Comparative Physiology,* 67(1):35-44, February 1966.

Eisinger, R.P. Influence of posture and diurnal rhythm on the renal excretion of acid - observations in normal and adrenalectomized subjects. Meeting of American Physiological Society, Illinois, April 1964. *Metabolism,* 15:76-87, January, 1966.

Eling, W. The circadian rhythms of nucleic acids. In: von Mayersbach, H., (ed.) *The Cellular Aspects of Biorhythms,* Berlin-Gottingen, Germany: Springer-Verlag, Inc., 1967.

Emlen, S.T. Bird Migration: Influence of physiological state upon celestial orientation. *Science,* 165:716-718, August 1969.

Enright, J.T. Entrainment of a tidal rhythm. *Science,* 147(3660): 864-867, February 1965.

Enright, J.T. Temperature and free-running circadian rhythm of the house finch. *Comparative Biochemistry and Physiology,* 18:463-475, 1966.

Feigin, R.D.; Dangerfield, H.G.; and Beisel, W.R. Circadian periodicity of blood amino-acids in normal and adrenalectomized mice. *Nature,* 221:94-95, 1969.

Feigin, R.D.; San Joaquin, V.H.; Haymond, M.W.; and Wyatt, R.G. Daily periodicity of the susceptibility of mice to pneumococcal infection. *Nature,* 224:379-380. 1969.

Friedman, S.B., and Ader, R. Adrenocortical response to novelty and noxious stimulation. *Neuroendocrinology,* 2:209-212, 1967.

Friedman, A.H., and Walker, C.A. Circadian rhythms: rat midbrain and caudate nucleus biogenic amine levels. *Journal of Physiology,* 197:77-86, 1968.

Galicich, J.H.; Halberg, F; French, L.A.; and Ungar, F. Effect of cerebral ablation on a circadian pituitary adrenocorticotropic rhythm in C mice. *Endocrinology,* 76:895-901, 1965.

Garcia, J.; Buchwald, N.A.; Feder, B.H.; and Koelling, R.A. Immediate detection of X-rays by the rat. *Nature,* 196(4858):1014-1015, 1962.

Garcia, J.; Buchwald, N.A.; et al. Electroencephalographic responses to ionizing radiation. *Science,* 140(3564):289-290, 1963.

Glick, D.; Ferguson, R.B.; Greenberg, L.J.; and Halberg, F. Circadian studies on succinic dehydrogenase, pantothenate and biotin of rodent adrenal. *American Journal of Physiology,* 200:811-814, 1961.

Goff, M.L.R., and Finger, F.W. Activity rhythms and diurnal light-dark control. *Science,* 154:1346-1348, 1966.

Halberg, F.; Albrecht, P.G.; and Barnum, C.P. Phase shifting of liver-glycogen rhythm in intact mice. *American Journal of Physiology,* 199:400, 1960.

Halberg, F.; Barnum, C.P.; and Vermund, H. Hepatic phospholipid metabolism and the adrenal. *Journal of Clinical Endocrinology and Metabolism,* 13:871, 1953.

Halberg, F.; Bittner, J.J.; and Gully, R.J. Twenty-four-hour periodic susceptibility to audiogenic convulsions in several stocks of mice. *Federation Proceedings,* 14:67-68, 1955.

Halberg, F.; Bittner, J.J.; and Smith, D. Mitotic rhythm in mice, mammary tumor milk agent, and breast cancer. *Proceedings of the American Association for Cancer Research,* 2:305, 1958.

Halberg, F.; Haus, E.; and Stephens, A. Susceptibility to ouabain and physiologic 24-hour periodicity. *Federation Proceedings,* 18:63,1959.

Halberg, F.; Jacobson, E.; Wadsworth, G.; and Bittner, J.J. Audiogenic abnormality spectra, 24-hour periodicity and lighting. *Science,* 128:657-658, 1958.

Halberg, F.; Peterson, R.E.; and Silber, R.H. Phase relations of 24-hour periodicities in blood cortiosterone. Mitoses in cortical adrenal parenchyma, and total body activity. *Endocrinology,* 64:222-230, 1959.

Halberg, F.; and Visscher, M.B. Temperature rhythms in blind mice. *Federation Proceedings,* 13:65, 1954.

Halberg, F.; Zander, H.R.; Houglum, M.W.; and Muhlemann, H.R. Daily variations in tissue mitoses, blood eosinophils and rectal temperature of rats. *American Journal of Physiology,* 177:361-366, 1954.

Haus, E.; Lakatua, D.; and Halberg, F. The internal timing of several circadian rhythms in the blinded mouse. *Experimental Medicine and Surgery,* 25:7-45, 1967.

Hayden, P., and Lindberg, R.G. Circadian rhythm in mammalian body temperature entrained by cyclic pressure changes. *Science,* 164:1288-1289, 1969.

Hayes, D.; Schechter, M.S.; and Sullivan, W.N. Biochemical look at insect diapause. *Symposium on Biological Rhythms, Entomological Society of America,* New York, N.Y., November 1967.

Hayward, J., and Baker, M. Role of Cerebral arterial blood in the regulation of brain temperature in the monkey. *American Journal of Physiology,* 215(2):389-403, 1968.

Holmquest, D.L., and Lipscomb, H.S. The response of thermal and activity rhythms in the rat to cyclic variations in adrenocortical function and environmental lighting. Proceedings of the International Union of Physiological Sciences. 7: 1968.

Honjo, S., Fujiwara, T., Takasaka, M., Suzuki, Y, and Imaizumi, K. Observations on the diurnal temperature variation of Cynomolgus monkeys (Macaca irus) and on the effect of changes in the routine of lighting upon this variation. *Japanese Journal of Medical Science and Biology,* 16:189-198, 1963.

Hoshizaki, T.; Adey, W.R.; Meehan, J.P.; Walter, D.O.; Berkout, J.I.; and Campeau, E. Central nervous, cardiovascular and metabolic data of a Macaca nemestrina during a 30-day experiment. In: Rohles, F.H. (ed.) *Circadian Rhythms in Non-Human Primates.* New York/Basel: Karger, 1969. pp. 8-38.

Jardetzky, C.D.; Barnum, C.P.; and Halberg, F. Physiologic 24-hour periodicity in nucleic acid metabolism and mitosis of inmature growing liver. *American Journal of Physiology,* 187:608. 1956.

Jerusalem, C. Circadian changes of the DNA-content in rat liver cells as revealed by histophotometric methods. In: von Mayersbach H., (ed.) *The Cellular Aspects of Biorhythms.* Berlin-Gottingen, Germany: Springer-Verlag, Inc., 1967.

Johnson, L. Diurnal patterns of metabolic variations in chick embryos. *Biological Bulletin* 131(2):308-322, 1966.

Kavanau, J.L., and Rischer, C.E. Program clocks in small mammals. *Science,* 161:1256-1259, September 1968.

Kuznetsova, S.S. Diurnal rhythm of radiosensitivity in mice and rats [Simpozium] *Biologicheskiye ritmy i voprosy razrabotski rezhimov truda i otdykha,* 1967. Materialy. Moscow, 1967. p. 44.

Manshardt, J., and Wurtman, R. Daily rhythm in the noradrenaline content of rat hypothalamus. *Nature,* 217(5128):574-575, 1968.

Menaker, M. Endogenous rhythms of body temperature in hibernating bats. *Nature,* 184:1251, 1959.

Minis, D.H., and Pittendrigh, C.S. Circadian oscillation controlling hatching; its ontogeny during embryogenesis of a moth. *Science,* 159:534-536, 1968.

Morley, A.; and Stohlman, Jr., F. Erythropoiesis in the dog: the periodic nature of the steady state. *Science,* 165:1025-1027, 1969.

Nishiitsutsuji-Uwo, J.; Petropulos, S.F.; and Pittendrigh, C.S. Central nervous system control of circadian rhythmicity in the cockroach: I. Role of the pars intercerebralis. *The Biological Bulletin,* 133(3):679-696, 1967.

Nishiitsutsuji-Uwo, J., and Pittendrigh, C.S. Central nervous system control of circadian rhythmicity in the cockroach. II. The pathway of light signals that entrain the rhythm. *Zeitschrift für Vergleichende Physiologies,* 58:1-13, 1968.

Nishiitsutsuji-Uwo, J., and Pittendrigh, C.S. Central nervous system control of circadian rhythmicity in the cockroach. III. The optic lobes, locus of the driving oscillation? *Zeitschrift für Vergleichende Physiologie.* 58:14-46. 1968.

Nishiitsutsuji-Uwo, J.; Townsend, R.N.; et al. Short communication, day-night variation, the enzymatic dephosphorylation of ATP in hamster liver fractions. *Comparative Biochemistry and Physiology,* 22:319-323, 1967.

Pauly, J.E., and Scheving, L.E. Daily leukocyte rhythms in normal and hypophysectomized rats exposed to different environmental light-dark schedules. *Anatomical Record,* 153(4):349-360, 1965.

Pauly, J.E., and Scheving, L.E. The innate rhythmic nature of several processes or events involved in the total mitotic cycle of dividing corneal epithelial cells in the rat. *Third International Congress of Histochemistry and Cytochemistry.* New York: Springer-Verlag, page 197, 1968.

Pittendrigh, C.S. Circadian systems: I. The driving oscillation and its assay in Drosphila Pseudoobscura. *Proceedings of the National Academy of Sciences,* 58(4):1762-1767, 1967.

Pittendrigh, C.S., and Minis, D.H. The entrainment of circadian oscillations by light and their role as photoperiodic clocks. *American Naturalist*, 98(902):261-294, 1964.
Pizzarello, D.J.; Isaak, D.; et al. Circadian rhythmicity in the sensitivity of two strains of mice to whole-body radiation. *Science*, 145:286-291, July, 1964.
Quay, W.B. Regional and Circadian differences in cerebral cortical serotonin concentration. *Life Sciences*, 4:379, 1965.
Rensing, L. Daily rhythmicity of corpus allatum and neurosecretory cells in drosophila melanogaster. *Science*, 144(3626):1586-1587, 1964.
Retiene, K.; Zimmermann, W.J.; Schindler, C.J.; and Lipscomb, H.S. A correlative study of resting endocrine rhythms in rats, *Acta Endocrinologica* 57:615-622, 1968.
Richards, A.G., and Halberg, F. Oxygen uptake rhythms in a cockroach gauged by variance spectra. *Experientia*, 20:40-42, 1964.
Richter, C.P. Biological foundation of personality differences. *American Journal of Orthopsychiatry*, 2:345-354, 1932.
Richter, C.P. Abnormal but regular cycles in behavior and metabolism in rats and catatonic-schizophrenics. In: Reiss, M., (ed.) *Psychoendocrinology.* New York: Grune & Stratton, 1958. pp. 168-181.
Richter, C.P. A hitherto unrecognized difference between man and other primates. *Science*, 154(3747):427, 1966.
Richter, C.P. Psychopathology of periodic behavior in animals and man. *Comparative Psychopathology.* New York: Grune & Stratton, 1967.
Richter, C.P. Inherent 24-hour and lunar clocks of a primate - the squirrel monkey. *Communications in Behavioral Biology.* Part A, (1): 305-332, 1968.
Scheving, L. Circadian rhythms in plasma organic phosphorous and sulfur of the rat; also susceptibility to strychnine. *Japanese Journal of Physiology*, 1970. In press.
Scheving, L.E. Circadian and ultradian rhythms in several physiological parameters of the rat. *Third International Congress of Histochemistry and Cytochemistry.* New York: Springer-Verlag, 1968.
Scheving, L.E.; Harrison, W.H.; Gordon, P.; and Pauly, J.E. Daily fluctuation (circadian and ultradian) in biogenic amines of the rat brain. *American Journal of Physiology*, 214:166-173, 1968.
Scheving, L.E., and Pauly, J.E. Circadian phase relationships of thymidine - H3 uptake, labeled nuclei, grain counts, and cell division rate in rat corneal epithelium. *The Journal of Cell Biology*, 32(3):677-683, 1967.
Scheving, L.E., and Pauly, J.E. Daily rhythmic variations in blood coagulation times in rats. *Anatomical Record*, 157:657-665, 1967.
Scheving, L.E.; Pauly, J.E.; and Tsai, Tien-Hu. Circadian fluctuation in plasma proteins in the rat. *American Journal of Physiology*, 215:1096-1101, 1968.
Skopik, S.D., and Pittendrigh, C.S. Circadian systems: II. The oscillation in the individual Drosophila Pupa: Its independence of developmental stage. *Proceedings of the National Academy of Sciences*, 58(5):1862-1869, 1967.
Sollberger, A. The control of circadian glycogen rhythms. *Annals of the New York Academy of Sciences*, 117(1):519-553, 1964.
Stacy, B.D., and Thorburn, G.D. Neurosecretory cells: Daily rhythmicity in leiobunum longpipes. *Science*, 152:1078-1079, 1966.
Stephens, G.J.; Halberg, F.; and Stephens, G.C. The blinded fiddler crab: An invertebrate model of circadian desynchronization. *Annals of the New York Academy of Sciences*, 117(1):386-406, 1964.
Stephens, G.J., and McGaugh, J.L. Periodicity and memory in mice: A supplementary report. *Communications in Behavioral Biology*, Part A, 2, (2):59-63, 1968.
Stephens, G.J., and McGaugh, J.L. Retrograde amnesia - effects of periodicity and degree of training. *Communications in Behavioral Biology*, Part A, 1(4):267-275, 1968.
Stephens, G.J.; McGaugh, J.L.; and Alpern, H.P. Periodicity and memory in mice. *Psychonomic Science*, 8(5):201-202, 1967.
Stroebel, C.F. Behavioral aspects of circadian rhythms. In: Zubin, J. and Hunt, H.F., (eds.), *Comparative Psychopathology.* New York: Grune and Stratton, 1967.
Stroebel, C.F. Biological rhythm correlates of disturbed behavior in the Rhesus monkey. In: Rohles, F.H. (ed.), *Circadian Rhythms in Nonhuman Primates.* New York: Karger, 1969. pp 91-105.
Strumwasser, F. The demonstration and manipulation of a circadian rhythm in a single neuron. In: Aschoff, J., (ed.), *Circadian Clocks:* Proceedings of Feldafing Summer School. Amsterdam: North-Holland Publishing Company, 1965.
Sullivan, W.N.; Cawley, B.M.; Oliver, M.; Hayes, D.K; and McGuire, J.U. Manipulating the photoperiod to damage insects. *Nature*, 221(5175): 60-61, 1969.
Swade, R.H., and Pittendrigh, C.S. Circadian locomotor rhythms of rodents in the artic. *The American Naturalist*, 101(922):431-466, 1967.
Thor, D.H.; and Hoats, D.L. A circadian variable in self-exposure to light by the rat. *Psychonomic Science*, 12(1), 1-2, 1969.
Ungar, F. and Halberg, F. Circadian rhythm in the in vitro response of mouse adrenal to adrenocorticotropic hormone. *Science*, 137:1058-1060, 1962.
Wahlstrom, G. The circadian rhythm of self-selected rest and activity in the canary and the effects of barbiturates, reserpine, monoamine oxidase inhibitors and enforced dark periods. *Acta Physiologica Scandinavia*, 65(supplementum 250), 1965, 67 pages.
Winget, C.M.; Card, D.H.; and Hetherington, N.W. Circadian oscillations of deep-body temperature and heart rate in a primate. (Cebus albafrons) *Aerospace Medicine*, 39:350-353, 1968.
Winget, C.M.; Rosenblatt, L.S., DeRoshia, C.W.; and Hetherington, N.W. Mechanisms of action of light on circadian rhythms in the monkey. *Life Sciences and Space Research*, 8, 1970. In press.
Wooley, D.E., and Timiras, P.S. Estrous and circadian periodicity and electroshock convulsions in rats. *American Journal of Physiology*, 202:379-382, 1962.

Circadian Rhythms: Human

Abernethy, J.D.; Farhi, L.E.; and Maurizi, J.J. Diurnal variations in urinary-alveolar N2 difference and effects of recumbency. *Journal of Applied Physiology*, 23:875-879, 1967.
Andrus, V.; Frank, G.; Gravem, H.; Halberg, F.; Harner, R.; Johnson, E.; and Matthews, J. Circadian periodicity, adrenal corticosteroids, and the EEG of normal man. *Journal of Psychiatric Research*, 4:73-86, November 1966.

Aschoff, J. Exogenous and endogenous components in circadian rhythms. *Symposium on Quantitative Biology*, Cold Spring Harbor, 25:11-27, 1960.

Aschoff, J. Human circadian rhythms in activity, body temperature and other functions. [Symposium] In: Brown, A.H. and Favorite, F.G., (eds.), *Life Science and Space Research V, International Space Science Symposium, 7th, Vienna, Austria, May 10-18, 1966, Papers.* Amsterdam: North-Holland Publishing Company, 1967, pp. 159-173.

Aschoff, J.C.; Giedke, H.; Pöppel. E. Tagesperiodische veranderungen der Reaktionszeit bei wahlreaktionen. *Zeitschrift für experimentelle und angewandte Psychologie.* In press, 1970.

Bartter, F.C.; Delea, C.S.; and Halberg, F. A map of blood and urinary changes related to circadian variations in adrenal cortical function in normal subjects. *Annals of the New York Academy of Sciences*, 98:969-983, 1962.

Beisel, W.R.; Feigin, R.D.; and Klainer, A.S. Factors affecting circadian periodicity of blood amino acids in man. *Metabolism*, 17:764-775, September 1968.

Berges, D. Investigation on the extent of diurnal variations in the electrocardiogram of healthy subjects. [Untersuchungen ufer das ausmass von tageschwankungen im ekg gesunder versuchspersonen]. *Zeitschrift für Kreislauforschung*, 54:35-49, 1965.

Bohlen, J.G. Circadian and circannual rhythms in Eskimos. *A Prospectus of Research Submitted to the Faculty of the Graduate School of the University of Wisconsin*, 1969.

Boriskin, V.V. Diurnal periodicity of basic physiological functions in personnel stationed in Antarctica. [Sympozium] *Biologicheskiye ritmy i voprosy razrabotki rezhimov truda i otdykha.* Materialy. Moscow, 1967, 18-19.

Cahn, A.A.; Folk G.E.; and Huston, P.E. Age comparison of human day-night physiological differences. *Aerospace Medicine*, 39(6):608-610, 1968.

Cranston, W.I. Diurnal variation in plasma volume in normal and hypertensive subjects. *American Heart Journal*, 68:427-428, 1964.

Curtis, G.C.; Fogel, M.L.; McEvoy, D.; and Zarate, C. Effects of weight, sex and diurnal variation on the excretion of 17-hydroxycorticosteroids. *Journal of Clinical Endocrinology and Metabolism*, 28(5):711-713, 1968.

Doctor, R.F., and Friedman, L.F. Thirty-day stability of spontaneous galvanic skin responses in man. *Psychophysiology*, 2:311-315, April 1966.

Doe, R.P.; Vennes, J.A.; and Flink, E.B. Diurnal variation of 17-hydroxycorticosteroids, sodium, potassium, magnesium and creatinine in normal subjects and in cases of treated adrenal insufficiency and Cushing's syndrome. *Journal of Clinical Endocrinology and Metabolism*, 20:253-265, February 1960.

Feigin, R.D.; Klainer, A.S.; and Beisel, W.R. Circadian periodicity of blood amino-acids in adult men. *Nature*, 215(5100):512-514, 1967.

Feigin, R.D.; Klainer, A.; and Beisel, W.R. Factors affecting circadian periodicity of blood amino acids in man. *Metabolism*, 17(9):764-775, 1968.

Fiorica, V.; Burr, M.J.; and Moses, R. Contribution of activity to the circadian rhythm in excretion of magnesium and calcium. *Aerospace Medicine*, 39(7):714-717, July 1968.

Frank, G.; Halberg, F.; Harner, R.; et al. Circadian periodicity, adrenal corticosteroids, and the EEG of normal man, *Journal of Psychiatric Research*, 4:73-86, 1966.

Frazier, T.W., Rummel, J.A., and Lipscomb, H.S. Circadian variability in vigilance performance. *Aerospace Medicine*, 39(4):383-395, 1968.

Gerd, M.A. Human work capacity at various periods of wakefulness. [Simpozium] *Biologicheskiye ritmy i voprosy razrabotki rezhimov truda i otdykha.* 1967. Materialy. Moscow, 1967, 20-21.

Gramvall, S., and Lundberg, U. Variations in pulse rate at a cycle ergometer test on Swedish Air Force Pilots. *Meddelanden Fran Flygoch Navalmedicinska Naemnden*, 12(1):4-5, 1963.

Halberg, F. Some physiological and clinical aspects of 24-hour periodicity. *The Journal-Lancet*, 73:20-32, 1953.

Halberg, F.; Anderson, J.A.; Ertel, R.; and Berendes, A. Circadian rhythm in serum 5-hydroxytryptamine in healthy men and male patients with mental retardation. *International Journal of Neuropsychiatry*, 3:4379-4386, 1967.

Halberg, F., Engeli, M.; and Hamburger, C. The 17-ketosteroid excretion of a healthy man on weekdays and weekends. *Experimental Medicine and Surgery*, 23:61-69, 1965.

Halberg, F.; Engeli, M.; Hamburger, C.; and Hillman, D. Spectral resolution of low-frequency, small-amplitude rhythms in excreted ketosteroid; probable androgen-induced circaseptan desynchronization. *Acta Endocrinologica Supplement*, 103:1-54, 1965.

Halberg, F.; Frank, G.; Harner, R.; Matthews, J.; Aaker, H.; Graven, H.; and Melby, J. The adrenal cycle in men on different schedules of motor and mental activity. *Experientia*, 17:282, 1961.

Haus, E., and Halberg, F. Circadian phase diagrams of oral temperature and urinary functions in a healthy man studied longitudinally. *Acta Endocrinologica*, 51:215-223, 1966.

Hersey, P. Emotional Cycles of man. *Journal of Mental Science*, 77:151-169, 1931.

Hildebrandt, Von G. Die koordination rhythmischer funktionen bei - menschen. *Verhandlungen der deutschen gesellschaft für innere medizin*, 73rd Kongress, 921-941, 1967.

Ivanov, D.I.; Malkin, V.B.; Popkof, V.L.; Ye,O.; and Cheryakov, I.N. Automatic Analysis of Diurnal Periodic Changes in Human EEG Rhythms. *Problems in Space Biology*, 4:642-645, Izd. vo, Nauka, Moscow, 1965. Clearinghouse for Federal Scientific and Technical Information, Springfield, Virginia.

Kaiser, I.H., and Halberg, F. Circadian aspects of birth. *Annals of the New York Academy of Sciences*, 98:1056-1068, 1962.

Kaneko, M., and Smith, R.E. Circadian variation in human peripheral blood flow levels and exercise responses. *Journal of Applied Physiology*, 25:109-114, August 1968.

Katz, F.H. Adrenal function during bed rest. *Aerospace Medicine*, 35:849-851, September 1964.

Konovalov, V.F.; Voronin, L.G.; and Konolalov, V. Electrographic data on the work of biological clocks. *Voprosy Psikhologii*, 12:87-94, November-December 1966.

Krylov, Yu.V. Circadian rhythm of hearing in humans after prolonged exposure to noise. *Aerospace Medicine*, 111-114, October 1968.

Laatikainen, T., and Vihko, R. Diurnal variation in the concentrations of solvolyzable steroids in human plasma. *Journal of Clinical Endocrinology and metabolism.* 28:1356-1360, September, 1968.

Lipscomb, H.S. Biological rhythms in man: a correlative study. *Excerpta Medica International Congress Series No. 99*, 6th Pan American Congress of Endocrinology, October 1965.

Lobban, M.C. Daily rhythms of renal excretion in Artic-dwelling Indians and Eskmos. *Quarterly Journal of Experimentall Physiology and Cognate Medical Sciences,* 52:401-410, October 1967.

Lobban, M.C., and Tredre, B.E. Renal diurnal rhythms in human subjects during bed-rest and limited activity. *Journal of Physiology,* 171(2):26-27, June 1964.

Meddis, R. Human circadian rhythms and the 48-hour day. *Nature,* 218:964-965, June, 1968.

Nichols, C.T., and Tyler, F.H. Diurnal variation in adrenal cortical function. *Annual Review of Medicine,* 18:313-324, 1967.

Ojemann, F.A., and Henkin, R.I. Steroid dependent changes in human visual evoked potentials. *Life Sciences,* 6:327-334, 1967.

Panferova, N.Ye. The diurnal rhythm of functions of humans in condition of limited mobility. *Fiziologicheskii Zhurnal SSSR,* 50:741-749, 1963.

Pöppel, E. Oszillatorische Komponenten in reaktionszeiten. *Die Naturwissenschaften,* 55:449-450, 1968.

Reinberg, A.; Halberg, F.; Ghata, J., et al. Rhythm circadien de diverses fonctions physiologiques de l'Homme adulte sain, actif et au repos (pouls, pression arterielle, excretions urinaires des 17OHCS des catecholamines et du potassium). *Journal de Physiologie,* 61:383, Paris, 1969.

Scheving, L.E. Mitotic activity in the human epidermis. *Anatomical Review,* 135:7-20, 1959.

Serge, G.; Turco, G.; and Ceresa, F., Assay of a compartmental analysis of blood glucose and insulin relationships in man. *Acta Diabetologica Latina,* 5:242-243, October 1968.

Southren, A.L., Todimoto, S.; Carmody, N.C.; and Isurugi, K. Plasma production rates of testosterone in normal adult men and women and in patients with the syndrome of feminizing testes. *Journal of Clinical Endocrinology and Metabolism,* 25:1441, 1965.

Stanbury, S.W., and Thomson, A.E. Diurnal variations in electrolyte excretion. *Journal of Clinical Endocrinology and Metabolism,* 11:267-293, 1961.

Vestergaard, P., and Leverett, R. Constancy of urinary creatinine excretion. *Journal of Laboratory and Clinical Medicine,* 51(2):211-218, 1958.

Voronin, L. G. Electrographic data on the work of biological clocks. *Voprosy Psikholgii,* 12:87-94, 1966.

Wadsworth, G.L.; Halberg, F.; Albrecht, P.; and Skaff, G. Peak urinary excretion of 5-hydroxyindoleacetic acid following arousal in human beings. *Physiologist,* 1:86, 1957.

Walsh, J.F., and Misiak, H. Diurnal variation of critical flicker frequency. *Journal of General Psychology,* 75:167-175, 1966.

Wever, R. Autonomous circadian rhythms in men as influenced by different light conditions, *Pflugers Archiv für Die Gesamte Physiologie,* 306:71-91, 1969.

Wurtman, R.J.; Rose, C.M.; Chou, C,; and Larin, F.F. Daily rhythms in the concentrations of various amino acids in human plasma. *New England Journal of Medicine,* 279(4):171-175, 1968.

Young, V.T.; Hussein, M.A.; Murray, E.; and Scrimshaw, N.S. Tryptophan intake, spacing of meals, and diurnal fluctuations of plasma tryptophan in men. *The American Journal of Clinical Nutrition,* 22(12):1563-1567, 1969.

Cycles: Ultradian and Infradian

Bohlen, J.G., Milan, F. A., and Halberg, F. Circumpolar Chronobiology. *Proceedings of the Ninth International Congress of Anatomists,* Leningrad, August 1970.

Bohlen, J. G. Circadian and circannual rhythms in Eskimos. *A Prospectus of Research Submitted to the Faculty of the Graduate School of the University of Wisconsin,* 1969.

Buchsbaum, M. Effects of cardiac and respiratory cycles on averaged visual evoked responses. *Electroencephalography and Clinical Neurophysiology,* 19:476-480, 1965.

Chance, B.; Pye, K.; and Higgins, J. Waveform generation by enzymatic oscillators. IEEE *Spectrum,* 4:79-86, 1967.

Dubois, F.S. Rhythms, cycles and periods in health and disease. *American Journal of Psychiatry,* 116: 114-119, 1959.

Fuchs, H. Seasonal and social factors in suicides. *Paper No. 96, IPU Conference,* 1961.

Globus, G.G. Observations on sub-circadian periodicity. (Abstract) Paper presented at the Association for the Psychophysiological Study of Sleep, March 1969. *Psychophysiology,* 1970, In press.

Globus, G.G. Rapid eye movement cycle in real time. *Archives of General Psychiatry,* 15:654-659, 1966.

Halberg, F.; Engeli, M.; and Hamburger, C. The 17-ketosteroid excretion of a healthy man on weekdays and weekends. *Experimental Medicine and Surgery,* 23:61-69, 1965.

Halberg, F.; Engeli, M.; Hamburger, C.; and Hillman, D. Spectral resolution of low-frequency, small-amplitude rhythms in excreted ketosteroid; probable androgen-induced circaseptan desynchronization. *Acta Endocrinologica Supplement,* 103: 1965.

Halberg, F., and Hamburger, C. 17-ketosteroid and volume of human urine: weekly and other changes with low frequency. *Minnesota Medicine,* 47:916-925, August 1964.

Halberg, F., and Reinberg, A. Rythmes circadiens et rythmes de basses frequences en physiologie humaine. *Journal de Physiologie,* 59(1):117-200, 1967.

Hartmann, E. The 90-minute sleep-dream cycle. *Archives of General Psychiatry,* 18:280-286, March 1968.

Haus, E., and Halberg, F. Circannual rhythm in level and timing of serum corticosterone in standardized inbred mature C-mice. *Journal of Environmental Research,* In press, 1970.

Hayashi, K.; Ota, M.; Kato, M.; Chiba, Y.; Narita, A. Seasonal and diurnal activities of biting insect attacking grazing cattle, with special reference to repellent-spraying. *Japanese Journal of Zootech. Science,* 38:376-84, 1967.

Hayes, D.; Schechter, M.S.; and Sullivan, W.N. Biochemical look at insect diapause. *Symposium on Biological Rhythms, Entomological Society of America,* New York, N.Y., November 1967.

Hersey, P. Emotional Cycles of man. *Journal of Mental Science,* 77:151-169, 1931.

Hildebrandt, Von G. Die bedeutung der unweltrezze für den tagesrhythmus des menschen. *Bader-Und Klimaheilkunde,* 13:626-644, December 1966.

Hildebrandt, Von G. Rhythmische koordination als ordnungsprinzip biologischer funktionen. *Die Umschau,* 19:592-596, 1962.

Hobson, A. Sleep and biorhythmicity. *Science,* 165(3896): 932-933, August 1969.

Jouvet, M. Insomnia and decrease of cerebral 5-hydroxytryptamine after destruction of the RAPHE system in the cat. *Advances in Pharmacology,* 6 B:265-279, 1968.

Kleitman, N. *Sleep and Wakefulness* (Rev. ed) Chicago: University of Chicago Press, 1963.

Kleitman, N. Basic rest-activity cycle. Abstract from paper presented at the Association for the Psychophysiological Study of Sleep, Boston, 1969. *Psychophysiology*, In press, 1970.

Lewis, J. A. Some observations on narcolepsy. Paper presented at the Association for the Psychophysiological Study of Sleep. Denver, 1968.

Malek, J.; Gleich, J.; and Malý, V. Characteristics of the daily rhythm of menstruation and labor. *Annals of the New York Academy of Sciences*, 98:1042-55, 1962.

Marotta, S.F., and Linwong, M. Excretion of urinary 17-ketosteroids and 17-ketogenic steroids. 1. Effects of age, time of day and season. *Chiengmai Medical Bulletin*, 5:167-181, 1966.

Othmer, E.; Hayden, M.P.; and Segelbaum, R. Encephalic cycles during sleep and wakefulness in humans: a 24-hour pattern. *Science*, 164(3878):447-449, April 1969.

Scheibel, M.E., and Scheibel, A.B. Activity cycles in neurons of the reticular formation. *Recent Advances in Biological Psychiatry*, 8:283-291, 1966.

Spoor, R.P., and Jackson, D.B. Circadian rhythms: Variation insensitivity of isolated rat atria to acetylcholine. *Science*, 154:782, 1966.

Stutte, K.H. and Hildebrandt, G. Untersu-hungen uber die koordination von herzschlag und atmung beim menschen. *Pflugers Archiv für die Gesamte Physiologie*, 2:289, 1966.

Thompson, M.; and Panella; G.; et al. Paleontological evidence of variations of length of synodic month since late Cambrian. *Science* 162:792-796, 1968.

Tromp, S.W. Blood sedimentation rate patterns in the Netherlands during the period 1955-1965. *International Journal of Biometeorology*, 11(1):105-117, March 1967.

Webster, J.H. The periodicity of the "sevens" in mind, man and nature: a neo-hippocratic study. *British Journal of Medical Psychology*, 24:277-282, 1951.

Yoshimuta, II. Seasonal changes in human body fluids. *Japanese Journal of Physiology*, 8:165-179, 1958.

Development: Infancy and Old Age

Ader, R. Early experiences accelerate maturation of the 24-hour adrenocortical rhythm. *Science*, 163:1225-1226, 1969.

Ader, R., and Grota, L.J. Rhythmicity in the maternal behavior of *Rattus Norvegicus. Animal Behavior*, 1970. In press.

Feinberg, I. The ontogenesis of Human Sleep and the Relationship of Sleep Variables to Intellectual Function in the Aged. *Comprehensive Psychiatry* 9(2):138-147, March 1968.

Feigin, R.D.; and Haymond, M.W. Circadian periodicity of blood amino acids in the neonate. *Pediatrics*, May, 1970.

Franks, R.C. Diurnal variation of plasma 17-hydroxycorticosteroids in children. *Journal of Clinical Endocrinology and Metabolism*, 27:75-78, 1967.

Grota, L.J.; and Ader, R. Continuous recording of maternal behavior in *Rattus norvegicus. Animal Behavior*, 17:722-729, 1969.

Hellbrügge, T. Ehrengut Lange, J., Rutenfranz, J., and Stehr, K. Circadian periodicity of physiological functions in different stages of infancy and childhood. *Annals of the New York Academy of Sciences*, 117:361-373, 1964.

Hellbrügge, T. Ontogènesé des rhythmes circadiens de l'enfant. In: de Ajuriaguerra, J., (ed.) *Cycles Biologiques et Psychiatrie*. Paris: Masson & Cie., 1968. pp. 159-183.

Honova, E.; Miller, S.A.; et al. Tyrosine transaminase: development of daily rhythm in liver of neonatal rat. *Science*, 162:999-1001, November 1968.

Lobban, M.C. Diurnal rhythms of renal excretion and of body temperature in aged subjects. *Journal of Physiology*, 188:48-49, 1967.

Marotta, S.F.; and Linwong, M. Excretion of urinary 17-ketosteroids and 17-ketogenic steroids. 1. Effects of age, time of day and season. *Chiengmai Medical Bulletin*, 5:167-181, 1966.

Montalbetti, N.; Bonanomi, L.; and Bonini, P. Il ritmo circadiano della funzione glicocorticoidea surrenalica nell 'eta senile. *Instituto Gerontologico, Bellora, Dell 'oppedale di gallarate*, 14:335-347, 1966.

Montalbetti, N.; Bonini, P.A.; and Ghiringhelli, F. I livelli nictemergli dei 17-idrossicorticosteroidi plasmatici nell 'angiosclerosi cerebrale senile. *Giornale Gerontologica*. 13:473-486, 1965.

Parmelee, A.H. Jr., Akiyama, Y.; Stern, E.; and Harris, M.A. A period of cerebral rhythm in newborn infants. *Experimental Neurology*, 25 (4):575-584, 1969.

Petren, T., and Sollberger, A. Developmental rhythms. In: Mayersbach, v.H., (ed.), *The Cellular Aspects of Biorhythms*. Berlin-Gottingen: Springer-Verlag Inc., 1967.

Roffwarg, H.P.; Muzio, J.N.; and Dement, W.C. Ontogenetic development of the human sleep-dream cycle. *Science*, 152:602-619, 1966.

Sollberger, A. The control of circadian glycogen rhythms. *Annals of the New York Academy of Sciences*, 117(1):519-553, 1964.

Sterman, M.B. Relationship of intrauterine fetal activity to maternal sleep stage. *Experimental Neurology*, Suppl. 4:98-106, 1967.

Sterman, M.B. and Hoppenbrowers, T. Sleep and activity rhythms in the human fetus, newborn, and adult. In: Sterman, M.B.; McGinty, D.J.; and Andinolfi, T. (eds.). *Neuro-Ontogeny and Behavior*. New York: Academic Press, In press, 1970.

Stern, E.; Parmelee, A.H.; Akiyama, Y.; Schultz, M.A.; and Wenner, W.H. Sleep cycle characteristics in infants. *Pediatrics*, 43(1): 1969.

Drugs, Toxins, and Stress

Albrecht, P.; Halberg, F.; and Bittner, J.J. Reserpine effects in the mouse and the adrenal. *Physiologist*, 1:6. 1957.

Baastrup, P.C., and Schou, M. Lithium as a prophylactic agent. Its effects against recurrent depressions and manic-depressive psychosis. *Archives of General Psychiatry*, 16:162-172, 1967.

Bruce, V.G., and Pittendrigh, C.S. An effect of heavy water on the phase and period of the circadian rhythm in Euglena. *Journal of Cellular and Comparative Physiology*, 56:25-31, 1960.

Christiaan, D.; van der Velde, M.D.; and Gordon, M.W. Manic-depressive illness, diabetes mellitus, and lithium carbonate. *Archives of General Psychiatry*, 21:478-485, 1969.

Cole, C.H., and Adkisson, P.H. A circadian rhythm, the susceptibility of an insect to an insecticide. In: Aschoff, J., (ed.), *Circadian Clocks. (Proceedings of the Feldafing Summer School)* Amsterdam: North-Holland Publishing Company. 1965.

Davis, W.M. Day-night periodicity in phenobarbital response of mice and the influence of socio-psychological conditions. *Experimentia,* 18:235-236, 1962.

Delayed-action drugs. *Science News Letter,* 89:87, February 5, 1966.

Dick, P.; Tissot, R.; and Pletscher, S. Influence des medicaments psychotropes sur les cycles biologiques et les cycles des psychoses, in *Cycles Biologigues et Psychiatrie,* ed. Ajuriaguerra, J. Paris: Masson and Cie, 1968.

Ertel, R.J.; Halberg, F. and Ungar, F. Circadian system-phase dependent toxicity and other effects of methopyrapone (SU-4885) in mice. *Journal of Pharmacology and Experimental Therapeutics,* 146:395-399, December 1964.

Everett, J.W., and Sawyer, C.H. A 24-hour periodicity in the "LH-Release Apparatus" of female rats, disclosed by barbiturate sedation. *Endocrinology,* 47:198-218, September 1950.

Feigin, R.D.: San Joaquin V.H.; Haymond, M.W.; and Wyatt, R.G. Daily periodicity of the susceptibility of mice to pneumococcal infection. *Nature,* 224:379-380, 1969.

Feldman, J. Lengthening the period of a biological clock in euglena by cycloheximide an inhibitor of protein synthesis. *Proceedings of the National Academy of Sciences,* 57(4):1080-1087, April 1967.

Friedman, A.H. and Walker, C. A. Circadian rhythms in central acetylcholine and the toxicity of cholinergic drugs. *Federation Proceedings,* 28:447, 1969.

Gattozzi, A. *Lithium in the treatment of mood disorders.* National Institute of Mental Health Monograph 5033. U.S. Dept. of Health, Education, & Welfare, Public Health Service, Wash., D.C., U.S. Govt. Print. Off., 1970. 99pp

Gosselink, J.G., and Standifer, L.C. Diurnal rhythm of sensitivity of cotton seedlings to herbicides. *Science,* 158:120-121, 1967.

Halberg, F.; Adkins, G.; and Marte, E. Reserpine effect upon the variance spectrum of human rectal temperature. *Federation Proceedings,* 21:347, 1962.

Halberg, F.; Haus, E.; and Stephens, A. Susceptibility to ouabain and physiologic 24-hour periodicity. *Federation Proceedings,* 18:63, 1959.

Halberg, F., Jacobson, E.; Wadsworth, G.; and Bittner, J.J. Audiogenic abnormality spectra, 24-hour periodicity and lighting. *Science,* 128:657-658, 1958.

Halberg, F.; and Stephens, A.N. Susceptibility to ouabain and physiologic circadian periodicity. *Proceedings of the Minnesota Academy of Sciences,* 27:139-143, 1959.

Hamberger, A.; Hyden, H.; and Lange, P.W. Enzyme changes in neurons and glia during barbituate sleep. *Science,* 151:1394-1395, March 1966.

Harner, R.N., and Halberg, F. Electrocorticographic difference in D_8 mice at times of daily high and low susceptibility to audiogenic convulsions. *Physiologist,* 1:34-35, 1958.

Hartmann, E., and Bernstein, J. Effect of drugs on sleep: long term human studies. Presented at the Association for the Psychophysiological Study of Sleep, Santa Fe, 1970. *Psychophysiology,* In press.

Haus, E.; Hanton, E.M.; and Halberg, F. 24-hour susceptibility rhythm to ethanol in fully fed, starved, and thirsted mice and the lighting regimen. *Physiologist,* 2:54, 1959.

Jones, F.; Haus, E.; and Halberg, F. Murine circadian susceptibility-resistance cycle to acetylcholine. *Proceedings of the Minnesota Academy of Sciences,* 31:61-62, 1963.

Kales, A.; Allen, C.; Scharf, M.; and Preston, T.A. Methodologic consideration and recommendation for sleep laboratory drug evaluation studies. Presented at Association for the Psychophysiological Study of Sleep, Santa Fe, 1970. *Psychophysiology,* In press.

Kales, A. (ed.) *Sleep: Physiology and Pathology* Philadelphia: Lippincott, 1969.

Koella, W.P.; Feldstein, A.; and Czicman, J.S. The effect of para chlorophenyl alanine on the sleep of cats. *Electroencephalography and Clinical Neurophysiology,* 25:481-490, 1968.

Krylov, Yu. V. Circadian rhythm of hearing in humans after prolonged exposure to noise. *Aerospace Medicine,* 111-114, October 1968.

Lincoln, R.G., and Hamner, K.C. An effect of gibberellic acid on the flowering of xanthium, a short day plant. *Plant Physiology,* 33(2): 101-104, 1958.

Marte, E., and Halberg, F. Circadian susceptibility rhythm to Librium. *Federation Proceedings,* 20:305, 1961.

Martin, M., and Hellman, D.E. Temporal variation in Su-4885 responsiveness in man: Evidence in support of circadian variation in ACTH secretion. *Journal of clinical endocrinology and metabolism,* 24:253-260, 1964.

Matthews, J.H.; Marte, E.; and Halberg, F. A circadian susceptibility resistance cycle to fluothane in made B_1 mice. *Canadian Anesthetists' Society Journal,* 11:280-290, 1964.

Morgan, M., and Mandell, A.J. Indole ethyl amine N-Methyltransferase in the brain. *Science,* 165:492, August, 1969.

Mouret, J.; Bobillier, P.; and Jouvet, M. Insomnia following parachlorphenylalanine in the rat. *European Journal of Pharmacology* 5:1, 1968-1969.

Mouret, J.R. Rhythme circadien de sommeil chez le rat modifications qar des agents pharmacologiques et physiques. *Journal Europeen de Toxicologie,* In Press.

Nanda, K.K.; and Hamner, K.C. The effect of temperature, auxins, anti-auxins, and some other chemicals on the endogenous rhythm affecting photoperiodic response of Biloxi soybean. *Planta* 53:53-68, 1959.

Pauly, J.E., and Scheving, L.E. Temporal variation in susceptibility of white rats to pentobarbitol sodium and tremorine. *International Journal of Neuropharmacology,* 3:651-658, 1964.

Pauly, J.E. and Scheving, L.E. Circadian susceptibility rhythms in response to various drugs in the rat. In Sayers, G., and Lunedei, A. (eds.), *Biorhythms in clinical and experimental endocrinology.* Proceedings First International Symposium, Florence, May, 1969. *Rassegna di Neurologica Vegetativa.* 234 pp. In press.

Pizzarello, D.J. Isaak, D.; et al. Circadian rhythmicity in the sensitivity of two strains of mice to whole-body radiation *Science,* 145:286-291, July 1964.

Radzialowski, F.M., and Bousquet, W.F. Circadian rhythm in hepatic drug metabolizing activity in the rat. *Life Sciences,* 6: 2545-2548, 1967.

Radzialowski, F.M., and Bousquet, W.F. Daily rhythmic variation in hepatic drug metabolism in the rat and mouse. *Journal of Pharmacology and Experimental Therapeutics,* 163:229-238, 1968.

Randrup, A., and Munkvad, I. Changes in urine volume and diurnal rhythm caused by reserpine treatment of schizophrenic patients. *British Journal of Psychiatry,* 112(483):173-176, February 1966.

Reinberg, A. The hours of changing responsiveness or susceptibility. *Perspectives in Biology and Medicine,* 11(1):111-128, 1967.

Reinberg, A. Les variations circadiennes de resistance ou de susceptibilite des organismes. In: de Ajuriaguerra, J. (ed.), *Cycles Biologiques et Psychiatrie.* Paris: Masson et Cie. 1968.

Reinberg, A., and Sidi, E. Circadian changes in the inhibitory effects of an antihistaminic drug in man. *Journal of Investigative Dermatology,* 46(4):415-419, April, 1966.

Reinberg, A.; Sidi, E.; and Ghata, J. Circadian reactivity rhythms of human skin to histamine or allergen and the adrenal cycle. *Journal of Allergy,* 36(3):273-283, 1965.

Reinberg, A.; Zagula-Mally, Z.W.; et al. Circadian rhythm in duration of salicylate excretion referred to phase of excretory rhythms and routine. *Proceedings of the Society for Experimental Biology and Medicine,* 124:826-832, 1967.

Reis, D.J.; Rifkin, M.; and Corvelli, A. Effects of morphine on cat brain norepinephrine in regions with daily monoamine rhythms. *European Journal of Pharmacology,* 8(1):149-152, 1969.

Richter, C.P. Lasting after-effects produced in rats by several commonly used drugs and hormones. *Proceedings of the National Academy of Sciences,* 45:1080, 1959.

Scheving. L. Circadian rhythms in plasma organic phosphorous and sulfur of the rat: also susceptibility to strychnine. *Japanese Journal of Physiology.* In press.

Scheving, L. Daily circadian rhythm in rats to D-Amphetamine sulphate: the effect of blinding and continuous illumination on the rhythm. *Nature,* 219 (5154):621-622, 1968.

Scheving, L.E.; Vedral, D.F.; and Pauly, J.E. A circadian susceptibility rhythm in rats to pentobarbital sodium. *Anatomical Record,* 160:741-750, 1968.

Schildkraut, J.J.; Schanberg, S.M.; and Kopin, I.J. The effects of lithium ion on H^3-norepinephrine metabolism in brain. *Life Sciences,* 5:1479-1483, 1966.

Stroebel, C.F. Biochemical, behavioral, and clinical models of drug interactions. Proceedings of the Fifth International Colloquium of Neuropsychopharmalogicum, Washington, D.C. *Excerpta Medica,* 1967.

Suter, R.B., and Rawson, K.S. Circadian activity rhythm of the deer mouse, Peromyscus: Effect of deuterium oxide. *Science,* 160:1011-1014, 1968.

Tsai, T.H.; Scheving, L.E.; and Pauly, J.E. Circadian rhythms in plasma inorganic phosphorus and sulfur in rat: also in susceptibility to strychnine. *Japanese Journal of Physiology.* In press.

Wahlstrom, G. The circadian rhythm of self-selected rest and activity in the canary and the effects of barbiturates, reserpine, monoamine oxidase inhibitors and enforced dark periods. *Acta Physiologica Scandinavia,* 65 (supplementum 250), 1965, 67.

Wahlstrom, G. Drugs which interfere with the metabolism of monoamines and biological cycles. In: de Ajuriaguerra, J., (ed.) *Cycles Biologiques et Psychiatrie.* Paris: Masson et Cie., 1968.

Walker, C.A. Pharmacological Implications of Circadian Rhythms: The relationship of circadian rhythms to specific CNS peripheral biological substances, physiological function and drug responsiveness. (Ph.D. Dissertation, Department of Pharmacology, Loyola University Medical Center, Chicago, Illinois), 1969.

Walker, C. A.; Speciale, S. G. Jr.; and Friedman, A. H. The influence of drug treatment on the ultrastructure of rat hypothalamus and caudate nucleus synaptic vesicles during a programmed light-dark cycle. *The International Journal of Neuropharmacology,* 1970. In press.

Wooley, D.E., and Timiras, P. S. Estrous and circadian periodicity and electroshock convulsions in rats. *American Journal of Physiology,* 202:379-382, 1962.

Wurtman, R.J.; Axelrod, J.; and Reis, D.J. Metabolic cycles of monoamines and their modification by drugs. In: de Ajuriaguerra, J. (ed.), *Cycles Biologiques et Psychiatrie* Paris: Masson et Cie., 1968. pp 373-381.

Zung, W.W.K. Antidepressant drugs and sleep. *Experimental Medicine and Surgery,* 27(1-2):124-137, 1969.

Zung, W.K. Effect of antidepressant drugs on sleeping and dreaming: II: on the adult male. *Excerpta Medica International Congress,* Series No. 150, pp. 1824-1826, 1968.

Zung, W.K. Effect of antidepressant drugs on sleeping and dreaming: III: on the depressed patient, *Biological Psychiatry,* 1:283-287, 1969.

Environment (Also See Light)

Adam, J.M.; Lobban, J.C.; and Tredre, B. Diurnal rhythms of renal excretion and of body temperature in Indian subjects after a sudden change of environment. *Journal of Physiology,* 177(1):18-19, March, 1965.

Adler, K. Extraoptic phase shifting of circadian locomotor rhythm in salamanders. *Science,* 164:1290-1291, 1969.

Aschoff, J. Exogenous and endogenous components in circadian rhythms. *Symposium on Quantitative Biology,* Cold Spring Harbor, 25:11-27, 1960.

Aschoff, J. Human circadian rhythms in activity, body temperature and other functions. [Symposium] In: Brown, A.H., and Favorite, F.G., (eds.), *Life Science and Space Research V, International Space Science Symposium, 7th, Vienna, Austria, May 10-18, 1966, Papers.* Amsterdam: North-Holland Publishing Company, 1967, pp. 159-173.

Aschoff, J. Time givers of 24-hour physiological cycles. In: Schaefer, K.E., (ed.), *Man's Dependence on the Earthly Atmosphere.* New York: Macmillan, 1962. 373-380 pp.

Aschoff, J., and Wever, R. Circadian rhythms of finches in light-dark cycles with interposed twilights. *Comparative Biochemistry and Physiology,* 16:507-514, 1965.

Barnwell, F.H. Daily and tidal patterns of activity in individual fiddler crab (Genus Uca) from the Woods Hole region. *Biological Bulletin,* 130(1):1-7, 1966.

Barnwell, F.H., and Brown, F.A. Response of planarians and snails. In: Barnothy, M.F., (ed.), *Biological Effects of Magnetic Fields.* New York: Plenum Press, Chapter I., 1964.

Boriskin, V.V. Diurnal periodicity of basic physiological functions in personnel stationed in Antarctica. [Symposium] *Biologicheskiye ritmy i voprosy razrabotki rezhimov truda i otdykha.* Materialy. Moscow, 1967, 18-19.

Bowers, W.S., and Blickenstaff, C.C. Hormonal termination of diapause in the alfalfa weevil. *Science,* 154:1673-1674, 1966.

Brown, F.A., Jr. Propensity for lunar periodicity in hamsters and its significance for biological clock theories. *Proceedings of the Society for Experimental Biology and Medicine,* 120:792-797, 1965.

Brown, F.A., Jr. Effects and after-effects on planarians of reversals of the horizontal magnetic vector. *Nature,* 209 (5022):533-535, 1966.

Brown, F.A., Jr. Endogenous biorhythmicity reviewed with new evidence. *Scientia,* 103 (V-VI): 245-259, 1968.

Brown, F.A., Jr. A Hypothesis for extrinsic timing of circadian rhythms. *Canadian Journal of Botany,* 47(2):287-298, February 1969.

Brown, F.A., Jr., and Park, Y.H. Duration of an after-effect in planarian following a reversed horizontal magnetic vector. *Biological Bulletin,* 128(3):347-355, 1965.

Brown, F.A., Jr., and Park, Y.H. Phase shifting a lunar rhythm in planarian by altering the horizontal magnetic vector. *Biological Bulletin,* 129(1):79-86, 1965.

Brown, F.A.; Park, Y.H.; and Zeno, J.R. Diurnal variation in organismic response to very weak gamma radiation. *Nature,* 211(5051):830-833, 1966.

Brown, F.A., and Park, Y.H. Synodic monthly modulation of the diurnal rhythm of hamsters. *Proceedings of the Society for Experimental Biology and Medicine,* 125:712-725, 1967.

Brown, F.A., Jr., and Park, Y.H. Association-formation between photic and subtle geophysical stimulus patterns – a new biological concept. *Biological Bulletin,* 132(3):311-319, 1967.

Bruce, V.G. Environmental entrainment of circadian rhythms. In: *Cold Spring Harbor Symposia on Quantitative Biology.* New York: Long Island Biological Association, 25:29-48, 1960.

Bruce, V.G., and Minis, D.H. Circadian clock action spectrum in a photoperiodic moth. *Science,* 163:583-585, 1969.

Bruce, V.G., and Pittendrigh, C.S. An effect of heavy water on the phase and period of the circadian rhythm in Euglena. *Journal of Cellular and Comparative Physiology,* 56:25-31, 1960.

Bruce, V.G., and Pittendrigh, C.S. Temperature independence in a unicellular "clock." *Proceedings of the National Academy of Sciences,* 42:676-682, 1956.

Bünning, E. *The Physiological Clock.* Revised, Second Edition. New York: Springer-Verlag, 1967.

Emlen, S.T. Bird Migration: influence of physiological state upon celestial orientation. *Science,* 165:716-718, August 1969.

Enright, J.T. Entrainment of a tidal rhythm. *Science,* 147:864-867, February 1965.

Farner, D.S. The photoperiodic control of reproductive cycles in birds. *American Scientist,* 52:137-56, 1964.

Garcia, J.; Buchwald, N.A.; et al. Electroencephalographic responses to ionizing radiation. *Science,* 140(3564):289-290, 1963.

Garcia, J.; Buchwald, N.A.; Feder, B.H.; and Koelling, R.A. Immediate detection of X-rays by the rat. *Nature,* 196(4858):1014-1015, 1962.

Grigoryev, Yu.G.; Darenskaya, N.G.; Druzhinin, Yu.P.; Kuznetsova, S.S.; and Serya, V.M. Diurnal rhythms and ionizing radiation effects. 12th Cospar Meeting: Prague, *Life Sciences,* 8:1969.

Halberg, F.; Nelson, W.; et al. Reproducibility of circadian temperature rhythm in the rat kept in continuous light of 30 lux intensity. *Physiologist,* 9(3):196, August 1966.

Hamner, K.C.; Finn, J.C. Jr.; Sirohi, G.S.; Hoshizaki, T.; and Carpenter, B.H. The Biological clock at the south pole. *Nature,* 195:476-480, 1962.

Hayden, P.; and Lindberg, R.G. Circadian rhythm in mammalian body temperature entrained by cyclic pressure changes. *Science,* 164:1288-1289, 1969.

Holmquest, D.L., Retiene K.; and Lipscomb, H.S. Circadian rhythms in rats: effects of random lighting. *Science,* 152:662-664, April 1966.

Kaiser, I.H. Effect of a 28-hour day on ovulation and reproduction in mice. *American Journal of Obstetrics and Gynecology,* 99:772-784, 1967.

Kerr, F.R., and Waisman, H.A. Environmental control of ovarian development in mosquitoes of the culex pipens complex. *Science,* 151: 824-825, February 1966.

Kovalchuk, A.V. Prospects of using certain biochemical reactions in the study of cosmic-ray variations. [o perspektivakh ispol zovaniia nekotorykh biologicheskikh reaktsii dlia izucheniia kolebanii kosmicheskoi radiatsii. In: Vernov, S.N., and Dorman, L.S., (eds.). Moscow: *Izdatel Stvo Nauka,* 8:206-208, 1967.

Lehman, A.; Magnes, J.; Samueloff, S., and Vider, E. Diurnal and seasonal metabolic changes in acid-base balance of blood from human subjects living in a hot climate. *Environmental Physiology and Psychology in Arid Conditions. XXIV. Proceedings of the Lucknow Symposium, 1962.* [Symposium sponsored by Central Drug Research Institute, India and UNESCO]. Paris, France: *UNESCO,* pp. 81-88, 1964.

Lobban, M.C. Daily rhythms of renal excretion in Arctic-dwelling Indians and Eskimos. *Quarterly Journal of Experimental Physiology and Cognate Medical Sciences,* 52:401-410, October 1967.

Lobban, M.C. Human renal diurnal rhythms in an Arctic mining community. *Journal of Physiology,* 165:75, 1966.

Lobban, M.C., and Simpson, H.W. Diurnal excretory rhythms in man at high altitudes. *Journal of Physiology,* 155:64-65, 1961.

Lowe, C.H.; Hinds, D.S.; et al. Natural free-running period in vertebrate animal population. *Science,* 156:531-534, 1967.

Matova, M.A. Studying shifts in biorhythms when there is an abrupt change in the geographic zone of habitation. [Simpozium] *Biologicheskiye ritmy i voprosy razrabotki rezhimov truda i otdyka,* 1967. Materialy. Moscow, 1967, 50-51.

Menaker, M. The free running period of the bat clock: seasonal variations at low body temperature. *Journal of Cellular and Comparative Physiology,* 57:81-86, 1961.

Menaker, M. Summer-winter differences in the circadian rhythms and the arousability of bats at low body temperatures. *Anatomical Record,* 137:381, 1960.

Pittendrigh, C.S., and Minis, D.H. The entrainment of circadian oscillations by light and their role as photoperiodic clocks. *American Naturalist,* 98(902):261-294, 1964.

Roberts, S.K. Circadian activity rhythms in cockroaches. I. The free-running rhythm in steady-state. *Journal of Cellular and Comparative Physiology,* 55:81-86, 1960.

Roberts, S.K. Circadian activity rhythms in cockroaches. II. Entrainment and phase shifting. *Journal of Cellular and Comparative Physiology,* 59:175-186, 1962.

Rocard, Y. Actions of a very weak magnetic gradient. The reflex of the dowser. Part IV, Chapter 2. In: Barnothy, M.F. (ed.), *Biological Effects of Magnetic Fields.* New York: Plenum Press, 1964.

Rosenthal, J.D.; Sullivan, W.N.; Adler, V.E.; and McGuire, J.U. Influence of temperatures and light regimens on diapause of samia cynthia pryeri. *Journal of Economic Entomology,* 61(2):578-579, 1968.

Swade, R.H., and Pittendrigh, C.S. Circadian locomotor rhythms of rodents in the arctic. *The American Naturalist,* 101(922):431-466, 1967.

Wever, R. The influence of weak electromagnetic fields on the circadian rhythm in man. *Zeitschrift Für Vergleichende Physiologie.* 56:111-128, 1967.

Wever, R. Principles of Circadian Rhythms in Men, Studied by the Effects of a Weak Alternating Electric Field. *Pflügers Archiv.,* 302:97-122, 1968.

Wever, R. The effects of electric fields on circadian rhythms in men. *Proceedings of Twelvth COSPAR Meetings.* In press.

Wever, R. Influence of electric fields on some parameters of circadian rhythms in man. *Proceedings of Friday Harbor Symposium on Chronobiology.* In press.

Zimmerman, W.F.; Pavlidis, T.; and Pittendrigh, C.S. Temperature compensation of the circadian oscillation in Drosophila pseudo obscura and its entrainment by temperature cycles, *Journal of Insect Physiology,* 14:669-684, 1968.

General and Review

Aschoff, J. Time givers of 24-hour physiological cycles. In: Schaefer, K.E. (ed.), *Man's Dependence on the Earthly Atmosphere.* New York: Macmillan, 1962. pp. 373-380.

Aschoff, J. Circadian Rhythms in Man. *Science,* 148:1427-1432, 1965.

Aschoff, J. The Biological Clock, *Abbott Tempo,* Book 1, pp. 14-17, 1968.

Biologic rhythms. *Therapeutic Notes,* 74:30-35, March-April 1967.

Biorhythms in clinical and experimental endocrinology. Sayers, G. and Lunedei, A., (eds.). Proceedings First International Symposium, Florence, May 1969. *Rassegna di Neurologica Vegetativa.* 234 pp. In press.

Brown, F.A., Jr. Biological Clocks. *Oceanology International,* July-August, 1967.

Brown, F.A., Jr. Periodicity in organisms. *McGraw-Hill Encyclopedia of Science and Technology.* New York: McGraw-Hill, 1967.

Brown, F.A., Jr. Rhythms, Biological. *Encyclopedia Britannica.* Benton, William, Publisher. Vol. 19:292-294, 1968.

Brown, F.A., Jr. A Hypothesis for extrinsic timing of circadian rhythms. *Canadian Journal of Botany,* 47(2):287-298, February, 1969.

Brozek, J. Psychorthythmics: a special review. *Psychophysiology,* 1(2):127-141, October 1964.

Dewan, E.M. Rhythms. *Science and Technology,* pp. 20-28, January 1969.

Halberg, F. Organisms as circadian systems; temporal analysis of their physiologic and pathologic responses, including injury and death. In: *Medical Aspects of Stress in the Military Climate.* Walter Reed Army Institute of Research, 1965. pp. 1-36.

Halberg, F. Chronobiology. *Annual Review of Physiology,* 31:675-725, 1969.

Halberg, F. (Investigator), and Luce, G. (Author). Techniques for assessing biological rhythms in medicine and psychiatry. *Mental Health Program Reports-3.* National Institute of Mental Health. Publication No. 1876, pp. 111-210, January 1969.

Harker, J.E. Diurnal rhythms and homeostatic mechanisms. *Symposia of the Society for Experimental Biology,* 18:283-300, 1964.

Mills, J.N. Human circadian rhythms. *Physiological Review,* 146(1):128-171, 1966.

Pittendrigh, C.S. Adaptation, natural selection and behavior. In: Roe, A., and Simpson, G.G., (eds.), *Behavior and Evolution.* Yale University Press, 1958. pp. 390-415.

Pittendrigh, C.S. Circadian rhythms and the circadian Organization of living systems. *Symposia on Quantitative Biology,* 25:159-182. New York: Long Island Biological Association, 1960.

Pittendrigh, C.S. Biological clocks, the functions, ancient and modern, of circadian oscillations. In: *Science in the Sixties. Proceedings of the 1965 Cloudcroft Symposium.* Air Force Office of Scientific Research, 1965. pp. 96-111.

Pittendrigh, C.S. The biologist in the solar system. *Bulletin of the Atomic Scientists,* 23:4-10, 1967.

Reinberg, A. Biorythmes et chronobiologie. *Presse Medicale,* 77:877, 1969.

Richter, C.P. (Investigator), and Luce, G.G. (Author). Role of Biological clocks in mental and physical health. *Mental Health Program Reports-3,* National Institute of Mental Health. PHS Publication Number 1876, 391-417, 1969.

Hormones and Metabolism

Ader, R. Early experiences accelerate maturation of the 24-hour adrenocortical rhythm. *Science,* 163:1225-1226, 1969.

Ader, R.; Friedman, B.; and Grota, L. J. "Emotionality" and adrenal cortical function: Effects of strain, test, and the 24-hour corticosterone rhythm. *Animal Behavior,* 15:37-44, 1967.

Ader, R., and Friedman, S. B. Plasma corticosterone response to environmental stimulation: Effects of duration of stimulation and the 24-hour adrenocortical rhythm. *Neuroendocrinology,* 3:378-386, 1968.

Abrams, R.L.; Parker, L; et al. Hypothalamic regulation of growth hormone secretion. *Endocrinology,* 78(3):605-613, 1966.

Agadzhanyan, N.A., and Rafikov, A.M. Diurnal variations in altitude tolerance and the role of the adrenocortical function. [Sympozium] *Biologicheskiye my i voprosy razrabotki rezhimov truda i otdyka,* 1967. Materialy. Moscow, 1967, 8-9.

Anton-Tay, F.; Chou, C.; and Anton, S. Brain serotonin concentration: Elevation following intraperitoneal administration of melatonin. *Science,* 162:277-278, 1968.

Barnum, C.P., and Halberg, F. Enhancement of DNA metabolism and mitotic activity by pituitary growth hormone in immature intact mouse liver. *Physiologist,* 1:3-4, 1958.

Barnum, C.P.; Jardetzky, C.D.; and Halberg, F. Nucleic acid synthesis in regenerating liver. *Texas Reports on Biology and Medicine,* 15:134-147, 1957.

Barnum, C.P.; Jardetzky, C.D.; and Halberg, F. Time relations among metabolic and morphologic 24-hour changes in mouse liver. *American Journal of Physiology,* 195:301-310, 1958.

Bartter, F.C.; Delea, C.S.; and Halberg, F. A map of blood and urinary changes related to circadian variations in adrenal cortical function in normal subjects. *Annals of the New York Academy of Sciences*, 98:969-983, 1962.

Beisel, W.R.; Feigin, R.D.; and Klainer, A.S. Factors affecting circadian periodicity of blood amino acids in man. *Metabolism,* 17:764-775, September 1968.

Black, I.B., and Axelrod, J. Elevation and depression of hepatic tyrosine transaminase activity by depletion and repletion of norepinephrine. *Proceedings of the National Academy of Sciences*, 59(4):1231-1234, 1968.

Black, I.B., and Axelrod, J. Inhibition of tyrosine transaminase activity by norepinephrine. *The Journal of Biological Chemistry*, 244:6124-6129, 1969.

Bunney Jr., W.E.; Fawcett, J.A.; Davis, J.M.; and Gifford, S. Further evaluation of urinary 17-hydroxycorticosteroids in suicidal patients. *Archives of General Psychiatry,* 21:138-150, 1969.

Bunney Jr., W.E., and Hartmann, E.L. Study of a patient with 48-hour manic depressive cycles, I & II. *Archives of General Psychology*, 12:611, 1965.

Canary, J.J.; Hellman, D.E.; and Mintz, D.H. Effect of altered thyroid function on calcium and phosphorus circadian rhythms. *Journal of Clinical Endocrinology and Metabolism,* 28:399-411, March 1968.

Clayton, G.W.; Librik, L; et al. Studies on the circadian rhythm of pituitary adrenocorticotropic release in man. *Journal of Clinical Endocrinology and Metabolism,* 23:975-980, 1963.

Curtis, G.C.; Fogel, M.L.; et. al. The effect of sustained affect on the diurnal rhythms of adrenal cortical activity. *Psychosomatic Medicine,* 28:696-713, 1966.

Curtis, G.C.; Fogel, M.L.; McEvoy, D; and Zarate, C. Effects of weight, sex and diurnal variation on the excretion of 17-hydroxycorticosteroids. *Journal of Clinical Endocrinology and Metabolism* 28(5):711-713, 1968.

Daughaday, W.H.; Othmer, E.; and Kipnis, D.M. Hypersecretion of growth hormone following REM deprivation. Report to the Fifty-first Annual Meeting of the Endocrine Society, New York, June 1969.

Deschamps, I,; Heilbronner, J.; and Canivet, J. Les variations de l'insuline chez le sujet normal au cours du nycthémere. *Annals Endocrinologie*, 30:589-596, 1969.

Dewar, H.A.; Menon, I.S.; Smith, P.A.; and White, R.W.B. Diurnal variations of fibrinolytic activity and plasma-11-hydroxycorticosteroid levels. *Lancet,* 2:531-532, September 1967.

Di Raimondo, C.; and Forsham, P.H. Some clinical implications of the spontaneous diurnal variation in adrenal and cortical secretory activity. *The American Journal of Medicine,* 21(3):321-323, 1956.

Doe, R.P.; Vennes, J.A.; and Flink, E.B. Diurnal variation of 17-hydroxycorticosteroids, sodium, potassium, magnesium and creatinine in normal subjects and in cases of treated adrenal insufficiency and Cushing's syndrome. *Journal of Clinical Endocrinology and Metabolism,* 20:253-265, February 1960.

Dossetor, J.B.; Gorman, H.M.; and Beck, J.C. The diurnal rhythm of urinary electrolyte excretion. *Metabolism,* 12(12):1083-1099, December 1963.

Durrell, J. Biological rhythms and psychiatry: psycho-endocrine mechanisms. In: de Ajuriaguerra, J., (ed.), *Cycles Biologiques et Psychiatrie.* Paris: Masson & Cie, pp. 321-328, 1968.

Durell, J.; Libow, L.S.; Kellam, S.G.; and Shader, R.I. Interrelationships between regulation of thyroid gland function and psychosis. *Endocrines and the Central Nervous System,* Association for Research in Nervous and Mental Diseases, Vol. 43, Chapter XXI, 387, 1966.

Eisinger, R.P. Influence of posture and diurnal rhythm on the renal excretion of acid - observations in normal and adrenalectomized subjects. *Metabolism,* 15:76-87, January 1966.

Eling, W. The circadian rhythms of nucleic acids. In: von Mayersbach, H.; (ed.), *The Cellular Aspects of Biorhythms.* Berlin-Gottingen, Germany: Springer-Verlag, Inc., 1967.

Engel, R.; Halberg, F.; Dassanyake, W.L.P.; and di Silva, G. Adrenal effects on time relations between rhythms of microfilariae and eosinophils in the blood. *American Journal of Tropical Medicine and Hygiene,* 11(5):653-663, September, 1962.

Engeli, M., and Halberg, F. Spectral analysis of steroid rhythms in data at equal or unequal intervals. *Federation Proceedings,* 23(2):p 89, March-April, 1964.

Eskin, I.A., and Mikhailova, N.V. Photoperiodicity and the function of the hypophysis and the adrenal cortex. *Bulletin of Experimental Biology and Medicine U.S.S.R.* (English translation), 46:999-1002, 1958.

Everett, J.W., and Sawyer, C.H. A 24-hour periodicity in the "LH-Release Apparatus" of female rats, disclosed by barbiturate sedation. *Endocrinology,* 47:198-218, September 1950.

Feigin, R.D.; Dangerfield, H.G.; and Beisel, W.R. Circadian periodicity of blood amino-acids in normal and adrenalectomized mice. *Nature,* 221:94-95, 1969.

Feigin, R.D.; Klainer, A.S.; and Beisel, W.R. Circadian periodicity of blood amino-acids in adult men. *Nature,* 215(5100):512-514, 1967.

Feigin, R.D.; Klainer, A.; and Beisel, W.R. Factors affecting circadian periodicity of blood amino acids in man. *Metabolism,* 17(9):764-775, 1968.

Forsham, P.H., and di Raimondo, V.C. Some clinical implications of the spontaneous diurnal variation in adrenocortical secretory activity. *American Journal of Medicine,* 21:321-323, 1956.

Franks, R.C. Diurnal variation of plasma 17-hydroxycorticosteroids in children, *Journal of Clinical Endocrinology and Metabolism,* 27:75-78, 1967.

Friedman, S.B., and Ader,R. Adrenocortical response to novelty and noxious stimulation. *Neuroendocrinology,* 2:209-212, 1967.

Fullerton, D.T. Circadian rhythm of adrenal cortical activity in depression. *Archives of General Psychiatry,* 19(6):674, 1968.

Galicich, J.H.; Halberg, F; French, L.A.; and Ungar, F. Effect of cerebral ablation on a circadian pituitary adrenocorticotropic rhythm in C mice. *Endocrinology,* 76:895-901, 1965.

George, J.M.; Delea, C.S.; and Bartter, F.C. The effect of sodium-retaining steroids on electrolyte concentration in human saliva. In: *Research on Pathogenesis of Cystic Fibrosis.* New York: Wickersham Printing Co., 1966.

Giusti, G.; Taccorondi, R.; et al. Nycthemeral rhythm of plasma cortisol levels in the Cushing syndrome associated with hyperplasia of the adrenal cortex. *Acta Neurovegetativa Rassegna,* 30:129-136, 1967.

Graber, A.L.; Cerchio, G.M.; and Abrams, R.L. *Circadian Variation of Intravenous Glucose Tolerance in Man, Metabolism.* In press.

Halberg, F. Discussion of glycogen metabolism. In: *Mechanisms of regulation of growth. Report of the Fortieth Ross Conference on Pediatric Research.* Columbus, Ohio: Ross Laboratories, 1962. pp. 31-32.

Halberg, F. Rhythmic interactions of steroidal and neural functions. In: Martini, L., et al. (eds.). *Hormonal Steroids.* Amsterdam: Excerpta Medica Foundation, pp. 966-979, 1967.

Halberg, F.; Barnum, C.P.; and Vermund, H. Hepatic phospholipid metabolism and the adrenal. *Journal of Clinical Endocrinology and Metabolism,* 13:871, 1953.

Halberg, F.; Frank, G.; Harner, R.; Matthews, J.; Aaker, H.; Graven, H.; and Melby, J. The adrenal cycle in men on different schedules of motor and mental activity. *Experientia,* 17:282, 1961.

Halberg, F., and Hamburger, C. 17-ketosteroid and volume of human urine, weekly and other changes with low frequency. *Medicine,* 47:916-925, August 1964.

Halberg, F.; Peterson, R.E.; and Silber, R.H. Phase relations of 24-hour periodicities in blood corticosterone. Mitoses in cortical adrenal parenchyma, and total body activity. *Endocrinology,* 64:222-230, 1959.

Halberg, F.; Vermund, H.; Halberg, E.; and Barnum, C.P. Adrenal hormones and phospholipid metabolism in liver cytoplasm of adrenalectomized mice. *Endocrinology,* 59:364-368, 1956.

Halberg, F.; Vestergaard, P.; and Sakai, M. Rhythmometry on urinary 17-ketosteroid excretion by healthy men and women and patients with chronic schizophrenia; possible chronopathology in depressive illness. *Archives d'Anatomie, d'Histologie et d'Embryologie Normales et Experimentales,* 51:301-311, 1968.

Hastings, J.W. Biochemical mechanisms involved in biological rhythms and cycles. In: de Ajuriaguerra, J., ed. *Cycles Biologiques et Psychiatrie.* Paris: Masson & Cie., 1968. pp. 127-140.

Hendricks, S.B. Metabolic control of Timing. *Science,* 141:21-27, 1963.

Henkin, R.I. Auditory detection and perception in normal man and in patients with adrenal cortical insufficiency: effect of adrenal cortical steroids. *Journal of Clinical Investigation,* 47(6):1269-1280, June 1968.

Henkin, R.I. Presence of corticosterone and cortisol in the central and peripheral nervous system of the cat. *Endocrinology,* 82(5):1058-1061, May 1968.

Henkin, R.I.; Gill, J.R.; and Bartter, F.C. Studies on taste thresholds in normal man and in patients with adrenal cortical insufficiency: The role of adrenal cortical steroids and of serum sodium concentration. *Journal of Clinical Investigations,* 42(5):727-735, 1963.

Henkin, R.I.; McGlone, R.E.; et al. Studies on auditory thresholds in normal man and in patients with adrenal cortical insufficiency: the role of adrenal cortical steroids. *Journal of Clinical Investigations,* 46(3):429-435, 1967.

Holmgren, H., and Swensson, A. Der Einfluss des Lichtesauf den 24-Stunden. Rhythmus der Aktivitat des Leberglykogens und der Korpertemeratur. *Acta Medica Scandinavica* (Supplement 278), 71-76, 1953.

Honova, E.; Miller, S.A.; et al. Tyrosine transaminase: development of daily rhythm in liver of neonatal rat. *Science,* 162:999-1001, November 1968.

Ishihara, I., and Komori, Y. Diurnal variations in the urinary excretion of corticosteroids in aging persons. *Annual Report on Research Institute Environmental Medicine, Nagoya, University,* 1957.

Jacey, M, and Schaefer, K.E. Circadian cycles of lactic dehydrogenase in urine and blood plasma. *Aerospace Medicine,* April 1968, p. 410.

Jardetzky, C.D.; Barnum, C.P.; and Halberg, F. Physiologic 24-hour periodicity in nucleic acid metabolism and mitosis of immature growing liver. *American Journal of Physiology,* 187:608, 1956.

Jerusalem, C. Circadian changes of the DNA-content in rat liver cells as revealed by histophotometric methods. In: von Mayersbach, H., (ed.) *The Cellular Aspects of Biorhythms.* Berlin-Gottingen, Germany: Springer-Verlag, Inc., 1967.

Johnson, L. Diurnal patterns of metabolic variations in chick embryos. *Biological Bulletin,* 131(2):308-322, 1966.

Kaempfer, R.; Meselson, M.; and Raskas, H.J. Cyclic dissociation into stable subunits and reformation of ribesomes during bacterial growth. *Journal of Molecular Biology,* 31:277-289, 1968.

Kahana, L.; Lebovitz, H.; et al. Endocrine manifestations of intracranial extrasellar lesions. *Journal of Clinical Endocrinology and Metabolism,* 22:304-324, March 1962.

Katz, F.H. Adrenal function during bed rest. *Aerospace Medicine,* 35:849-851, September 1964.

Kirshner, N.; Sage, H.J.; and Kirschner, A.G. Release of catecholamines and specific protein from adrenal glands. *Science,* 162:529-531, October 1968.

Klevecz, R.R., and Ruddle, F.H. Cyclic changes in enzyme activity in synchronized mammalian cell cultures. *Science,* 159:634-636, February 1968.

Knapp, M.S.; Keane, P.M.; and Wright, J.G. Circadian rhythm of plasma 11-hydroxycorticosteroids in depressive illness, congestive heart failure and Cushing's syndrome. *British Journal of Psychiatry,* 2:27-30, 1967.

Krieger, D.T., and Krieger, H.P. The effects of intrahypothalamic injection of drugs on ACTH release in the cat. *Exerpta Medica International Congress Series,* 83:640-645, 1964.

Krieger, D.T., and Krieger, H.P. Adrenal function in central nervous system disease. *Endocrines and the Central Nervous System,* 43:400-417, 1966.

Krieger, D.T., and Krieger, H.P. The circadian variation of the plasma 17-OHCS in central nervous system disease. *Journal of Clinical Endocrinology and Metabolism,* 26:939, 1966.

Krieger D.T., and Krieger, H.P. The effect of short-term administration of CNS acting drugs on the circadian variation of the plasma 17-OHCS in normal subjects. *Neuroendocrinology,* 2:232, 1967.

Krieger, D.T., and Krieger, H.P. Circadian pattern of plasma 17-hydroxycorticosteroid: alteration by anticholinergic agents. *Science,* 155:1421-1422, March 1967.

Krieger, D.T., and Rizzo, F. Circadian periodicity of plasma 17-OHCS: mediation by serotonin dependent pathways, *American Journal of Physiology,* 217:1703, 1969.

Krieger, D.T.; Silverberg, A.I.; Rizzo, F.; and Krieger, H.P. Abolition of circadian periodicity of plasma 17-OHCS levels in the cat. *American Journal of Physiology,* 215:915, 1968.

Laatikainen, T., and Vihko, R. Diurnal variation in the concentrations of solvolyzable steroids in human plasma. *Journal of Clinical Endocrinology and Metabolism,* 28:1356-1360, September 1968.

Lohrenz, F.N.; Fullerton, D.T.; et al. Adrenocortical function in depressive states, study of circadian variation in plasma and urinary steroids. *International Journal of Neuropsychiatry,* 4: 21-25, 1968.

Malek, J.; Gleich, J.; and Maly, V. Characteristics of the daily rhythm of menstruation and labor. *Annals of the New York Academy of Sciences,* 98:1042-55, 1962.

Marotta, S.F., and Linwong, M. Excretion of urinary 17-ketosteroids and 17-ketogenic steroids: 1. Effects of age, time of day and season. *Chiengmai Medical Bulletin,* 5:167-181, 1966.

Meites, J. Direct studies of the secretion of hypothalamic hypophysiotropic hormones (HHH). In: Meites, J. (ed.), *Hypophysiotropic hormones of the Hypothalamus.* Baltimore, Md: Williams and Wilkins Co., 1970.

Meites, J. Releasing factors in hypothalamic control of the anterior pituitary in mammals. (Symposium on Hypothalamic Control of the Anterior Pituitary). American Association for the Advancement of Science, December 1965.

Mintz, D.H.; Hellman, D.E.; and Canary, J.J. Effects of Altered thyroid function on calcium and phosphorus circadian rhythms. *Journal of Clinical Endocrinology and Metabolism,* 28:399-411, 1968.

Montalbetti, N.; Bonini, P.A.; and Ghiringhelli, F. I livelli nictemerali dei 17-idrossicorticosteroidi plasmatici nell' angiosclerosi cerebrale senile. *Giornale di Gerontologia,* 13:473-486, 1965.

Montalbetti, N.; Ghiringhelli, G.; Bonini, P.A.; and Bonanomi, L. Adrenal Rhythms during human senescence. *Acta Endocrinologica,* Supplement 119:44, 1967.

Nelson, W. Aspects of circadian periodic changes in phosphorus metabolism in mice. *American Journal of Physiology,* 206(3): 589-598, 1964.

Nichols, T.; Nugent, C., and Tyler, F.H. Diurnal variation in suppression of adrenal function by glucocorticoids. *Journal of Clinical Endocrinology and Metabolism,* 25:343-349, 1965.

Nichols, C.T., and Tyler, F.H. Diurnal variation in adrenal cortical function. *Annual Review of Medicine,* 18:313-324, 1967.

Ojemann, F.A., and Henkin, R.I. Steroid dependent changes in human visual evoked potentials. *Life Sciences,* 6:327-334, 1967.

Oppenheimer, J.H.; Fisher, L.V.; and Jailer, J.W. Disturbance of the pituitary-adrenal interrelationship in diseases of the central nervous system. *Journal of Clinical Endocrinology and Metabolism,* 21(a):1023-1036, 1961.

Orth, D.N.; Island, D.P.; and Liddle, G.W. Experimental alteration of the circadian rhythm in plasma cortisol (17-OHCS) concentration in man. *Journal of Clinical Endocrinology and Metabolism,* 27:549-555, 1967.

Orth, D.N., and Island, D.P. Light synchronization of the circadian rhythm in plasma cortisol (17-OHCS) concentration in man. *Journal of Clinical Endocrinology,* 29:479-486, 1969.

Parker, D.C.; Sassin, J.F.; Mace, J.W.; Gotlin, R.W.; and Rossman, L.G. Human growth hormone release during sleep: electroencephalographic correlation. *Journal of Clinical Endocrinology and Metabolism,* 29(6):871-874, 1969.

Pauly, J.E., and Scheving, L.E. Daily leukocyte rhythms in normal and hypophysectomized rats exposed to different environmental light-dark schedules. *The Anatomical Record,* 153(4):349-360, 1965.

Pauly, J.E., and Scheving, L.E. Circadian rhythms in the blood glucose and the effect of different lighting schedules, hypophysectomy, adrenal medullectomy and starvation. *Journal of Anatomy,* 120:627-636, 1967.

Perkoff, G.T.; Eik-Nes, K.; et al. Studies of the diurnal variation of plasma 17-hydroxycorticosteroids in man. *Journal of Clinical Endocrinology and Metabolism,* 19:432-443, 1959.

Pincus, G. A diurnal rhythm in the excretion of urinary ketosteroids by young men. *Journal of Clinical Endocrinology,* 3(4):195-199, 1943.

Rapoport, M.; Feigin, R.D.; Bruton, J.; and Beisel, W.R. Circadian rhythm for tryptophan pyrrolase activity and its circulating substrate. *Science,* 153:1642-44, 1966.

Reinberg, A. Rythmes circadiens du metabolisme du potassium chez l'Homme. *Le Potassium et La Vie.* Paris: Presses Universitaires de France, 1969, pp. 103-104.

Retiene, K.; Zimmermann, W.J.; Schindler, C.J.; and Lipscomb, H.S. A correlative study of resting endocrine rhythms in rats, *Acta Endocrinologica,* 57:615-622, 1968.

Richter, C.P. Hormones and rhythms in man and animals. In: Pincus, G., (ed.), *Recent Progress in Hormone Research* (Vol. 13). New York: Academic Press, 1957. pp. 105-159.

Rust, C.C., and Meyer, R.K. Hair color, molt and testis size in male, short-tailed weasels treated with melatonin. *Science,* 165:921-922, 1969.

Scheving, L.E., and Pauly, J.E. Effect of adrenalectomy, adrenal medullectomy and hypophysectomy on the daily mitotic rhythm in the corneal epithelium of the rat. In: von Mayersbach, H.; (ed.), *The Cellular Aspects of Biorhythms,* Berlin: Springer-Verlag, 1967.

Scheving, L.E., and Pauly, J.E. Circadian phase relationships of thymidine -^{3}H uptake, labeled nuclei, grain counts, and cell division rate in rat corneal epithelium. *The Journal of Cell Biology,* 23:(3)677-683, 1967.

Schumacher, G.F.B. Acute phase protein in serum of women using hormonal contraceptives. *Science,* 153:901-902, 1966.

Scrimshaw, N.S.; Habicht, J.P.; Pellet, P.; Piché, M.L.; and Cholakos, B. Effects of Sleep Deprivation and reversal of diurnal activity on protein metabolism of young men. *American Journal of Clinical Nutrition,* 19:313-319, 1966.

Serio, M.; Tarquini, B; Contini, P; Bucalossi, A.; and Toccafondi, R. Plasma cortisol response to insulin and circadian rhythm in diabetic subjects. *Diabetes,* 17(3):124-127, 1968.

Sharp, G.W.G.; Slorach, S.A.; and Vipond, H.J. Diurnal rhythms of keto and ketogenic steroid excretion and the adaptation to changes of the activity-sleep routine. *Journal of Endocrinology,* 22:377-385, 1961.

Shenkin, H. The effect of pain on the diurnal pattern of plasma corticoid levels. *Neurology,* 14(12):1112-1117, 1964.

Sholitan, L.J.; Werk, E.E., Jr.; and Marnell, R.T. Diurnal variation of adrenocortical function in non-endocrine disease states. *Metabolism,* 10:632-646, 1961.

Silverberg, A.I.; Rizzo, F.; and Krieger, D.T. Nycthemeral periodicity of plasma 17-OHCS levels in elderly subjects. *Journal of Clinical Endocrinology,* 28:1666, 1968.

Simpson, H.W., and Lobban, M.C. Effects of a 21-hour day on the human circadian excretory rhythms of 17-hydroxycorticosteroids and electrolytes. *Aerospace Medicine,* 38:1205-1213, 1967.

Simpson, H.W.; Lobban, M.C.; and Halberg, F. Near 24-hour rhythms in subjects living on a 21-hour routine in the artic summer at 78°N - revealed by circadian amplitude ratios. *Rassegna di Neurologia Vegetativa.* In press.

Smith, W.R.; Ulvedal, F.; and Welch, B.E. Steroid and catecholamine studies in pilots during prolonged experiments in a space cabin simulator. *Journal of Applied Physiology,* 18:1257-1263, 1963.

Southren, A.L.; Gordon, G.G.; et al. Mean plasma concentration, metabolic clearance and basal plasma production rates of testosterone in normal young men and women using a constant infusion procedure: Effect of time of day and plasma concentration on the metabolic clearance rate of testosterone. *Journal of Clinical Endocrinology and Metabolism,* 27(5):686-694, 1967.

Southren, A.L.; Todimoto, S.; Carmody, N.C.; and Isurugo, K. Plasma production rates of testosterone in normal adult men and women and in patients with the syndrome of feminizing testes. *Journal of Clinical Endocrinology and Metabolism,* 25:1441, 1965.

Szczepanska, E.; Preibisz, J.; Drzewiecki, K.; and Kozlowski, S. Studies on the circadian rhythm of variations of the blood antidiuretic hormone in humans. *Polish Medical Journal,* 7:517-523, 1968.

Takashi, Y.; Kipnis, D.M.; and Daughaday, W.H. Growth hormone accretion during sleep. *Journal of Clinical Investigations,* 47(9):2079-2090, 1968.

Ungar, F., and Halberg, F. Circadian rhythm in the in vitro response of mouse adrenal to adreno corticotropic hormone. *Science,* 137:1058-1060, 1962.

Vestergaard, P.; Leverett, R.; and Douglas, W.R. Spontaneous variability in the excretion of combined neutral 17-ketosteroids in the urine of chronic schizophrenic patients. *Psychiatric Research Reports, No. 6,* American Psychiatric Association, 74-89, October 1956.

Vestergaard, P., and Leverett, R. Excretion of combined neutral urinary 17-ketosteroids in short-term collection periods. *Acta Endocrinologica,* 25(1):45-53, 1957.

Weitzman, E.D.; Goldmacher, D.; Kripke, D.; MacGregor, P.; Kream, J.; and Hellman, L. Reversal of sleep-waking cycle-effect on sleep stage pattern and certain neuroendocrine rhythms. *Trans American Neurological Association,* 93:153-157, 1968.

Weitzman, E.D. (Investigator), Luce, G. (Author). Biological Rhythms-Indices of pain, adrenal hormones, sleep, and sleep reversal. *Mental Health Program Reports-3,* National Institute of Mental Health, PHS Publication No. 1876, 319-332, 1969.

Weller, L.A.; Margen, S.; and Calloway, D.H. Variation in fasting and postprandial amino acids of men fed adequate or protein-free diets. *American Journal of Clinical Nutrition,* 1577-1583, 1969.

Wurtman, R.J. Effects of light and visual stimuli on endocrine function. In: Martini L., and Ganong W.F., (eds.), *Neuroendocrinology,* 2:20-59, 1967. New York: Academic Press.

Wurtman, R.J., and Axelrod, J. The physiologic effects of melatonin and the control of its biosynthesis. *Problems Actuels d'Endocrinologie et de Nutrition,* 10:189-200, 1966.

Wurtman, R.J., and Axelrod, J. Daily rhythmic changes in tyrosine transaminase activity of the rat liver. *Proceedings of the National Academy of Science,* 57(6):1594-1599, 1967.

Wurtman, R.J.; Chou, C.; and Rose, C.M. Daily rhythm in tyrosine concentration in human plasma: persistence on low-protein diets. *Science,* 158:660-663, 1967.

Wurtman, R.J.; Rose, C.M.; Chou, C.; and Larin, F.F. Daily rhythms in the concentrations of various amino acids in human plasma. *New England Journal of Medicine,* 279(4):171-175, 1968.

Wurtman, R.J.; Shoemaker, W.J.; and Larin, F.F. Mechanism of the daily rhythm in hepatic tyrosine transaminase activity: role of dietary tryptophan. *Proceedings of the National Academy of Sciences,* 59(3):800-807, 1968.

Wurtman, R.J.; Shoemaker, W.J.; Larin, F.F.; and Zigmond, M. Failure of brain norepinephrine depletion to extinguish the daily rhythm in hepatic tyrosine transaminase activity. *Nature,* 219:1049-1050, 1968.

Young, V.T.; Hussein, M.A.; Murray, E.; and Scrimshaw, N.S. Tryptophan intake, spacing of meals, and diurnal fluctuations of plasma tryptophan in men. *The American Journal of Clinical Nutrition,* 22(12):1563-1567, 1969.

Illness

Ader, R. Behavioral and physiological rhythms and the development of gastric erosions in the rat. *Psychosomatic Medicine,* 29(4):345-353, 1967.

Biorhythms in clinical annd experimental endocrinology, In: Sayers, G., and Lunedei, A. (eds.) Proceedings First International Symposium, Florence, May 1969. *Rassegna di Neurologia Vegetativa,* 234 pp. In press.

Chipman, D.M., and Sharon, N. Mechanism of lysozyme action. *Science,* 165(3892):454-465, August 1969.

Di Raimondo, V.C., and Forsham, P.H. Some clinical implications of the spontaneous diurnal variation in adrenal and cortical secretory activity. *The American Journal of Medicine,* 21(3):321-323, 1956.

Engel, R.; Halberg, F.; Dassanayake, W.L.P.; and De Silva, J. Adrenal effects on time relations between rhythms of blood microfilariae and eosinophils. *American Journal of Tropical Medicine and Hygiene,* 2(5):653-663, September 1962.

Engel, R.; Halberg, F.; and Gully, R.J. The diurnal rhythm in EEG discharge and in circulating eosinophilis in certain types of epilepsy. *Electroencephalography and Clinical Neurophysiology,* 4:115-116, 1952.

Engel, R.; Halberg, R.; Tichy, F.Y.; and Dow, R. Electrocerebral activity and epileptic attacks at various blood sugar levels (with a case report). *Acta Neurovegetativa,* 147-167, 1954.

Engel, R.; Halberg, F.; Ziegler, M.; and McQuarrie, I. Observations on two children with diabetes mellitus and epilepsy. *Lancet,* 72:242-248, 1952.

Falliers, C.; Purcell, K.; and Hahn, W. (Investigators); and Luce, G. (Author). Psychodynamics of asthmatic children. Mental Health Program Reports - 2, National Institute of Mental Health, Chevy Chase, Md., 1966. PHS Publication 1743. pp. 149-162.

Feigin, R.D.; San Joaquin, V.H.; Haymond, M.W.; and Wyatt, R.G. Daily periodicity of susceptibility of mice to pneumococcal infection. *Nature.* 224:379-380, 1969.

Forsham, P.H., and di Raimondo, V.C. Some clinical implications of the spontaneous diurnal variation in adrenocortical secretory activity. *American Journal of Medicine,* 21:321-323, 1956.

Fremont-Smith, D.; Harter, J.G.; and Halberg, F. Circadian rhythmicity of proximal interphalangeal joint circumference of patients with rheumatoid arthritis. *Rheumatism and Arthritis,* 12:294, 1969.

Garcia-Sainz, M., and Halberg, F. Mitotic rhythms in human cancer reevaluated by electronic computer programs - evidence for temporal pathology, *Journal of the National Cancer Institute,* 37:279-292, 1966.

Giusti, G.; Taccorondi, R.; et al. Nycthemeral rhythm of plasma cortisol levels in the Cushing syndrome associated with hyperplasia of the adrenal cortex. *Acta Neurovegetativa Rassegna,* 30:129-136, 1967.

Gjessing, L.R. A review of the biochemistry of periodic catatonia. *Excerpta Medica International Congress Series,* No. 150, 1966.

Gjessing, L.R.; Jenner, F.A.; and Harding, G.F.A. The EEG in three cases of periodic catatonia. *British Journal of Psychiatry,* 113(504):1271-1282, 1967.

Goldman, R. Studies in diurnal variation of water and electrolyte excretion: Nocturnal diuresis of water and sodium in congestive cardiac failure and cirrhosis of the liver. *Journal of Clinical Investigation,* 30:1192-1199, 1951.

Graber, A.L.; Cerchio, G.M.; and Abrams, R.L. Circadian variation of intravenous glucose tolerance in man. *Metabolism.* In press.

Halberg, F. Organisms as circadian systems; temporal analysis of their physiologic and pathologic responses, including injury and death. *Medical Aspects of Stress in the Military Climate,* Walter Reed Army Institute of Research, April 1964, pp. 1-36.

Halberg, F. Some physiological and clinical aspects of 24-hour periodicity. *The Journal-Lancet,* 73:20-32, 1953.

Halberg, F.; Bittner, J.J.; and Smith, D. Mitotic rhythm in mice, mammary tumor milk agent, and breast cancer. *Proceedings of the American Association for Cancer Research,* 2:305, 1958.

Halberg, F., and Falliers, C.J. Variability of physiologic circadian crests in groups of children studied "transversely." *Pediatrics,* 68:741-746, 1966.

Halberg, F.; Good, R.A.; and Levine, H. Some aspects of the cardiovascular and renal circadian systems. *Circulation* 34:715-717, 1966.

Hamerston, O.; Elveback, L.; Halberg, F.; and Gully, R.J. Correlation of absolute basophil and eosinophil counts in blood from institutionalized human subjects. *Journal of Applied Physiology,* 9:205-207, 1956.

Henkin, R.I. Auditory detection and perception in normal man and in patients with adrenal cortical insufficiency: effect of adrenal cortical steroids. *Journal of Clinical Investigation,* 47(6):1269-1280, June 1968.

Henkin, R.I.; Gill, J.R.; and Bartter, F.C. Studies on taste thresholds in normal man and in patients with adrenal cortical insufficiency: The role of adrenal cortical steroids and of serum sodium concentration. *Journal of Clinical Investigations,* 42(5):727-735, 1963.

Henkin, R.I.; McGlone, R.E.; et al. Studies on auditory thresholds in normal man and in patients with adrenal corticol insufficiency: the role of adrenal cortical steroids. *Journal of Clinical Investigations,* 46(3):429-435, 1967.

Hildebrandt, Von G. Balneologie und rhythmusforschung. *Allgemeine Therapeutik,* 6(7), July-Stepmber 1961.

Hildebrandt, Von G. Storungen der rhythmischen koordination und ihre balneotherapeutische beeinflussing. *Bader-Und Klimaheilkunde,* 4, August 1963.

Kahana, L.; Lebovitz, H.; et al. Endocrine manifestations of intracranial extrasellar lesions. *Journal of Clinical Endocrinology and Metabolism,* 22:304-324, March 1962.

Krieger, D.T.; and Krieger, H.P. Adrenal function in central nervous system disease. *Endocrine and the Central Nervous System,* 43:400-417, 1966.

Libow, L.S., and Durell, J. Interrelationships between thyroid function and psychosis: I. A case of periodic psychosis with coupled alterations in thyroid function. *Psychosomatic Medicine,* 27,369, 1965.

Linquette, M.; Fossati, P.; Racadot, A.; Hubschman, B.; and Decoulx, M. Les variations circadiennes de la cortisolemie dans la maladie de Cushing et les etats frontieres. *Annals of Endocrinology,* 29:67-76, 1967.

Lunedei, A. In: Pozzy, E.L. (ed.), *Importanza dei fattori psichici nella patologia umana considetta internistica.* Roma: SEU, 1961. pp. 267-279.

McDaniel, W.B. The moon, werewolves and medicine. *College of Physicians of Philadelphia, Transactions and Studies,* 18:113-122, 1950.

Mead, R. A discourse concerning the action of the sun and moon on animal bodies; and the influence which this may have in many diseases. Published in London in 1708.

Menzel, W. Langwellige Rhythmen bei inneren Krankeiten. *Verhandlungen der Deutschen Gesellschaft für innere Medizin,* 73:962-973, 1967.

Menzel, W. Pertubations des rhythmes circadiares chez l'homme, y compris aspect psychosomatique. In: de Ajuriaguerra, J. ed. *Cycles Biologiques et Psychiatrie.* Paris: Masson et Cie, 1968., pp. 205-221.

Möllerstrom, J., and Sollberger, A. Fundamental concepts underlying the metabolic periodicity in diabetes. *Annals of the New York Academy of Sciences,* 98(4):984-994, 1962.

Möllerstrom, J., and Sollberger, A. The 24-hour rhythm of metabolic processes in diabetes; citric acid in the urine. *Acta Medica Scandinavica,* 160:25, 1958.

Oppenheimer, J.H.; Fisher, L.V.; and Jailer, J.W. Disturbance of the pituitary-adrenal interrelationship in diseases of the central nervous system. *Journal of Clinical Endocrinology and Metabolism,* 21(a):1023-1036, 1961.

Reimann, H.A. *Periodic Disease,* Oxford, England: Blackwell Scientific Publications, 1963, 189 pp.

Reinberg, A. The hours of changing responsiveness or susceptibility. *Perspectives in Biology and Medicine,* 11(1):111-128, 1967.

Reinberg, A.; Sidi, E.; and Ghata, J. Circadian reactivity rhythms of human skin to histamine or allergen and the adrenal cycle. *Journal of Allergy,* 36(3):273-283, 1965.

Serge, G.; Turco, G.; and Ceresa, F. Assay of a compartmental analysis of blood glucose and insulin relationships in man. *Acta Diabetologica Latina,* 5:242-243, October 1968.

Serio, M.; Tarquini, B.; Contini, P.; Bucalossi, A.; and Toccafondi, R. Plasma cortisol response to insulin and circadian rhythm in diabetic subjects. *Diabetes,* 17(3):124-127, 1968.

Shilov, V.M., and Kozar, M.I. Changes in the immunological reactivity of man exposed to various day schedules in a sealed chamber. [Simpozium] *Biologicheskiye ritmy i voprosy razrabotki rezhimov truda i otdykha, Materialy.* Moscow, 1967, 71-72.

Sholitan, L.J.; Werk, E.E., Jr.; and Marnell, R.T. Diurnal variation of adrenocortical function in non-endocrine disease states. *Metabolism,* 10:632-646, 1961.

Southren, A.L.; Gordon, G.G.; et al. Mean plasma concentration, metabolic clearance and basal plasma production rates of testosterone in normal young men and women using a constant infusion procedure: Effect of time of day and plasma concentration on the metabolic clearance rate of testosterone. *Journal of Clinical Endocrinology and Metabolism,* 27(5):686-694, 1967.

Tarquini, B.; Della Corte, M.; and Orzalesi, R. Circadian studies on plasma cortisol in subjects with peptic ulcer. *Journal of Endocrinology,* 38:475-476, 1967.

Isolation: Free-Running

Angiboust, R., and Saumande, P. Speleologic expedition in the vitarelles caverns in August 1964 - Biological and psychophysiological results. [Expedition speleologique a la cavite des vitarelles en Aout 1964 - Resultats biologiques et psycho-physiologiques]. *Revue de Medecine Aeronautique,* 4:22-24, 1965.

Apfelbaum, M., and Nillus, P. Evolution de la conductance physiologique chez les femmes vivant a 11°C pendant quinze jours. *Revue Francaise Etudes Clin. et Biol.,* 12:80-85, 1969.

Apfelbaum, M.; Reinberg, A.; Nillus, P.; and Halberg, F. Rhythmes circadiens de l'alternance veille-sommeil pendant l'isolement souterrain de sept jeunes femmes. *Presse Medicale,* 77:879-882, 1969.

Aschoff, J. Circadian Rhythms in Man. *Science,* 148:1427-1432, 1965.

Aschoff, J. Time givers of 24-hour physiological cycles. In: Schaefer, K.E., (ed.), *Man's Dependence on the Earthly Atmosphere.* New York: Macmillan, 1962, 373-380.

Aschoff, J. Human circadian rhythms in activity, body temperature and other functions. [Symposium] In: Brown, A.H. and Favorite, F.G., (eds.), *Life Science and Space Research V, International Space Science Symposium, 7th, Vienna, Austria.* Amsterdam: North-Holland Publishing Company, 1967, pp. 159-173.

Aschoff, J.; Gerecke, U.; and Wever, R. Desynchronization of human circadian rhythms. *Japanese Journal of Physiology,* 17:450-457, 1967.

Bayevskiy, R.M.; Chernyayeva, S.A.; et al. Investigation of the physiological condition of men exposed to isolation and hypodynamia during the shift to a new work and rest schedule. [Sympozium] *Biologicheskiye ritmy i voprosy razrabotki rezhimov truda i otdykha,* Materialy. Moscow, 1967, 12-13.

Cherepakhin, M.A. Normalization of physiological function under conditions of hypokinesia caused by a regime of motor activity. [Simpozium] *Biologicheskiye ritmy i voprosy razrabotki rezhimov truda i otdykha,* 1967. Materialy. Moscow, 1967, 69-70.

Chipman, D.M., and Sharon, N. Mechanism of lysozyme action. *Science,* 1-5(3892):454-465, August 1969.

Clogg, B.R., and Schaefer, K.E. Studies of circadian cycles in human subjects during prolonged isolation in a constant environment using 8-channel telemetry systems. Memorandum Report No. 66-4(NASA) February 19, 1966.

Colin, J.; Noudas, Y.; Boutelier, C.; Timbal, J.; and Siffre, M. Etude du Rhythm Circadien de la temperature Centrale d'un Sujet au cours d'Isolement Souterrain de 6 mois. *XVI International Congress of Aviation Space Medicine,* Lisbon, September 11-15, 1967.

Colin, J.; Timbal, J.; et al. Rhythm of the rectal temperature during a 6-month free-running experiment. *Journal of Applied Physiology,* 25(2):170-176, August 1968.

Dushkov, B.A. Change in rhythmic activity and in motor coordination during confinement to a small-capacity closed chamber. [Simpozium] *Biologisheskiye ritmy i razrabotki rezhimov truda i otdykha,* 1967. Materialy. Moscow, 1967, 27.

Ghata, J.; Halberg, F.; et al. Rhythmes circadiens desynchronises du cycle (17-hydroxycorticosteroides, temperature rectale, veille-sommeil) chez deux sujets adultes sains. *Annales d'Endocrinologie,* 30(2):245-260, 1969.

Gorbov, F.D.; Novikov, M.A.; et al. Group activity under stress conditions (during prolonged group isolation). [Sympozium] *Biologicheskiye ritmy i voprosy razrabotki rezhimov truda i otdykha,* 1967. Materialy. Moscow, 1967, 24-26.

Halberg, F.; Nelson, W.; et al. Reproducibility of circadian temperature rhythm in the rat kept in continuous light of 30 lux intensity. *Physiologist,* 9(3):196, August 1966.

Halberg, F.; Siffre, M.; Engeli, M.; Hillman, D.; and Reinberg, A. Etude en libre-cours des rythmes circadiens du pouls de l'alternance veilles-sommeil et de l'estimation du temps pendant les deux mois de dejour souterrain d'un homme adulte jeune. *Comptes-rendus de l'Acadamie de Science,* 260:1259-1262, 1965.

Korotayev, M.M.; Mikhaylovskiy, G.P.; Tsyganova, N.I. Changes in the human organism's reactivity during a prolonged stay in a small chamber. [Simpozium] *Biologicheskiye ritmy i voprosy razrabotki rezhimov truda i otdykha,* 1967. Materialy. Moscow. 1967, 35-36.

Litsov, A.N. Daily dynamics of some physiological functions and human work capacity in isolation. *Space Biology and Medicine,* 2(4):142-148, November 1968.

Lowe, C.H.; Hinds, D.S.; et al. Natural free-running period in vertebrate animal population. *Science,* 156:531-534, 1967.

Meddis, R. Human circadian rhythms and the 48-hour day. *Nature,* 218:964-965, June 1968.

Miasnikov, V.I. Electroenphalographic changes in persons isolated for long periods. *Kosmicheskie Issledovaniia,* 2:154-161, January-February 1964; and in *Cosmic Research,* 2:133-138, January-February 1964.

Mills, J.N. Circadian rhythms during and after three months in solitude underground. *Journal of Physiology,* 174:217-231, 1964.

Mills, J.N. Sleeping habits during four months in solitude. *Journal of Physiology,* 189:30-31, March 1967.

Myasinkov, V.N. Circadian rhythm of physiological functions in man under conditions of isolation. *Aerospace Medicine,* 73-78, October 1968.

NASA. The effects of confinement on long duration manned space flights. Proceedings of the NASA Symposium, November 17, 1966.

Pöppel, E. Desynchronisationen circadianer rhythmen innerhalbeiner isolierten gruppe. *Pflugers Archiv für die Gesammelte Physiologie,* 299:11-12, 1968.

Pöppel, E. Desynchronization of circadian rhythms within an isolated group. *Pflugers Archiv fur Die Gesammelte Physiologie,* 299:364-370, March 1968.

Reinberg, A. Halberg, F.; Ghata, J., and Siffre, M. Spectre thermique (rhythmes de la temperature rectale) d'une femme adulte saine avant, pendant, apress on isolement souterraine de trois mois. *Comptes Rendus Academie Sciences,* 262:782-785, 1966.

Reinberg, A. Eclairement et cycle menstruel de la femme. Rapport au Collogue International du CRNS "La photo-regulation de la reproduction chez les Oiseaux et les maumiferes", *Montpellier,* 1967.

Romashkin-Timanov, V.I., and Peskov, N.N. Combined effect of hypodynamia and certain extreme factors on the functional conditions of human central nervous and cardiovascular systems. *Simpozium Biologicheskiye ritmy i voprosy razrabotki rezhimov truda i otdykha.* Materialy. Moscow, Moscow, 1967, 58-59.

Siffre, M; Reinberg, A.; Halberg, F.; Ghata, J.; Perdriel, G.; and Slind, R. Studies on two healthy human subjects isolated for several months underground. *La Presse Medicale,* 74:915-919, 1966.

Simpson, H.W., and Lobban, M.C. Effects of a 21-hour day on the human circadian excretory rhythms of 17-hydroxycorticosteroids and electrolytes. *Aerospace Medicine,* 38(12): 1205-1213, 1967.

Simpson, H.W.; Lobban, M.C.; and Halberg, F. Near 24-hour rhythms in subjects living on a 21-hour routine in the arctic summer at 78 degrees N - revealed by circadian amplitude ratios. *Ressegna di Neurologia Vegetativa.* In press.

Smith, W.R.; Ulvedal, F.; and Welch, B.E. Steroid and catecholamine studies on pilots during prolonged experiments in a space cabin simulator. *Journal of Applied Physiologyy,* 18:1257-1263, 1963.

Wever, R. The Duration of Re-entrainment of Circadian Rhythms after Phase Shifts of the Zeitgeber. *Journal of Theoretical Biology,* 13:187-201, 1966.

Wever, R. The influence of weak electromagnetic fields on the circadian rhythm in man. *Zeitschrift Für Vergleichende Physiologie,* 56:111-128, 1967.

Wever, R. Autonomous Circadian Rhythms in Men as Influenced by Different Light Conditions. *Pflügers Archiv,* 306:71-91, 1969.

Wever, R. Influence of electric fields on some parameters of circadian rhythms in man. *Proceedings of Friday Harbor Symposium on Chronobiology.* In press.

Light: Photoperiodism, Reproduction, and Circadian Rhythms

Axelrod, J.; Snyder, S.H.; Heller, A.; and Moore, R.Y. Light induced changes in pineal hydroxyindole-O-methyltransferase: abolition by lateral hypothalamic lesions. *Science,* 154:989-90, 1966.

Axelrod, J.; Wurtman, R.J.; and Winget, C.M. Melatonin synthesis in the hen pineal gland and its control by light. *Nature,* 201(4924):1134, 1964.

Axelrod, J.; Wurtman, R.J.; and Snyder, S.H. Control of hydroxyindole O-methyltransferase activity in the rat pineal gland by environmental lighting. *Journal of Biological Chemistry,* 240(2):949-954, 1965.

Benoit, J., and Assenmacher, I. The control by visible radiations of the gonadotropic activity of the duck hypophysis. *Recent Progress in Hormone Research No. 15*:143-164, 1959.

Blaney, L.T.; and Hamner, K.C. Interrelations among effects of temperature, photoperiod, and dark period on floral initiation of Biloxi soybean. *Botanical Gazette,* 119:(1)10-24, 1957.

Bruce, V.G., and Minis, D.H. Circadian clock action spectrum in a photoperiodic moth. *Science,* 163:583-585, 1969.

Bruce, V.G., and Pittendrigh, C.S. Resetting the euglena clock with a single light stimulus. *American Naturalist,* 92:295-305, 1958.

Bruce, V.G.; Weight, F.; and Pittendrigh, C.S. Resetting the sporulation rhythm in Pilobolus with short light flashes of high intensity. *Science,* 131(3402):728-730, 1960.

Bünning, E. Known and Unknown Principles of Biological Chronometry. In: Fischer, R., (ed.) *Interdisciplinary Perspectives of Time. Annals of the New York Academy of Sciences,* 138:515-524, 1967.

Bünning, E. *The Physiological Clock.* Revised, Second Edition, New York: Springer-Verlag, 1967.

Coulter, M.W. and Hamner, K.C. Quantitative assay of photoperiodic floral inhibition and stimulation in biloxi soybean. *Plant Physiology,* 40(5):873-881, September 1965.

Dewan, E.M. On the possibility of a perfect rhythm method of birth control by periodic light stimulation. *American Journal of Obstetrics and Gynecology,* 99(7):1016-1019, December 1967.

Dewan, E.M. Rhythms. *Science and Technology,* pp.20-28, January 1969.

Emlen, S.T. Bird Migration: Influence of physiological state upon celestial orientation. *Science,* 165:716-718, August 1969.

Eskin, I.A., and Mikhailova, N.V. Photoperiodicity and the function of the hypophysis and the adrenal cortex. *Bulletin of Experimental Biology and Medicine U.S.S.R.,* (English translation), 46:999-1002, 1958.

Everett, J.W., and Sawyer, C.H. A 24-hour periodicity in the "LH-Release Apparatus" of female rats, disclosed by barbiturate sedation. *Endocrinology* 47:198-218, September 1950.

Farner, D.S. The photoperiodic control of reproductive cycles in birds. *American Scientistt,* 52:137-56, 1964.

Fiske, V.M., and Huppert, L.C. Melatonin action on pineal varies with photoperiod. *Science,* 162:279, October 1968.

Fox, H.M. Lunar periodicity in reproduction. *Proceedings of the Royal Society of Britain,* 95, 1923.

Gerritzen, F. Influence of light on human circadian rhythms. *Aerospace Medicine,* 37:66-70, 1966.

Gerritzen, F.; Strengers, T; et al. Studies on the influence of fast transportation on the circadian excretion pattern of the kidney in humans. *Aerospace Medicine,* 40:264-271, March 1969.

Goff, M.L.R., and Finger, F.W. Activity rhythms and diurnal light-dark control. *Science,* 154:1346-1348, 1966.

Halberg, F.; Halberg, E.; Barnum, C.P.; and Bittner, J.J. Physiologic 24-hour periodicity in human beings and mice, the lighting regimen and daily routine. In: Withrow, R.B., (ed.) *Photoperiodism and Related Phenomena in Plants and Animals.* Washington, D.C.: Publication of the American Association for the Advancement of Science, 1959. pp. 803-878.

Halberg, F.; and Visscher, M.B. Temperature rhythms in blind mice. *Federation Proceedings,* 13:65 1954.

Hamner, K.C., and Takimoto, A. Circadian rhythms and plant photoperiodism. *The American Naturalist,* 98(902):295-322, September-October 1964.

Harker, J.E. The effect of photoperiod on the developmental rate of Drosophila pupae. *Journal of Experimental Biology,* 43:411-423, 1965.

Haus, E.; Lakatua, D.; and Halberg, F. The internal timing of several circadian rhythms in the blinded mouse. *Experimental Medicine and Surgery*, 25:7-45, 1967.

Hoffman, K. Overt circadian frequencies and circadian rule. In: Aschoff, J., (ed.), *Circadian Clocks.* Proceedings of the Feldafing Summer School. Amsterdam: North-Holland Publishing Company, 1965.

Hollwich, F. Von. Der Einfluth des Augenlichtes auf Stoffwechselvorgange, *Acta Neurovegetativa,* 30(1-4):201-212, 1967.

Hollwich, F.Von. The influence of light via the eye on the endocrine system. *Anales del Instituto Barraquer* IX(1, 2):133-142, 1969.

Hollwich, F. Von, and Dieckhues, B. Augenlicht und Nebennierenrindenfunktion, *Deutsche Medizinische Wochenschrift,* 1-17, December 22, 1967.

Hollwich, F. Von, and Dieckhues, B. Eosinopeniereaktion und Sehvermogen. *Sonderdruck aus Klinische Monatblatter fur Augenheilkunde,* 152(1)11-16, 1968.

Hollwich, F. Von and Tilgner, S. The influence of light via the eyes on the thyroid and testes. *Deutsche Medizinische Wochenschrift,* 87:2674, 1962.

Holmgren, H., and Swensson, A. Der Einfluss des lichtes auf den 24-Stunden. Rhythmus der Activitat des Leberglykogens und der Korpertemperatur. *Acta Medica Scandinavica* (Supplement 278):71-76, 1953.

Holmquest, D.L., and Lipscomb, H.S. The response of thermal and activity rhythms in the rat to cyclic variations in adrenocortical function and environmental lighting. *Proceedings of the International Union of Physiological Sciences,* 7, 1968.

Holmquest, D.L.; Retiene K.; and Lipscomb, H.S. Circadian rhythms in rats: effects of random lighting. *Science,* 152:662-664, April 1966.

Honjo, S.; Fujiwara, T.; Takasaka, M.; Suzuki, Y; and Imaizumi, K. Observations on the diurnal temperature variation of cynomolgus monkeys (Macaca irus) and on the effect of changes in the routine of lighting upon this variation. *Japanese Journal of Medical Science and Biology,* 16:189-198, 1963.

Hoshizaki, T.; Brest, D.E.; and Hamner, K.C. Xanthium leaf movements in light and dark. *Plant Physiology,* 44:151-152, 1969.

Hoshizaki, T., and Hamner, K.C. Circadian leaf movements: persistence in bean plants grown in continuous high-intensity light. *Science,* 144:1240-1241, 1964.

Hoshizaki, T., Hamner, K.C. Interactions between light and circadian rhythms in plant photoperiodism. *Photochemistry and Photobiology,* 10:87-97, 1969.

Kaiser, I.H. Effect of a 28-hour day on ovulation and reproduction in mice. *American Journal of Obstetrics and Gynecology,* 99:772-784, 1967.

Kaiser, I.H., and Halberg, F. Circadian aspects of birth. *Annals of the New York Academy of Sciences,* 98:1056-1060, 1962.

Kerenyi, N.A., and Sarkar, K. The postnatal transformation of the pineal gland. *Acta Morphologia Academia Scientia Hungaria,* 16:223, 1968.

Kopell, B.S.; Lunde, D.T.; Clayton, R.B., and Moos, R.H. Variations in some measures of arousal during the menstrual cycle. *Journal of Nervous and Mental Disease,* 148:180-187, 1969.

Lincoln, R.G., and Hamner, K.C. An effect of gibberellic acid on the flowering of xanthium, a short day plant. *Plant Physiology,* 33(2):101-104, 1958.

Lisk, R.D. Direct effects of light on the hypothalamus. (Symposium on Hypothalamic Control of the Anterior Pituitary). American Association for the Advancement of Science, December 27, 1965.

Lisk, R.D.; and Sawyer, C.H. Induction of Paradoxical Sleep by Lights-off Stimulation. *Proceedings of the Society of Experimental Biology and Medicine,* 123:664-667, 1966.

Lobban, M.C., and Tredre, B.E. Perception of light and the maintenance of human renal diurnal rhythms. *Journal of Physiology,* 89(1):32-33, March 6, 1967.

Lobban, M.C., and Tredre, B. Renal diurnal rhythms in blind subjects. *Journal of Physiology,* 170:29-30, 1964.

Makarova, L.G. Changes of EEG in a healthy individual with trigger light stimulation. *Bulletin of USSR, Academy of Medical Science, Institute of Neurology, Moscow,* 62:6-11, November 1966.

Meites, J. Direct studies of the secretion of hypothalamic hypophysiotropic hormones (HHH), In: Meites, J. (ed.), *Hypophysiotropic Hormones of the Hypothalamus.* Baltimore: Williams and Wilkins Co., 1970, pp. 261.

Meites, J. Releasing factors in hypothalamic control of the anterior pituitary in mammals. (Symposium on Hypothalamic Control of the Anterior Pituitary). American Association for the Advancement of Science, December, 1965.

Menaker, M., and Eskin, A. Circadian clock in photoperiodic time measurement: A test of the Bünning Hypothesis. *Science,* 157:1182-1184, 1967.

Menaker, M. Lunar periodicity with reference to live births. *American Journal of Obstetrics and Gynecology,* 98(7):1002-1004, 1967.

Minis, D.H., and Pittendrigh, C.S. Circadian oscillation controlling hatching: its ontogeny during embryogenesis of a moth. *Science,* 159:534-536, 1968.

Moore, R.Y.; Heller, A.; Wurtman, R.J.; and Axelrod, J. Visual pathway mediating pineal response to environmental light. *Science,* 155:220-223, 1967.

Nanda, K.K., and Hamner, K.C. Studies on the nature of the endogenous rhythm affecting photoperiodic response of Biloxi soybean. *Botanical Gazette,* 120(1):14-25, 1958.

Nanda, K.K., and Hamner, K.C. The effect of temperature, auxins, anti-auxins, and some other chemicals on the endogenous rhythm affecting photoperiodic response of Biloxi soybean. *Planta,* 53:53-68, 1959.

Oishi, T.; and Kato, M. The pineal organ as a possible photoreceptor in photoperiodic testicular responses in Japanese quail. Memoirs of the Faculty of Science; Kyoto University, 2:Series D:12-18, 1968.

Orth, D.N., and Island, D.P. Light synchronization of the circadian rhythm in plasma cortisol (17-OHCS) concentration in man. *Journal of Clinical Endocrinology,* 29:479-486, 1969.

Pittendrigh, C.S. On the mechanism of the entrainment of a circadian rhythm by light cycles. In: Aschoff, J., (ed.) *Circadian Clocks.* Proceedings of the Feldafing Summer School. Amsterdam: North-Holland Publishing Company, 1965.

Pittendrigh, C.S. The circadian oscillation in Drosophila pseudoobscura pupae: A model for the photoperiodic clock. *Zeitschrift fur Planzenphysiologie,* 54:275-307, 1966.

Pittendrigh, C.S., and Minis, D.H. The entrainment of circadian oscillations by light and their role as photoperiodic clocks. *American Naturalist,* 98(902):261-294, 1964.

Reid, H.B., Moore, P.H.; and Hamner, K.C. Control of flowering on xanthium pensylvanicum by red and far-red light. *Plant Physiology,* 42:532-540, 1967.

Reinberg, A. Eclairement et cycle menstruel de la femme. Rapport au Collogue International du CRNS, *La photo-regulation de la reproduction chez les Oiseaux et les maumiferes,* Montpellier, 1967.

Rosenthal, J.D.; Sullivan, W.N.; Adler, V.E.; and McGuire, J.U. Influence of temperatures and light regimens on diapause of samia cynthia pryeri. *Journal of Economic Entomology,* 61(2):578-579, 1968.

Rust, C.C., and Meyer, R.K. Hair color, molt and testis size in male, short-tailed weasels treated with melatonin. *Science,* 165:921-922, 1969.

Saunders, D.S. Time measurements in insect photoperiodism: Reversal of a photoperiodic effect by chilling. *Science,* 156:1126-1127, 1967.

Scheving, L.E.; Verdal, D.F.; and Pauly, J.E. Daily circadian rhythm in rats to D-Amphetamine sulphate: effect of blinding and continuous illumination on the rhythm. *Nature,* 219(5154):621-622, 1968.

Shumate, W.H.; Reid, H.B.; and Hamner, K.C. Floral inhibition of biloxi soybean during a 72-hour cycle. *Plant Physiology,* 42:1511-1518, 1967.

Skopik, S.D., and Pittendrigh, C.S. Circadian systems. II. The oscillation in the individual Drosophila Pupa: Its independence of developmental stage. *Proceedings of the National Academy of Sciences,* 58(5):1862-1869, 1967.

Snyder, S.H.; Ganong, W.; Axelrod, J; and Fischer, J.E. Control of the circadian rhythm in serotonin content of the rat pineal gland. *Proceedings of the National Academy of Sciences,* 53:301-305, 1965.

Stephens, G.J.; Halberg, F.; and Stephens, G.C. The blinded fiddler crab: An invertebrate model of circadian desynchronization. *Annals of the New York Academy of Sciences,* 117(1):386-406, 1964.

Sullivan, W.N.; Cawley, B.M.; Oliver, M.; Hayes, D.K.; and McGuire, J.U. Manipulating the photoperiod to damage insects. *Nature,* 221(5175):60-61, 1969.

Takimoto, A., and Hamner, K.C. Studies on red light interruption in relation to timing mechanism involved in the photoperiodic response of Pharbitis nil. *Plant Physiology,* 40:852-854, 1965.

Thor, D.H.; and Hoats, D.L. A circadian variable in self-exposure to light by the rat. *Psychonomic Science,* 12(1):1-2, 1968.

Timonen, S.; Franzas, B.; and Wichmann, K. Photosensibilty of the Human Pituitary. *Annales Chirurgiae et Gynaecologiae Femniae,* 165-172, 1964.

Underwood, H., and Menaker, M. Photoperiodically significant photoreception in sparrows: is the retina involved? *Science,* 167:301, 1970.

Wever, R. Autonomous Circadian Rhythms in Men as Influenced by Different Light Conditions. *Pflügers Archiv,* 306:71-91, 1969.

Winget, C.M., and Card, D.H. Daily rhythm changes associated with variations in light intensity and color. *Life Sciences and Space Research,* 5:148-158, 1967.

Winget, C.M.; Rahlman, D.F.; and Pace, N. Phase relationship between circadian rhythms and photoperiodism (Cebus albifrons). In: Rohles, F.; (ed.), *Circadian Rhythms in Non-Human Primates.* New York: Karger, 1968.

Winget, C.M.; Rosenblatt, L.S., DeRoshia, C.W.; and Hetherington, N.W. Mechanisms of action of light on circadian rhythms in the monkey. *Life Sciences and Space Research,* 8. In press.

Wurtman, R.J. Effects of light and visual stimulii on endocrine function. In: Martini, L; and Ganong, W.F., (eds.), *Neuroendocrinology,* Chapter 13, New York: Academic Press, 1967.

Wurtman, R.J.; Axelrod, J.; Chu, E.W.; and Fischer, J.E. Mediation of some effects of illumination on the rat estrous cycle by the sympathetic nervous system. *Endocrinology,* 75:366, 1964.

Wurtman, R.J.; Axelrod, J.; Sedvall, G.; and Moore, R.Y. Photic and neural control of the 24-hour norepinephrine rhythm in the rat pineal gland. *Journal of Pharmacology and Experimental Therapeutics,* 157(3):487-492, 1967.

Menstrual Syndrome

Dalton, K. *The Premenstrual Syndrome.* Springfield, Illinois: Charles C. Thomas, 1964.

Dewan, E.M. On the Possibility of a perfect rhythm method of birth control by periodic light stimulation. *American Journal of Obstetrics and Gynecology,* 99(7):1016-1019, December 1967.

Janowsky, D.S.; Gorney, R; Castelnuovo-Tedesco, P.; and Stone, C.B. Premenstrual-menstrual increases in psychiatric hospital admission rates. *American Journal of Obstetrics and Gynecology,* 103:189-191, 1969.

Kopell, B.S.; Lunde, D.T.; Clayton, R.B., and Moos, R.H. Variations in some measures of arousal during the menstrual cycle. *Journal of Nervous and Mental Disease,* 148:180-187, 1969.

Malck, J.; Gleich, J.; and Maly, V. Characteristics of the daily rhythm of menstruation and labor. *Annals of the New York Academy of Sciences,* 98:1042-55, 1962.

Moos, R.H., Typology of menstrual cycle symptoms. *American Journal of Obstetrics and Gynecology,* 103(3):390-402, 1969.

Moos, R.H.; Kopell B.S.; Melges, F.T.; Yalom, I.D.; Lunde, D.T.; Clayton, R.B.; and Hamburg. D.A. Fluctuations in symptoms and moods during the menstrual cycle. *Journal of Psychosomatic Research,* 13:37-44, 1969.

Reinberg, A. Eclairement et cycle menstruel de la femme. Rapport au Collogue International du CRNS "La photo-regulation de la reproduction chez les Oiseaux et les maumiferes," *Montpellier,* 1967.

Smith, S.L.; and Sauder, C. Food cravings, depression, and premenstrual problems. *Psychosomatic Medicine,* 31:281-287, 1969.

Swanson, E.M., and Foulkes, D. Dream content and the menstrual cycle. *Journal of Nervous and Mental Disease,* 145(5):358-363, 1968.

University of Colorado Medical School, *Symposium on Menstrual Mechanisms, Denver, Colorado:* December 6-7, 1965.

Zacharias, L., and Wurtman, R.J. Blindness: its relation to age of menarche. *Science,* 144(3622):1154-1155. 1964.

Mental Illness

Berendes, H.W.; Marte, E.; Ertel, R.J.; McCarthy, J.A.; Anderson, J.A.; and Halberg, F. Circadian physiologic rhythm and lowered blood 5-hydroxytryptamine in human subjects with defective mentality. *Physiologist,* 3:20, 1960.

Bunney, Jr., W.E.; Fawcett, J.A.; Davis, J.M.; and Gifford, S. Further evaluation of urinary 17-hydroxycorticosteroids in suicidal patients. *Archives of General Psychiatry,* 21:138-150, 1969.

Bunney, Jr., W.E., and Hartmann, E.L. Study of a patient with 48-hour manic depressive cycles: I. and II. *Archives of General Psychology,* 12:611-625, 1965.

Cahn, A.A.; Folk, G.E.; and Huston. P.E. Age comparison of human day-night physiological differences. *Aerospace Medicine,* 39(6):608-610, 1968.

Cookson, B.A.; Quarrington, B.; and Huszka, L. Longitudinal study of periodic catatonia. *Journal of Psychiatric Research,* 5:15-38, 1967.

Cranmer, J.L. Rapid weight-changes in mental patients. *Lancet,* 259-262, August 10, 1957.

Curtis, G.C.; Fogel, M.L.; et al. The effect of sustained affect on the diurnal rhythms of adrenal cortical activity. *Psychosomatic Medicine,* 26:696-713, 1966.

Curtis, G.C.; Fogel, M.L.; McEvoy, D. and Zarate, C. Effects of weight, sex and diurnal variation on the excretion of 17-hydroxycorticosteroids. *Journal of Clinical Endocrinology and metabolism,* 28(5):711-713, 1968.

Dick, P.; Tissot, R.; and Pletscher, S. Influence des medicaments psychotropes sur les cycles biologiques et les cycles des psychoses. In: Ajuriaguerra, J., (ed.), *Cycles Biologiques et Psychiatrie.* Paris: Masson and Cie, 1968.

Durrell, J.; Libow, L.S.; Kellam, S.G.; and Shader, R.I. Interrelationships between regulation of thyroid gland function and psychosis. *Endocrines and the Central Nervous System,* Association for Research in Nervous and Mental Diseases, Vol. 43, Chapter XXI, 3872, 1966.

Durrell, J. Biological rhythms and psychiatry: psycho-endocrine mechanisms. In: de Ajuriaguerra, J., (ed.), *Cycles Biologiques et Psychiatrie.* Paris: Masson & Cie., pp. 321-328, 1968.

Elithorn, A.; Bridges, P,K.; et al. Observations on some diurnal rhythms in depressive illness. *British Medical Journal,* 1620-1623, December 1966.

Engel, R.; Halberg, F.; and Gully, R.J. The diurnal rhythm in EEG discharge and in circulating eosinophils in certain types of epilepsy. *Electroencephalography and Clinical Neurophysiology,* 4:115-116, 1952.

Feinberg, I. Eye movement activity during sleep and intellectual function in mental retardation. *Science,* 159:1256, 1968.

Fleeson, W.; Glueck, B.C., Jr.; and Halberg, F. Persistence of daily rhythms in eosinophil count and rectal temperature during "regression" induced by intensive electroshock therapy. *Physiologist,* 1:28, 1957.

Fullerton, D.T. Circadian rhythm of adrenal cortical activity in depression. *Archives of General Psychiatry,* 19(6):674, 1968.

Gjessing, L.R. Studies of periodic catatonia, II: The urinary excretion of phenolic amines and acids with and without loads of different drugs. *Journal of Psychiatric Research,* 2(3):149-162, 1964.

Gjessing, L.R. A review of the biochemistry of periodic catatonia. *Excerpta Medica International Congress Series,* No. 150, 1966.

Gjessing, R., and Gjessing, L.R. Some main trends in the clinical aspects of periodic catatonia. *Acta Psychiatrica Scandinavica,* 37(1):1-13, 1961.

Gjessing, L.R.; Jenner, F.A.; Harding, G.F.A.; and Johannessen, N.B. The EEG in three cases of periodic catatonia. *British Journal of Psychiatry,* 113(504):1271-1282, 1967.

Glueck, B.C. The use of computers in patient care. *Mental Hospitals,* 16(4):117-120, 1965.

Glueck, B.C., (Investigator) and Luce, G. (Writer) The Computer as psychiatric aid and research tool. *Mental Health Program Reports–2,* National Institute of Mental Health, Chevy Chase, Md., February 1968, PHS Publication 1743. pp. 353-363.

Goodwin, J.C.; Jenner, F.A.; Lobban, M.C.; and Sheridan, M. Renal Rhythms in a patient with a 48-hour cycle of psychosis during a period of life on an abnormal time routine. *Journal of Physiology,* 17:16-17, 1964.

Halberg, F. Physiologic considerations underlying rhythmometry with special reference to emotional illness. In: de Ajuriaguerra, J., (ed.) *Cycles Biologiques et Psychiatrie.* Paris: Masson & Cie., 1968. pp. 73-126.

Halberg, F.; Engel, R.; Halberg, E.; and Gully, R.J. Diurnal variations in amount of electroencephalographic paroxysmal discharge and diurnal eosinophil rhythm of epileptics on days with clinical seizures. *Federation Proceedings,* 11:63, 1952.

Halberg, E.; Halberg, F.; and Bittner, J.J. Daily periodicity of convulsions in man and in mice. Report from the 5th Conference of the Society for the Study of Biologic Rhythms, Stockholm, p. 97, 1967.

Halberg, F.; Jacobson, E.; Wadsworth, G.; and Bittner, J.J. Audiogenic abnormality spectra, 24-hour periodicity and lighting. *Science,* 128:657-658, 1958.

Halberg, F.; Vestergaard, P.; and Sakai, M. Rhythmometry on urinary 17-ketosteroid excretion by healthy men and women and patients with chronic schizophrenia; possible chronopathology in depressive illness. *Archives d'Anatomie, d'Histologie et d'Embryologie Normales et Experimentales,* 51:301-311, 1968.

Harding, G.F.A., and Jenner, F.A. The electroencephalogram in three cases of periodic psychosis. *Electroencephalography and Clinical Neurophysiology,* 21:59-66, 1966.

Hartmann, E. Longitudinal studies of sleep and dream patterns in manic-depressive patients. *Archives of General Psychiatry,* 19:312-329, September 1968.

Hatotani, N.; Ishida, C.; Yura, R.; Maeda, M.; Kato, Y.; and Nomura, J. Psycho-pysiological studies of atypical psychoses-endocrinological aspect of periodic psychoses. *Folia Psychiatrica et Neurologica Japonica,* 16(3): 248-292, 1962.

Jenner, F.A. Periodic psychoses in the light of biological rhythm research. *International Review of Neurobiology,* 11:129-169, 1968.

Jenner, F.A. Studies of recurrent and predictable behavior. Proceedings: Leeds Symposium on Behavioral Disorders, 1965.

Jenner, F.A.; Gjessing, L.R.; Cox, J.R.; Davies-Jones, A.; Hullin, R.P.; and Hanna, S.M. A manic-depressive psychotic with a persistent 48-hour cycle. *British Journal of Psychiatry,* 113(501):895-910, 1967.

Jenner, F.A.; Goodwin, J.C.; Sheridan, M.; Tauber, I.J.; and Lobban M.C. The effect of an altered time regime on biological rhythms in a 48-hour periodic psychosis. *British Journal of Psychiatry,* 114:213, 1968.

Knapp, M.S.; Keane, P.M. and Wright, J.G. Circadian rhythm of plasma 11-hydroxycorticosteroids in depressive illness, congestive heart failure and Cushing's syndrome. *British Journal of Psychiatry,* 2:27-30, 1967.

Krieger, D.T.; Kolodny, H.D.; and Warner, R.R.P. Serum serotonin in nervous system disease. *Neurology,* 14(6):578-580, June 1964.

Lobban, M.C., and Tredre, B.E. Diurnal rhythms of electrolyte excretion in depressive illness. *Nature,* 199(4894):667-669, 1963.

Lohrenz, F.N.; Fullerton, D.T.; et al. Adrenocortical function in depressive states, study of circadian variation in plasma and urinary steroids. *International Journal of Neuropsychiatry,* 4: 21-25, 1968.

Melges, F.T., and Fougerousse, C.E., Jr. Time sense, emotions, and acute mental illness. *Journal of Psychiatric Research,* 4:127-140, 1966.

Richter, C.P. Biological approach to manic-depressive insanity. *Proceedings of the Association for Research in Nervous and Mental Disease,* 11:611, 1930.

Richter, C.P. Abnormal but regular cycles in behavior and metabolism in rats and catatonic-schizophrenics. In: Reiss, M., (ed.), *Psychoendocrinology.* New York: Grune & Stratton, 1958, pp. 168-181.

Richter, C.P. Psychopathology of periodic behavior in animals and man. *Comparative Psychopathology.* New York: Grune & Stratton, 1967.

Rubin, R.T.; Young, W.M.; and Clark, B.R. 17-hydroxycorticosteroid and vanillylmandelic acid excretion in a rapidly cycling manic-depressive. *Psychosomatic Medicine,* 30:162-171, 1968.

Stroebel, C.F. Behavioral aspects of circadian rhythms. In: Zubin, J., and Hunt, H.F., (eds.), *Comparative Psychopathology.* New York: Grune and Stratton, 1967.

Stroebel, C.F. A biologic rhythm approach to psychiatric treatment. In: *Proceedings of the Seventh Medical Symposium.* New York: Yorktown Heights, IBM Press, 1967, pp. 215-241.

Stroebel, C.F. Biologic rhythm correlates of disturbed behavior in the Rhesus monkey. In: Rohles, F.H. (ed.), *Circadian Rhythms in Nonhuman Primates.* New York: Karger, 1969, pp 91-105.

Stroebel, C.F. (Investigator) and Luce, G. (Writer). The Importance of biological clocks in mental health. Mental Health Program Reports—2. National Institute of Mental Health, Chevy Chase, Md., February 1968. PHS Publication 1743. pp 323-351.

Sueda, T. Schwankungen im 24-Stunden Rhythmus bei manisch-depressiven. Kranksein - Pathophysiologie des manisch-depressiven Krenkseins. *Psychiatria et Neurologia Japonica,* 9:1449-1485, 1962.

Tamura, A. Changes of diurnal rhythm of Na-, K. and Ca-excretion in urine with disorders in brain. *Psychiatria et Neurologica Japonica,* 5:405-23, 1965.

Vestergaard, P.; Leverett, R.; and Douglas, W.R., Spontaneous variability in the excretion of combined neutral 17-ketosteroids in the urine of chronic schizophrenic patients. *Psychiatric Research Reports, No. 6,* American Psychiatric Association, pp. 74-89, October 1956.

Wakoh, T. Endocrinological studies on periodic psychosis. *Mie Medical Journal,* 9(2):351-390, 1959.

Wakoh, T.; Takekoshi, A.; Yoshimoto, S.; Yoshimoto, K.; Hiramoto, K.; and Kurosawa, R. Pathophysiological study of the periodic psychosis (atypical endogeneous psychosis) with special reference to the comparison with the chronic schizophrenic. *Mie Medical Journal,* 10(3):317-396, 1960.

Yamashita, I; Shinohara, S.; Nakazawa, A.; Yoshimura, Y.; Ito, K.; and Takasugi, K. Endocrinological study of atypical psychosis. *Japonica,* 16(3):293-298, 1962.

Methods: Data Analysis

Adey, W.R.; Kado, R.T.; and Walter, D.O. Computer analysis of EEG data from Gemini flight G-T 7. *Aerospace Medicine,* 38(4):345-359, April, 1967.

Cole, L.C. Biological Clock in the Unicorn. *Science,* 125:874-876, 1957.

Engeli, M., and Halberg, F. Spectral analysis of steroid rhythms in data at equal or unequal intervals. *Federation Proceedings,* 23(2):897 March-April, 1964.

Enright, J.T. The search for rhythmicity in biological time-series. *Journal of Theoretical Biology,* 8:426-468, 1965.

Fischer, R.; and Sollberger, A. Interdisciplinary perspectives of time. In: Weyer, E.M. and Fischer, R. (eds.) *Annals of the New York Academy of Sciences,* 138(2):367-415, 1967.

Gabrieli, E.R., and Sollberger, A. In: Gabrieli, E.R.; Albertson, P.D.; and Krauss, M., (eds.). Conference on the use of data mechanization and computers in clinical medicine. *Annals of New York Academy of Sciences,* 161(2):371, 1969.

Garcia-Sainz, M., and Halberg, F. Mitotic rhythms in human cancer reevaluated by electronic computer programs - evidence for temporal pathology. *Journal of the National Cancer Institute,* 37:279-292, 1966.

Globus, G.G. Quantification of the sleep cycle as a rhythm. *Psychophysiology.* In press, 1970.

Glueck, B.C. The use of computers in patient care. *Mental Hospitals,* 16(4):117-120, 1965.

Glueck, B.C. (Investigator), and Luce, G. (Author). The Computer as psychiatric aid and research tool. *Mental Health Program Reports-2,* National Institute of Mental Health, Chevy Chase, Md., February 1968, PHS Publication 1743. pp. 353-363.

Gordon, R.; Spinks, J.; Dulmanis, A.; Hudson P.; Halberg, F.; and Bartter, F. Amplitude and phase relations of several circadian rhythms in human plasma and urine resolved by cosinor: demonstration of rhythm for tetrahydrocortisol and tetrahydrocorticosterone. *Clinical Science,* 35:307-324, 1968.

Halberg, F. Circadian temporal organization and experimental pathology. VII conference Internazionale della Societa per lo studio dei Ritimi Biologici, Siena, pp. 52-69. September 1969.

Halberg, F. Physiologic considerations underlying rhythmometry with special reference to emotional illness. In: de Ajuriaguerra, J., (ed.) *Cycles Biologiques et Psychiatrie.* Paris: Masson & Cie., 1968, pp. 73-126.

Halberg, F. Physiologic rhythms. *Physiological Problems in Space Travel.* Springfield, Illinois: Charles C. Thomas, 1964, pp. 298-322.

Halberg, F. Physiologic rhythms and bioastronautics. In: Schafer, K.E., (ed.), *Bioastronautics.* New York: Macmillan, 1964, pp. 181-195.

Halberg, F. Temporal coordination of physiological function. In: *Cold Spring Harbor Symposia on Quantitative Biology*. New York: Long Island Biological Association, 1960. pp. 289-310.

Halberg, F.; Cohen, S.L.; and Flink, E.B. The new tools for the diagnosis of adrenal dysfunction. *Journal of Laboratory and Clinical Medicine,* 38:817, 1951.

Halberg, F.; Engeli, M.; Hamburger, C.; and Hillman, D. Spectral resolution of low-frequency, small-amplitude rhythms in excreted ketosteroid; probable androgen-induced circaseptan desynchronization. *Acta Endocrinologica Supplement,* 103:1-54, 1965.

Halberg, F., and Falliers, C.J. Variability of physiologic circadian crests in groups of children studied "transversely." *Pediatrics,* 68:741-746, 1966.

Halberg, F.; Reinhardt, J.; et al. Agreement in endpoints from circadian rhythmometry on healthy human beings living on different continents. *Experimenta,* 25:107-112, 1969.

Halberg, F.; Stein, M.; Diffley, M.; Panofsky, H.; and Adkins, G. Computer techniques in the study of biologic rhythms. *Annals of the New York Academy of Sciences,* 115:695-720, 1964.

Hoffman, K. Overt circadian frequencies and circadian rule. In: Aschoff, J., (ed.), *Circadian clocks.* Proceedings of the Feldafing Summer School. Amsterdam: North-Holland Publishing Company, 1965.

Hoshizaki, T., and Hamner, K.C. Computer analysis of the leaf movements of pinto beans. *Plant Physiology,* 44:1045-1050, 1969.

Iberall, A.S., and Cardon, S.Z. Control in biological systems-physical review. *Annals of the New York Academy of Sciences,* 117:445-515, 1964.

Kripke, D.F.; Crowley, T.J.; Pegram, G.V.; and Halberg, F. Circadian rhythmic amplitude modulation of Berger-Region frequencies in electroencephalograms from Macaca Mulatta. *Rassegna di Neurologia Vegetativa,* 22:519-525, 1968.

Möllerstrom, J., and Sollberger, A. Fundamental concepts underlying the metabolic periodicity in diabetes. *Annals of the New York Academy of Sciences,* 98(4):984-994, 1962.

Pavlidis, T. A model for circadian clocks. *Bulletin of Mathematical Biophysics,* 29:781-791, 1967.

Pittendrigh, C.S. and Bruce, V.G. An oscillator model for biological clocks. In: Rudnick A., (ed.), *Rhythmic and Synthetic Processes in Growth.* Princeton: Princeton University Press, 1957.

Richards, A.G., and Halberg, F. Oxygen uptake rhythms in a cockroach gauged by variance spectra. *Experientia,* 20:40-42, 1964.

Sollberger, A. Collapsible three-dimensional diagram and model. *Medical and Biological Illustration,* 15(1):12-18, 1965.

Sollberger, A. Biological measurements in time: with special reference to synchronization mechanisms. *Annals of the New York Academy of Sciences,* 138(2):567-599, 1967.

Sollberger, A. Rhythms and biological cycles. In: Ajurieagurra, J. de, (ed.), *Cycles Biologiques et Psychiatrie,* Paris: Masson & Cie, 1968, pp. 43-56.

Sollberger, A. How are biological rhythms related to the normal values concept. Conference on the use of data mechanization and computers in clinical medicine. *Annals of New York Academy of Sciences,* 161(2):602, 1969.

Sollberger, A., Methods for identifying environmental factors controlling rhythmic processes in biology. *Scientia,* 104:683-684, 1969.

Sollberger, A.; Apple, H.P.; Greenway, R.M.; King, P.H.; Lindan, O.; and Reswick, J.B. Automation in biological rhythm research with special reference to Homo Sapiens. *Annals of the New York Academy of Sciences,* 161(2):184, 1969.

Sollberger, A. Problems in the statistical analysis of short periodic time series, *Journal of Interdisciplinary Cycle Research,* 1, 1970. In press.

Wever, R. A mathematical model for circadian rhythms. In: Aschoff, J. (ed.), *Circadian Clocks.* Amsterdam: North Holland Publishing Company, 1965, 47-63.

Wever, R. Mathematical Models of Circadian Rhythms and their Applicability to Men. In: *Cycles biologiques et psychiatrie.* Paris: Masson & Cie. 1968. pp. 61-72.

Phase Shifts and Work-Rest Schedules

Adams, O.S.; Levine, R.B.; and Chiles, W.D. Research to investigate factors affecting multiple task psychomotor performance. USAir Force WADC Technical Report, No. 59-120, 1959.

Adams, O.S., and Chiles, W.D. Human performance as a function of the work-rest cycle. U.S. Air Force WADC Technical Report, No. 60-248, 1960.

Adams, O.S., and Chiles, W.D. Human performance as a function of the work-rest ratio during prolonged confinement. U.S. Air Force WADC Technical Report, No. 61-720, 1961.

Adler, K. Extraoptic phase shifting of circadian locomotor rhythm in salamanders. *Science,* 164:1290-1291, 1969.

Alluisi, E.A. Research in Performance assessment and enhancement. U.S. Army Behavioral Science Research Laboratory and Performance Research Laboratory, University of Louisville, Louisville, Kentucky, 1969.

Alluisi, E.A. Sustained performance (Chapter 3). In: Bilodau and Bilodau, (eds.), *Principles of Skill Acquisition,* New York: Academic Press, 1969.

Alluisi, E.A.; Chiles, W.D.; et al. Human group performance during confinement. AMRL Technical Documentary Report, 63-87 Aerospace Medical Division, Wright-Patterson Air Force Base, November 1963.

Alluisi, E.A., and Chiles, W.D. Sustained performance, work-rest scheduling, and diurnal rhythms in man. *Acta Psychologica,* 27:436-442, 1967.

Alluisi, E.A.; Chiles, W.D.; and Hall, T.J. Combined effects of sleep loss and demanding work-rest schedules on crew performance. AMRL Technical Documentary Report, 64-63 Aerospace Medical Division, Wright-Patterson Air Force Base, June 1964.

Andlauer, P., and Metz, B. Variations nychtemerales de la frequence des accidents du travail continu. *Archives de maladies professionnelles,* 14(6):613, 1953.

Andrezheyuk, N.I. Effects of different work and rest routines on subjects kept in relative isolation. *Medicine,* 52: 631, October, 1968.

Bayevskiy, R.M.; Chernyayeva, S.A.; et al. Investigation of the physiological condition of men exposed to isolation and hypodynamia during the shift to a new work and rest schedule. [Simpozium] *Biologicheskiye ritmy i voprosy razrabotki rezhimov truda i otdykha.* Materialy. Moscow, 1967, 12-12.

Blake, M.J.F. Relationship between circadian rhythm of body temperature and introversion-extroversion. *Nature,* 215(5103):896-897, 1967.

Blake, M.J.F.; Colquhoun, W.P.; and Edwards, R.S. Experimental studies of shift work, a comparison of rotating and stabilized four-hour shift systems. *Ergonomics,* 2:437-453, September 1968.

Bonjer, F.H. Physiological aspects of shiftwork. *Proceedings of the International Consortium on Occupation Health,* 13:848-849, 1960.

Cantrell, G.K., and Hartman, B.O. Crew performance on demanding work/rest schedules compounded by sleep deprivation. Brooks AFB, Texas: School of Aerospace Medicine, November 1967.

Chiles, W.D.; Alluisi, E.A.; and Adams, O.S. Work schedules and performance during confinement. *Human Factors,* 10:143-146, 1968.

Chipman, D.M., and Sharon, N. Mechanism of Lysozyme Action. *Science,* 165(3892):454-465, August 1969.

Colquhoun, W.P. Biological rhythms and shift work. *Spectrum,* December 1967.

Colquhoun, W.P.; Blake, M.J.F.; and Edwards, R.S. Experimental studies of shift-work I: A comparison of "rotating and stabilized" 4-hour shift systems. *Ergonomics,* 11(5)437-453, 1968.

Colquhoun, W.P.; Blake, M.J.F.; and Edwards, R.S. Experimental studies of shift-work. II: stabilized 8-hour shift systems. *Ergonomics,* 11 (6):527-546, 1968.

Colquhoun, W.P.; Blake, M.J.F.; and Edwards, R.S. Experimental studies of shift-work. III: stabilized 12-hour shift systems. *Ergonomics* 12(6) : 865-882, 1969.

Dehart, R.L. Work-rest cycle in aircrewmen fatigue. *Aerospace Medicine,* 38(2):1174-1179, November 1967.

Dirken, J.M. Industrial shift work: Decrease in well-being and specific effects. *Ergonomics,* 9(2):115-124, 1966.

Dushkov, B.A. and Kosmolinskii, F.P. Rational establishment of cosmonaut work schedules. In: Gurovskii, N.N. (ed.), *Papers on the Psychophysiology of the Labor of Astronauts* (a collection of Russian articles). Foreign Translation Division, Clearinghouse, Department of Commerce, Springfield, Virginia, 22151: AD-684-690. 1968. English pages: 182.

Eranko, O. 25-hour day: one solution to the shift-work problem. *International Congress on Occupation and Health,* 3:134 1967.

Flink, E.B., and Doe, R.P. Effect of sudden time displacement by air travel on synchronization of adrenal function. *Proceedings of the Society for Experimental Biology and Medicine,* 100:494-501, 1959.

Galler, S.R. Relevance of biological orientation research to the field of bioastronautics. *Aerospace Medicine,* 32:535, 1961.

Gambashidze, G.M. Possibilities of body adaptation to day and night shift work. [Vozmozhnosti prisposobleniaa organizma k smennym i nochnym rabotam]. *Gigenia Truda I Professional Nye Zabolevaniia,* 1:12-17, January 1965.

Gerritzen, F. The diurnal rhythm in water, chloride, sodium and potassium excretion during a rapid displacement from East to West and vice versa. *Aerospace Medicine,* 33:697-701, 1962.

Gerritzen, F. Influence of light on human circadian rhythms. *Aerospace Medicine,* 37:66-70, 1966.

Gerritzen, F.; Strengers, T.; and Esser, S. The behavior of the circadian rhythm in water and electrolyte excretion before, during and after a flight from Amsterdam to Anchorage and Tokyo, and on secondary influences on circadian kidney function. XVI International Congress on Aviation and Space Medicine, Lisbon, 11-15, September 1967.

Gerritzen, F.; Strengers, T.; et al., Studies on the influence of fast transportation on the circadian excretion pattern of the kidney in humans. *Aerospace Medicine,* 40:264-271, March 1969.

Ghata, J.; Fourn, P.; and Borrey, F. Application de l'Etude des Variations Circadiennes a L'Analyse des Vols Comportant le Passage de Fuseaux Horaires. XVI International Congress on Aviation and Space Medicine, Lisbon, 11-15, September 1967.

Halberg, F.; Albrecht, P.G.; and Barnum, C.P. Phase shifting of liver-glycogen rhythm in intact mice. *American Journal of Physiology,* 199:400, 1960.

Hartman, B.O., and Cantrell, G.K. Sustained pilot performance requires more than skill. *Aerospace Medicine,* 8:(7)801-803 August 1967.

Hauty, G.T. Individual differences in phase shifts of the human circadian system and performance deficit. In: Brown, A.H. and Favorite, F.G. eds. *Life Sciences and Space Research* V. Amsterdam: North-Holland Publishing Company, 1967. pp. 135-147.

Hautv, G.T. Relationship between operator proficiency and effected changes in biological circadian periodicity. *Aerospace Medicine,* 34:100-105, February 1966.

Hauty, G.T., and Adams, T. Phase shifts of the human circadian system and performance deficit during the periods of transition: I. East-West flight. *Aerospace Medicine,* 37(7):668-674, July, 1963.

Hauty, G.T., and Adams T. Phase shifts of the human circadian system and performance deficit during the periods of transition: III. North-South flight. *Aerospace Medicine,* 37(12):1257-1262, 1966.

Hauty, G.T., and Adams T. Phase shifts of the human circadian system and performance deficit during the periods of transition: II. West-East flight. *Aerospace Medicine,* 37:1027-33, 1966.

Hauty, G.T.; Steinkamp, G.R.; Hawkins, W.R.; and Halberg, F. Circadian performance rhythms in men adapting to an 8-hour day. *Federation Proceedings,* 19:54, 1960.

Holmquest, D.L.; Retiene K.; and Lipscomb, H.S. Circadian rhythms in rats: effects of random lighting. *Science,* 152:662-664, April 1966.

Howitt, J.S.; Balkwill, J.S.; Whiteside, T.C.D.; and Whittingham, P.D.G.V. Flight deck work loads in civil air transport aircraft. *FPRC Reports* 1240, August 1965, and 1264, December 1966. Ministry of Defense, Great Britain.

Kakurin, L.I., and Tokarev, Iu. Investigation of astronaut efficiency in an experiment applicable to space flight problems. [k roprosu issledovaniia rabotosposobnosti zadacham kosmicheskogo poleta] In: Sisakian, N.M., and Iazovskii, V.I., (eds.), *Problems of Space Biology,* 3:226-234, 1964.

Kharabuga, S.G. Study of the effect of the daily schedule on daily periodics. [Simpozium] *Biologicheskiye ritmy i voprosy razrabotki rezhimov truda i otdykha,* 1967. Materialy. Moscow, 1967, 67-68.

Klein, K.E.; Brüner, H.; Rehme, H.; Stolze, J.; Steinhoff, W.D.; and Wegman, H.M. Circadian rythms of pilots' efficiency and effects of multiple time zone travel. *Aerospace Medicine* 41 (1970) In press.

Klein, K.E.; Brüner, H.; and Ruff, S. Investigations on stress imposed on air crew in civil jet aircraft during long-range flight: report on results on the northern Atlantic route. German Experimental Institute for Aeronautics and Astronautics, Research Report 65-44, October 1965.

Klein, K.E.; Brüner, H.; and Ruff, S. Investigation regarding stress on flying personnel in long-distance jet flights. *Zeitschrift für Flugwissenschaften,* 14:109, 1966.

Klein, K.E.; Wegmann, H.M.; and Brüner, H. Circadian rhythm in indices of human performance, physical fitness and stress resistance. *Aerospace Medicine*, 39:512-518, May 1968.

Kleitman, N. Sleep wakefulness cycle in submarine personnel. *Human Factors in Undersea Warfare.* Baltimore: Waverly Press, 1949, pp. 329-341.

Kosmolinskaya, F.P. Biological rhythms and development of work and rest regimes for cosmonauts. *Aerospace Medicine,* 1(5):136-141, February 1967.

Kosmolinskiy, F.P., and Kozar, M.I. Some indices of the stress reaction during chamber experiments with various work and rest schedules. [Simpozium] *Biologicheskiye ritmy i voprosy razrabotki rezhimov truda i otdykha,* 1967. Materialy. Moscow, 1967, 37-39.

Kripke, D.F.; Cook, B.; and Lewis O.F. Sleep in nightworkers: Electroencephalographic recordings. *Psychophysiology,* 1970. In press.

Kuznetsov, O.N.; Lebedev, O.N.; and Litsov, A.N. Concerning individual psychological features of human adaptation to altered daily regimes. [Simpozium] *Biologicheskiye ritmy i voprosy razrabotki rezhimov truda i otkykha,* 1967. Materialy. Moscow, 1967, 40-41.

Lafontaine, E.; Lavernhe, J.; Courillon, J; Medvedoff, M.; and Ghata, J. Influence of air travel on circadian rhythms. *Aerospace Medicine,* 38(9):944-947, September 1967.

Lafontaine, E.; Lavernhe, J.; and Pasquet, J. Subjective and objective responses to disruptions in the circadian rhythms during long distance commercial flights East-West and vice versa. *Revue De Medecine Aeronautique et Spatiale,* 7(26):121-123, 1968.

Lipscomb, H.S.; Rummel, J.A.; et al. Circadian rhythms in simulated and manned orbital space flight. Paper presented at the 37th Annual Aerospace Medical Meeting, Las Vegas, Nevada, April 1966.

Litsov, A.N. Daily dynamics of some physiological functions and human work capacity in isolation. *Space Biology and Medicine,* 2(4):142-148, November 1968.

Litsov, A.N. Experimental investigation of the dynamics of certain physiological functions and work capacity of man under the condition of regular or altered diurnal schedules. [Simpozium] *Biologicheskiye ritmy i voprosy razrabotki rezhimov truda i otdykha,* 1967. Materialy. Moscow, 1967, 45-46.

Meddis, R. Human circadian rhythms and the 48-hour day. *Nature,* 218: 964-965, June 1968.

Medvedeff, E.; Barre, Y.; and Laverne, J. Les rythmes circadiens des activites cortico surrenale et medullo surrenale influence des decalages horaires. Service Medical d'Air France, l, Square Max Hymans, Paris, 15 e.

Menzel, W. Perturbations des rhythmes circadiares chez l'homme, y compris aspect psychosomatique. In: de Ajuriaguerra, J. (ed.), *Cycles Biologiques et Psychiatrie.* Paris: Masson et Cie., 1968. pp. 205-221.

Mills, J.N. Circadian rhythms and shift workers. *Transactions of the Society of Occupational Medicine,* 17(1):5-7, 1967.

Mohler, S.R. Fatigue in aviation activities. *Aerospace Medicine,* 37 :722-732, 1966.

Mohler, S.R.; Dille, J.R.; and Gibbons, H.L. The Time Zone and circadian rhythms in relation to aircraft occupants taking long-distance flights. *American Journal of Public Health,* 58 (8):1404-1409, 1968.

Myasnikov, V.I. The importance of sleep in organizing the cosmonaut's schedule of daily activity. [Simpozium] *Biologicheskiye ritmy i voprosy razrabotki reshimov truda i otdykha,* 1967. Materialy. Moscow, 1967, 52-53.

Ray, J.T.; Martin, O.E., Jr.; and Alluisi, E.A. Human performance as a function of the work-rest cycle: a review of selected studies. *National Academy of Sciences Research Council Publication No. 882,* 1961.

Reinberg, A. Evaluation of circadian dyschronism during transmeridan flights. *Life Sciences and Space Research,* 12th Cospar, Prague. In Press.

Roberts, S.K. Circadian activity rhythms in cockroaches. II. Entrainment and phase shifting. *Journal of Cellular and Comparative Physiology,* 59:175-186, 1962.

Saito, H. Some considerations on reduction of working hours in Japan from the view-point of science of labour. *Journal of the Science of Labour,* 40: 469-486, 1964.

Sasaki, T. Effect of rapid transposition around the earth on diurnal variation in body temperature. *Proceedings of the Society for Experimental Biology and Medicine,* 115:1129-1131, April 1964.

Sharp, G.W.G.; Slorach, S.A.; and Vipond, H.J. Diurnal rhythms of keto- and ketogenic steroid excretion and the adaptation to changes of the activity-sleep routine. *Journal of Endocrinology,* 22:377-385, 1961.

Shift of the work and rest schedule as a functional test of the resistance of regulatory mechanisms of physiological functions. [Simpozium] *Biologicheskiye ritmy i voprosy razrabotki rezhimov truda i otdykha,* 1967. Materialy. Moscow, 1967, 14-15.

Shilov, V.M., and Kozar, M.I. Changes in the immunological reactivity of man exposed to various day schedules in a sealed chamber. [Simpozium] *Biologicheskiye ritmy i voprosy razrabotki rezhimov truda i otdykha,* 1967. Materialy. Moscow, 1967, 71-72.

Siegel, P.V.; Gerathewohl, S.J.; and Mohler, S.R. Time zone effects: disruption of circadian rhythms poses a stress on the long-distance air traveller. *Science*, 164:1249-1255, 1969.

Simpson, H.W., and Lobban, M.C. Effects of a 21-hour day on the human circadian excretory rhythms of 17-hydroxycorticosteroids and electrolytes. *Aerospace Medicine,* 38(11):1205-1212, 1967.

Simpson, H.W.; Lobban, M.C.; and Halberg, F. Near 24-hour rhythms in subjects living on a 21-hour routine in the arctic summer at 78°N - revealed by circadian amplitude ratios. *Rassegna di Neurologia Vegetativa.* In press.

Strughold, H. Physiologic day-night cycle after global flight. *Journal of Aviation Medicine,* 23:464, 1952.

Strughold, H. The physiological clock across time zones and beyond. *Air U. Review,* 19:28-33, July-August 1968.

Strughold, H. The physiological clock in aeronautics and astronautics. *Annals of New York Academy of Sciences,* 134:413-422, November 1965.

Thiis-Evensen, E. Shift-work and health. *Industrial Medicine and Surgery,* 27(10): 493-494, 1957.

Thomas, L. Keep an eye on your inner clock. *Reader's Digest,* 61-64, August 1966.

Tune, G.S. A note on the sleep of shift workers. *Ergonomics,* 11(2):183-184, 1968.

Wever, R. The duration of re-entrainment of circadian rhythms after phase shifts of the Zeitgeber. *Journal of Theoretical Biology.* 13:187-201, 1966.

Wever, R. The influence of weak electromagnetic fields on the circadian rhythm in man. *Zeitschrift Für Vergleichende Physiologie,* 56:111-128, 1967.

Wilkinson, R.T., and Edwards, R.S. Stable hours and varied work as aids to efficiency. *Psychonomic Science,* 13(4):205-206, 1968.

Yurugi, R.; Totsuka, T.; Naba, K.; Iizuka, M.; Akiyama, T.; and Kimara, K. Experimental studies on the effects of reverse day-night cycle upon the twenty-four-hour rhythm of living functions. *Aerospace Medicine* 32(3), March 1961.

Pineal Gland

Axelrod, J. Enzymatic synthesis of the skin-lightening agent, melatonin, in amphibians. *Nature,* 208(5008):306, 1965.

Axelrod, J.; Shein, H.M.; and Wurtman, R.J. Stimulation of C^{14} melatonin synthesis from C^{14} tryptophan by noradrenaline in rat pineal organ culture. *Proceedings of the National Academy of Sciences* 62:544-549, 1969.

Axelrod, J., and Wurtman, R.J. Photic and neural control of indolamine metabolism in the rat pineal gland. In: Garolstein, S., and Moore, P.A. (eds.), *Advances in Pharmacology.* New York: Academic Press, 1968, 157 pp.

Axelrod, J.; Wurtman, R. J.; and Snyder, S.H. Control of hydroxyindole O-methyltransferase activity in the rat pineal gland by environmental lighting. *Journal of Biological Chemistry,* 240(2):949-954, 1965.

Axelrod, J.; Wurtman, R.J.; and Winget, C.M. Melatonin synthesis in the hen pineal gland and its control by light. *Nature,* 201(4924):1134, 1964.

Cohen, R.A. Some Clinical Biochemical and Physiological Actions of the Pineal Gland. *Annals of Internal Medicine,* 61(6):1144-1161, 1964.

Fiske, V.M., and Huppert, L.C. Melatonin action on pineal varies with photoperiod. *Science,* 162:279, October 1968.

Gaston, S., and Menaker, M. Pineal function: the biological clock in the sparrow? *Science,* 160:1125-1127, 1968.

Kastin, A.J.; Redding, T.W.; and Schally, A.V. MSH activity in rat pituitaries after pinealectomy. *Proceedings of Society for Experimental Biology and Medicine,* 24:1275-1277, 1967.

Kerenyi, N.A., and Sarker, K. The postnatal transformation of the pineal gland. *Acta Morphologia Academia Scientia Hungaria,* 16: 223, 1968.

Moore, R.Y.; Heller, A.; Wurtman, R.J.; and Axelrod, J. Visual pathway mediating pineal response to environmental light. *Science,* 155:220-223, 1967.

Oishi, T., and Kato, M. The pineal organ as a possible photoreceptor in photoperiodic testicular responses in Japanese quail. *Memoirs of the Faculty of Science, Kyoto University,* 2: Series D: 12-18, 1968.

Quay, W.B. Circadian and estrous rhythms in pineal and brain serotonin. *Progress in Brain Research,* 8:61-63, 1964.

Shein, H.M., and Wurtman, R.J. Cyclic adenosine monophosphate: stimulation of melatonin and serotonin synthesis in cultured rat pineals. *Science,* 166: 519-520, October, 1969.

Snyder, S.H., and Axelrod, J. Circadian rhythms in pineal serotonin: Effect of monoamine oxidase inhibition and reserpine. *Science,* 149(3683):542-544, 1965.

Snyder, S.H.; Axelrod, J.; Wurtman, R.; and Fischer, J. Control of 5-hydroxytryptophan decarboxylase activity in the rat pineal gland by sympathetic nerves. *Journal of Pharmacology and Experimental Therapeutics,* 147(3):371-375, 1965.

Snyder, S.H.; Ganong, W.; Axelrod, J.; and Fischer, J.E. Control of the circadian rhythm in seratonin content of the rat pineal gland. *Proceedings of the National Academy of Sciences,* 53: 301-305, 1965.

Snyder, S.; Zweig, M.; Axelrod, J.; and Fischer, J.E. Control of the circadian rhythms in serotonin content of the rat pineal gland. *Proceedings of the National Academy of Sciences,* 53(2):301-305, 1965.

Wetterberg, L.; Geller, E.; and Yuwiler, A. Harderian Gland: An extraretinal photoreceptor influencing the Pineal Gland in neonatal rats? *Science,* 167:884-885, February 6, 1970.

Wurtman, R.J., and Axelrod, J. The physiologic effects of melatonin and the control of its biosynthesis. *Problems Actuels D'Endocrinologie et de Nutrition,* 10:189-200, 1966.

Wurtman, R.J., and Axelrod, J. A 24-hour rhythm in the content of norepinephrine in the pineal and salivary glands of the rat. *Life Sciences,* 5:665-669, 1966.

Wurtman, R.J.; Axelrod, J.; and Fischer, J.E. Melatonin synethesis in the pineal gland. Effect of light mediated by the sympathetic nervous system. *Science,* 143(3612); 1328-1330, 1964.

Wurtman, R.J.; Axelrod, J.; Sedvall, G.; and Moore, R.Y. Photic and neural control of the 24-hour norepinephrine rhythm in the rat pineal gland. *Journal of Pharmacology and Experimental Therapeutics,* 157(3):487-492, 1967.

Wurtman, R.J.; Axelrod, J.; Snyder, S.H.; and Chu, E.W. Changes in the enzymatic synthesis of melatonin in the pineal during the estrous cycle. *Endocrinology,* 76:798-800, 1965.

Wurtman, R.J.; Chu, E.W.; and Axelrod, J. An inhibitory effect of melatonin on the estrous phase of the estrous cycle of the rodent. *Endocrinology,* 75(2):238-242, 1964.

Sleep

Abernethy, J.D.; Farhi, L.E.; and Maurizi, J.J. Diurnal variations in urinary-alveolar N2 difference and effects of recumbency. *Journal of Applied Physiology,* 23:875-879, 1967.

Abrams, R.L.; Parker, L.; et al. Hypothalamic regulation of growth hormone secretion. *Endocrinology,* 78(3):605-613, 1966.

Angeleri, F.; Bergonzi, P.; and Ferroni, A. Sleep stages and cycles in epileptics. A statistical study on 87 nocturnal polygraphic records. [Le fasi de i cicli del sonno nocturno negli epilettici. Studio statistico in 87 registrazioni poligrafiche]. *Riv Pat Nerv Ment* (Ital.), 13(4):953-955, 1967.

Brebbia, D.R., and Altshuler, K.Z. Stage related patterns and nightly trends of energy exchange during sleep. In: Kline and Lask, (eds.), *Computers and Electronic Devices in Psychiatry,* New York: Grune and Stratton. 1968.

Brebbia, D.R., Altshuler, K.Z., and Kline, N.S. Lithium and the electroencephalogram during sleep. *Diseases of the Nervous Systems,* 30:541-546, 1969.

Brodan, V.; Brodanova, M.; Friedman, B.; and Kuhn, E. Influence of sleep deprivation on iron metabolism. *Nature,* 213:1041-1042, 1967.

Broughton, R.J. Sleep disorders: disorders of arousal? *Science,* 159:1070-1078, 1968.

Cantrell, G.K., and Hartman, B.O. Crew performance on demanding work/rest schedules compounded by sleep deprivation. Brooks AFB, Texas: School of Aerospace Medicine, November 1967.

Capek, R.; Babej, M.; and Radil-Weiss T. Drug-induced sleep cycle impairment. A possible indicator of central nervous system side-effects. In: Baker, S.B. de C., and Tripod, J., (eds.), *Sensitization to Drugs.* Proceedings of the European Society for the Study of Drug Toxicity, vol. 10, and Excerpta Medica International Congress Series, 181:47-50, 1969.

Crowley, T.J.; Kripke, D.F.; Halberg, F.; Pegram, G.V.; and Schildkraut, J.J. Circadian rhythms in monkeys: sleep, EEG, EMG, body and eye movement and temperature. *Psychophysiology,* 6:242-243, 1969.

Daughaday, W.H.; Othmer, E.; and Kipnis, D.M. Hypersecretion of growth hormone following REM deprivation. Report to the Fifty-first Annual Meeting of the Endocrine Society, New York, June 1969.

Dement, W.C., and Kleitman, N. Cyclic variations in EEG during sleep and their relation to eye movements, body motility and dreaming. *Electroencephalography and Clinical Neurophysiology,* 9(4):673-690, 1957.

Dewan, E.M. The P(programming) hypothesis for REMS. *Psychophysiology,* 4: 365, 1968.

Feinberg, I. Eye movement activity during sleep and intellectual function in mental retardation. *Science,* 159:1256, 1968.

Feinberg, I. The ontogenesis of Human Sleep and the Relationship of Sleep Variables to Intellectual Function in the Aged. *Comprehensive Psychiatry,* 9(2):138-147, March 1968.

Frank, G.; Halberg, F.; Harner, R.; Matthew, J.; Johnson, E.; Gravem, H.; and Andrus, V. Circadian periodicity, adrenal corticosteroids, sleep deprivation and the EEG in normal men. *Journal of Psychiatric Research,* 4:73-86, 1966.

Friedman, A.H., and Walker, C.A. Rat brain amines, blood histamine and glucose levels in relationship to circadian changes in sleep induced by pentobarbital sodium. *Journal of Physiology* (London), *202*:133-147, 1968.

Globus, G.G. Observations on sub-circadian periodicity (Abstract). Paper presented at the Association for the Psychophysiological Study of Sleep, March 1969. *Psychophysiology.* In press.

Greenberg, R., and Dewan, E.M. Aphasia and rapid eye movement sleep. *Nature* 223:183-184, 1969.

Hansen, N.E. Sleep-related plasma hemoglobin levels. *Acta Medica Scandinavica,* 184(6):547-549, 1968.

Hartmann, E. Longitudinal studies of sleep and dream patterns in manic-depressive patients. *Archives of General Psychiatry,* 19:312-329, September 1968.

Hartmann, E. The 90-minute sleep-dream cycle. *Archives of General Psychiatry,* 18:280-286, March 1968.

Hartmann, E. (Ed.). *Sleep and Dreaming,* Boston: Little Brown, 1970. 435 pp.

Hayashi T. The relationship between circadian sleep and gamma-hydroxybutyrate (4HB) in brain. *Experimental Medical Surgery,* 25(2-4):148-155, 1968.

Hayward, J. Brain temperature regulation during sleep and arousal in the dog. *Experimental Neurology,* 21(2):201-212, 1968.

Hobson, A. Sleep and biorhythmicity. *Science,* 165(3896): 932-933, August 1969.

Huertas, J., and McMillin, J.K. Paradoxical sleep: effect of low partial pressure of atmospheric oxygen. *Science,* 159:745-746. February 1968.

Jouvet, M. Biogenic Amines and the states of sleep. *Science,* 163 (3862):32-41, 1969.

Jouvet, M. Insomnia and decrease of cerebral 5-hydroxytryptamine after destruction of the RAPHE system in the cat. *Advances in Pharmacology,* 6 B: 265-279, 1968.

Jouvet, M. The states of sleep. *Scientific American,* 216:62-72, 1967.

Karacan, I; Finley, W.W.; Williams, R.L.; and Hursch, C.J. Changes in stage 1-REM and Stage-4 sleep during naps. *Biological Psychiatry.* In press, 1970.

Kardzic, V.; and Marsulla, B. Deprivation of paradoxical sleep and brain glycogen. *Journal of Neurochemistry,* 16(1):29-33, 1969.

Kleitman, N. *Sleep and Wakefulness* (Rev. ed.) Chicago: University of Chicago Press, 1963.

Kleitman, N. Basic rest-activity cycle. Symposium on Sleep and Biorhythmicity. Association for the Psychophysiological Study of Sleep, Boston, March 1969.

Koukkou, M., and Lehmann, D. EEG and memory storage in sleep experiments with humans. *Electroencephalography and Clinical Neurophysiology,* 25:455-462, November 1968.

Kripke, D.F., and O'Donoghue, J.P. Perceptual deprivation, REM sleep, and an ultradian biological rhythm. *Psychophysiology,* 5:231-232, 1968.

Lewis, J.A. Some observations on narcolepsy. Paper presented at the Association for the Psychophysiological Study of Sleep, Denver, 1968. *Psychophysiology,* 1969.

Luce, G. *Current research on sleep and dreams.* U.S. Department of Health, Education, and Welfare, Public Health Service, National Institute of Mental Health, Public Health Service Publication No. 1389, 1965.

Luce, G., and Segal, J. *Sleep.* New York: Coward-McCann, 1966.

Mandell, A.J.; Spooner, C.E.; and Brunet, D. Whither the "Sleep Transmitter". *Biological Psychiatry,* 1(1):13-30, New York; Plenum Publishing Corporation, 1969.

Mink, W.D.; Best, J.; and Olds, J. Neurons in paradoxical sleep and motivated behavior. *Science,* 158:1335-1337, 1967.

Monroe, L.J. Psychological and physiological differences between good and poor sleepers. *Journal of Abnormal Psychology,* 72:255-264, 1967.

Mouret, J.R., and Bobillier, R. Circadian rhythms of sleep and brain biogenic amines in the rat. Alterations independent of light. *Life Sciences.* In press.

Mouret, J.; Bobillier, P.; and Jouvet, M. Insomnia following parachlorphenylalanine in the rat. *European Journal of Pharmacology,* 5:1, 1968-1969.

Parker, D.C.; Sassin, J.F.; Mace, J.W.; Gotlin, R.W.; and Rossman, L.G. Human growth hormone release during sleep: electroencephalographic correlation. *Journal of Clinical Endocrinology and Metabolism,* 29(6):871-874, 1969.

Passouant P. Périodicité Nycthémérale du sommeil rapide au cours de la narcolepsie. In: de Ajuriaguerra, J. (ed.). *Cycles Biologiques et Psychiatrie.* Paris: Masson & Cie, 1968. pp 223-225.

Passouant, P.; Halberg, F.; Genicot, R.; Popoviciu, L.; and Baldy-Moulinier, M. La périodicité des accés narcoleptiques et le rythme ultradien du sommeil rapide. *Revue Neurologique,* 121 (2):155-164, 1969.

Passouant, P.; Popiciu, L.; Velok, G.; Baldy-Moulinier, M. Etude polygraphique des narcolepsies aux cours du nycthemere. *Revue Neurologie,* 118 (6):431-441, 1968.

Pujol, J.F.; Hery, F.; Durand, M.; and Glowinski, J. Increase in serotonin synthesis in the brainstem of the rat after selective deprivation of paradoxical sleep. [Augmentation de la synthese de la serotonine dans le tronc cerebral chez le rat apres privation selective du sommeil paradoxal]. *D R Acad Sci,* (d) (Paris), 267(3):371-372, 1968.

Rechtschaffen, A., and Kales, A. Manual of standardized terminology, techniques and scoring system for sleep stages of human subjects. National Institutes of Health Publication No. 204. Washington: Superintendent of Documents, U.S. Government Printing Office, Book 1-62, 1968.

Reis, D.J.; Corvelli, A.; and Conners, J. Circadian and ultradian rhythms of serotonin regionally in cat brain. *The Journal of Pharmacology and Experimental Therapeutics,* 167(2):328-333, 1969.

Richter, C.P. Sleep and activity: Their relation to the 24-hour clock. Sleep and altered states of consciousness. *Association for Research in Nervous and Mental Disease,* 45:8-29, 1967.

Roffwarg, H.P.; Muzio, J.N.; and Dement, W.C. Ontogenetic development of the human sleep-dream cycle. *Science,* 152:602-619, 1966.

Steriade, M., and Iosif, G. Opposite changes in responsiveness of the motor and somaesthetic cortex during natural sleep and arousal. [Abstract] *Electroencephalography and Clinical Neurophysiology,* 25(3):299, 1968.

Sterman, M.B.; Howe, R.C.; and Macdonald, L.R. Facilitation of Spindleburst sleep by instrumental conditioning of waking EEG activity. *Science,* 167:1146-1148, February 20, 1970.

Sterman, M.B.; Wyrwicka, W.; and Roth, S. Electrophysiological correlates and neural substrates of alimentary behavior in the cat. *Annals of the New York Academy of Sciences,* 157 (2):723-729, 1969.

Stern, E.; Parmelee, A.H.; Akiyama, Y.; Schultz, M.A.; and Wenner, W.H. Sleep cycle characteristics in infants. *Pediatrics,* 43(1), 1969.

Storch, J., and Hildebrandt, G. Methodische grundlagen zur bestimmung der puls-atem-koppelung beim menschen und ihr verhalten im schlaf. *Pflugers Archiv fur Die Gesamte Physiology,* 289(2), 1966.

Stoyva, J.; Zimmerman, J.; and Metcalf, D. Distorted visual feedback and augmented REM sleep. *Association for the Psychophysiological Study of Sleep-Santa Fe,* 1970.

Takashi, Y.; Kipnis, D.M.; and Daughaday, W.H. Growth hormone secretion during sleep. *Journal of Clinical Investigations,* 47(9):2079-2090, 1968.

Tepas, D.I. Evoked brain response as a measure of human sleep and wakefulness. *Aerospace Medicine,* 38:148-153, 1967.

Webb, W.B., and Agnew, Jr., H.W. Sleep cycling within twenty-four hour periods. *Journal of Experimental Psychology,* 64(2):158-160, 1967.

Weitzman, E.D.; Goldmacher, D.; Kripke, D.; MacGregor, P.; Kream, J.; and Hellman, L. Reversal of sleep-waking cycle-effect on sleep stage pattern and certain neuroendocrine rhythms. *Trans American Neurological Association,* 93: 153-157, 1968.

Weitzman, E.D.; Kripke, D.; Goldmacher, D.; McGregor, P.; and Nogeire, C.: Acute Reversal of the Sleep-Waking Cycle in Man: Effect on Sleep Stage Patterns. *Archives of Neurology.* In Press, 1970.

Weitzman, E.D. (Investigator), and Luce, G. (Author). Biological Rhythms—Indices of pain, adrenal hormones, sleep, and sleep reversal. *Mental Health Program Reports-3,* National Institute of Mental Health, PHS Publication No. 1876, 319-332, 1969.

Weitzman, E.D.; Schaumburg, H.; and Fishbein, W. Plasma 17-Hydroxycorticosteroid levels during sleep in man. *Journal of Endocrinology and Metabolsim,* 26:121-127, 1966.

Wyrwicka, W.; and Sterman, M.B. Instrumental conditioning of sensorimotor cortx EEG spindles in the waking cat. *Physiology and Behavior,* 3:703-707, 1968.

Zung, W.W.K. Antidepressant drugs and sleep. *Experimental Medicine and Surgery,* 27(1-2); 124-137, 1969.

Zung, W.K. Effect of anti-depressant drugs on sleeping and dreaming. III: on the depressed patient. *Biological Psychiatry,* 1:283-287, 1969.

Space

Altukhov, G.V.; Belai, V.E.; Egorov, A.D.; and Vasilev, P.V. Federation of American Societies for experimental biology, Washington, D.C. Diurnal rhythm of autonomic functions in cosmic flight. *Izv Akademiia Nauk SSSR, Ser. Biol.,* (Moscow), 30(2):182-187, 1965.

Altukhov, G.V.; Belai, P.E.; Egorov, A.D.; and Vasilev, P.V. Diurnal rhythm of sympathetic functions during space flight [Sutochnaia ritmika vegetativnykh funktsii v kosmicheskom polete] In: Parin, V.V., and Kasian, I.I., (eds.), Seri- in-Medico-Biological Studies of Weightlessness. [Medico-Biologicheskie Issledovaniia V Nevesomosti. Moscow: Stuo Meditsina, 1968, pp. 201-205.

Alyakrinsiy, B.S. The ways and principles of the development of biorhythm studies and its importance in the organization of spaceflights. [Sympozium] *Biologicheskiye ritmy i voprosy razrobotki rezhimov truda i otdykha,* 1967. Materialy. Moscow, 1967, 3-7.

Analysis of Crew Performance in the Apollo Command Module, Phase II. Vol. II Appendix. Baltimore, Maryland: Martin Company, October 1966, 109 pp.

Aschoff, J. Significance of circadian rhythms for space flight. In: Bedwell, T.C., and Strughold, H. (eds.), *Proceedings of Third International Symposium on Bioastronautics and the Exploration of Space.* San Antonio, 1964.

Berry, C.A.; Coons, D.O.; Catterson, A.D.; and Kelly, G.F. Man's response to long-duration flight in the Gemini spacecraft. Gemini Midprogram Conference, including experiment results, NASA, Washington, D.C., 235-261, 1966.

Brown, F.A. Jr. The biological rhythm problem and its bearing on space biology. In: Kaufman, W.C., (ed.), *Advances in the Astronautical Sciences.* North Hollywood, California; Western Periodicals Company, 1963.

Dietlein, L.F., and Vallbona, C. Experiment M-4, inflight phonocardiogram measurements of the duration of the cardiac cycle and its phases during the orbital flight of Gemini V. Gemini Midprogram Conference, including experiment results, NASA Washington, D.C., 397-402, 1966.

Duke, M.B. Biosatellite III: preliminary findings. *Science,* 166:492-493, 1969.

Dushkov, B.A., and Kosmolinskii, F.P. Rational establishment of cosmonaut work schedules. (Ratsional nom postroenii rezhma truda kosmonavtov). In: Gurovskii, N.N., (ed.), *Papers on the Psychophysiology of the Labor of Astronauts.* (a collection of Russian articles). Foreign Translation Division, Clearinghouse, Department of Commerce, Springfield, Virginia, 22151: AD-684-690, 1968. English pages: 182.

Fedotov, Y., and Yudin, Y. Biological rhythms and astronautics. *Krasnaya Zvezda,* 212(11817):6, September 1962. Translated into English at Foreign Technology Division, Air Force Systems Command, Wright Field, Patterson AFB, Ohio, March 1964.

Halberg, F. Physiologic rhythms. In: *Physiological Problems in Space Travel.* Springfield, Illinois: Charles C. Thomas Company, 1964.

Halberg, F. Physiologic rhythms and bioastronautics. In: Schafer, K.E., (ed.), *Bioastronautics.* New York: Macmillan, 1964. pp. 181-195.

Halberg, F.; Vallbona, C.; Dietlein, L.F.; Rummel, J.A.; Berry, C.E.; Pitts, G.; and Nunnely, S. Human circadian circulatory rhythms during weightlessness in extraterrestrial flight or bedrest with and without exercise. United States Public Health Service, NASA, Texas Medical Center, 1969.

Hoshizaki, T.; Adey, W.R.; Meehan, J.P.; Walter, D.O.; Berkout, J.I.; and Campeau, E. Central nervous, cardiovascular and metabolic data of a Macaca nemestrina during a 30-day experiment. In: Rohles, F.H., (ed.), *Circadian Rhythms in Non-Human Primates.* New York: Karger, 1969. pp. 8-38.

Katkovskiy, B.S., and Pilyavskiy, O.A. Effect of prolonged hypokinesia on human resistance to physical labor. [Simpozium] *Biologicheskiye ritmy i voprosy razrabotki rezhimov truda i otdykha,* 1967. Materialy. Moscow, 1967, 32-33.

Lebedev, V.I. Scientist reviews problems of space psychology. *Science and Life* (Russian), 3:25-29, September 1968.

Pittendrigh, C.S. Circadian rhythms, space research, and space flight. *Life Sciences and Space Research.* Amsterdam; North-Holland Publishing Company, 1967, pp. 122-134.

Rummel, J.; Sallin, E.; and Lipscomb, H. Circadian rhythms in simulated and manned orbital space flight. *Rassegna di Neurologia Vegetativa,* 21(1-2): 41-56, 1967.

Strughold, H. Solved and unsolved space medical problems. *Aerospace Medicine,* 38(5):520-535, 1967.

Wheden, G.D.; Lutwak, L.; Neuman, W.F.; and LaChance, P.A. Experiment M-7, calcium and nitrogen balance. Gemini Midprogram Conference, including Experiment results, NASA, Washington, D.C., 1966, 405-415.

Time Sense and Time Estimation

Baldwin, R.O., and Thor, D.H. Time of day estimates at six times of day under normal conditions. *Perceptual and Motor Skills,* 21:904-906, 1965.

Crawford, M.L.J., and Thor, D.H. Circadian activity and noise comparisons of two confined groups with and without reference to clock time. *Perceptual and Motor Skills,* 19:211-216, 1964.

Denner, B.; Wapner, S.; and Werner, H. Rhythmic activity and the discrimination of stimuli in time. *Perceptual and Motor Skills,* 19:723-729, December, 1964.

Erickson, M.H., and Cooper, L.F. *Time Distortion in Hypnosis.* Baltimore: Williams and Wilkins, 1959.

Fischer, R. The biological fabric of time. In: Fischer, R., (ed.), *Interdisciplinary Perspectives of Time.* Annals of the New York Academy of Sciences,138:440-488, 1967.

Fraisse, P. *The Psychology of Time.* London: Eyre and Spottiswoode, 1964.

Fraisse, P.; Siffre, M.; Oleron, G.; and Zuili, N. Le rhythme vielles-sommeil et l'estimation du temps. In: de Ajuriaguerra, J., (ed.), *Cycles Biologiques et Psychiatrie.* Paris: Masson & Cie, 1968. pp. 257-265.

Grossman, J.S., and Hallenbeck, C.E. Importance of time and its subjective speed. *Perceptual and Motor Skills,* 20:1161-1166, (part 2), June 1965.

Levinson, J.Z. Flicker fusion phenomena. *Science,* 160:21-28, 1968.

Lockhart, J.M. Ambient temperature and time estimation. *Journal of Experimental Psychology,* 73:286-291, February 1967.

Melges, F.T., and Fougerousse, C.E., Jr. Time sense, emotions, and acute mental illness. *Journal of Psychiatric Research,* 4:127-140, 1966.

Ornstein, R.E. *On the Experience of Time.* Harmondsworth, England: Penguin Books Ltd. 1969. 115 pp.

Pfaff, D. Effects of temperature and time of day on time judgments. *Journal of Experimental Psychology,* 76:419-422, March 1968.

Stephens, G.J., and Halberg, F. Human time estimation. *Nursing Research,* 14(4):310--317, 1965.

Surwillo, W.W. Time perception and the "internal clock": Some observations on the role of the electroencephalogram. *Brain Research,* 2:390-392, 1966.

Tepas, D.I. Evoked brain response as a measure of human sleep and wakefulness. *Aerospace Medicine,* 38:148-153, 1967.

Thor, D.H. Diurnal variability in time estimation. *Perceptual and Motor Skills,* 15:451-454, 1962.

Thor, D.H. Time perspective and time of day. *The Psychological Record,* 12(4): 417-422, 1962.

Treisman, M. The psychology of time. *Discovery,* 26:40-45, October 1965.

Walsh, J.F., and Misiak, H. Diurnal variation of critical flicker frequency. *Journal of General Psychology,* 75:167-175, 1966.

Wilkinson, R.T. Evoked response and reaction time. *Acta Psychologica,* 27:235-245, 1967.